Proceedings

Ein stetig steigender Fundus an Informationen ist heute notwendig, um die immer komplexer werdende Technik heutiger Kraftfahrzeuge zu verstehen. Funktionen, Arbeitsweise, Komponenten und Systeme entwickeln sich rasant. In immer schnelleren Zyklen verbreitet sich aktuelles Wissen gerade aus Konferenzen, Tagungen und Symposien in die Fachwelt. Den raschen Zugriff auf diese Informationen bietet diese Reihe Proceedings, die sich zur Aufgabe gestellt hat, das zum Verständnis topaktueller Technik rund um das Automobil erforderliche spezielle Wissen in der Systematik aus Konferenzen und Tagungen zusammen zu stellen und als Buch in Springer.com wie auch elektronisch in Springer Link und Springer Professional bereit zu stellen. Die Reihe wendet sich an Fahrzeug- und Motoreningenieure sowie Studierende, die aktuelles Fachwissen im Zusammenhang mit Fragestellungen ihres Arbeitsfeldes suchen. Professoren und Dozenten an Universitäten und Hochschulen mit Schwerpunkt Kraftfahrzeug- und Motorentechnik finden hier die Zusammenstellung von Veranstaltungen, die sie selber nicht besuchen konnten. Gutachtern, Forschern und Entwicklungsingenieuren in der Automobil- und Zulieferindustrie sowie Dienstleistern können die Proceedings wertvolle Antworten auf topaktuelle Fragen geben.

Today, a steadily growing store of information is called for in order to understand the increasingly complex technologies used in modern automobiles. Functions, modes of operation, components and systems are rapidly evolving, while at the same time the latest expertise is disseminated directly from conferences, congresses and symposia to the professional world in ever-faster cycles. This series of proceedings offers rapid access to this information, gathering the specific knowledge needed to keep up with cutting-edge advances in automotive technologies, employing the same systematic approach used at conferences and congresses and presenting it in print (available at Springer.com) and electronic (at Springer Link and Springer Professional) formats. The series addresses the needs of automotive engineers, motor design engineers and students looking for the latest expertise in connection with key questions in their field, while professors and instructors working in the areas of automotive and motor design engineering will also find summaries of industry events they weren't able to attend. The proceedings also offer valuable answers to the topical questions that concern assessors, researchers and developmental engineers in the automotive and supplier industry, as well as service providers.

Alexander Heintzel
(Hrsg.)

Internationaler Motorenkongress 2024

Springer Vieweg

Hrsg.
Alexander Heintzel
Springer Fachmedien Wiesbaden
Wiesbaden, Deutschland

ISSN 2198-7432 ISSN 2198-7440 (electronic)
Proceedings
ISBN 978-3-658-47965-7 ISBN 978-3-658-47966-4 (eBook)
https://doi.org/10.1007/978-3-658-47966-4

Die Deutsche Nationalbibliothek verzeichnet diese Publikation in der Deutschen Nationalbibliografie; detaillierte bibliografische Daten sind im Internet über https://portal.dnb.de abrufbar.

Planung/Lektorat: Stella Müller
Springer Vieweg ist ein Imprint der eingetragenen Gesellschaft Springer Fachmedien Wiesbaden GmbH und ist ein Teil von Springer Nature.
Die Anschrift der Gesellschaft ist: Abraham-Lincoln-Str. 46, 65189 Wiesbaden, Germany

Geleitwort

Herzlich willkommen, liebe Kongressteilnehmer!
Die Sorgen zum weltweiten Klimawandel nehmen weiter zu. In Europa überwiegt politisch geprägt noch die Diskussion um ein Verbot des Verbrennungsmotors über definierte CO_2-Flottengrenzwerte und Emissionsgesetze. Interessant ist, wann sich die einseitige batterieorientierte Schwerpunktlösung bei Pkw und Nfz eher den Zielorientierungen anderer Regionen anpassen wird.

Nicht nur führende asiatische Länder setzen gemeinsam mit Industrie und Wissenschaft auf technologieoffene Innovationen für die Energieerzeugung und Nutzung in der Mobilität. Die zur Zielerreichung unabdingbare Defossilisierung der Bestandsflotten kann nur so volkswirtschaftlich vertretbar und rasch eingebunden werden. Eine Schlüsselrolle kommt bei der komplexen Anwendungsvielfalt dabei der Schaffung von kreislauforientierten Energie-, Rohstoff-, Liefer- und Produktionsketten zu.

Der Kongress wird die zukünftige Rolle des Verbrennungsmotors und seiner nichtfossilen Kraftstoffe aus globaler Sicht führender Experten beleuchten. Sichern Sie sich Ihren Wissensvorsprung und profitieren Sie!

- Es erwarten Sie internationale Referenten, hochkarätige Vorträge und Diskussionsrunden.
- Nutzen Sie den Kongress zum Netzwerken – der Abend der Motoren-Community bietet interessante Gespräche in ungezwungener Atmosphäre.
- Eine begleitende Fachausstellung informiert über innovative Produkte und Dienstleistungen.

Wir freuen uns auf Ihre Teilnahme!
Im Namen der Programmbeiräte

Prof. Dr. Peter Gutzmer
Wissenschaftlicher Leiter des Kongresses
Herausgeber ATZ | MTZ-Gruppe

Foreword

Warm welcome to all participants
Concern about global climate change continues to increase. In Europe, discussions are still being dominated for political reasons by a ban on internal combustion engines through defined CO_2 fleet limit values and emissions standards. It will be interesting to see when the proposed one-sided, battery-oriented solution for passenger cars and commercial vehicles will become more adapted to the target orientations of other regions.

Leading Asian countries and others, together with industry and science, are relying on innovations for the production and use of energy in mobility that are open to all technologies. Only in this way can the defossilization of existing vehicle fleets – which is absolutely essential if we are to achieve our goals – be implemented quickly and in an economically justifiable manner. Due to the complex diversity of applications, a key role will be played by the creation of circular energy, raw material, supply, and production chains.

The congress will examine the future role of the internal combustion engine and its non-fossil fuels from the global perspective of leading experts.

Stay abreast of current trends and benefit from a lead in knowledge!

- You can expect international speakers as well as top-level presentations and panel discussions.
- The congress is a great opportunity to "network" – the evening event for the engine community offers stimulating discussions in an informal atmosphere.
- The trade exhibition, held in parallel, provides ample information about innovative products and services.

We look forward to your participation.

On behalf of the program advisory boards

Prof. Dr. Peter Gutzmer
Scientific Director of the Congress
Editor-in-Charge ATZ | MTZ Group

Tagungsbericht MoKo 24

Verbrennungsmotoren für den Klimaschutz

Die 11. Auflage des Internationalen Motorenkongresses findet am 27. und 28. Februar 2024 in Baden-Baden statt. Mit 280 Teilnehmern vor Ort und weiteren 50, die sich online zugeschaltet haben.

In seinen Grußworten hob Tagungsleiter und Herausgeber Prof. Peter Gutzmer die dringende Notwendigkeit von technologieoffener Entwicklung erneut hervor und verwies auf die laufenden Verhandlungen auf europäischer Ebene. Gutzmer betonte die weltweite Relevanz des Verbrennungsmotors und dessen Rolle für einen effektiven Klimaschutz. Auch zeige sich, dass der Verbraucher den eingeschlagenen Weg hin zu vollelektrischen Fahrzeugen nicht bereit ist, mitzugehen. „Es hat sich noch nie eine Technologie aufgrund von Zwang durchgesetzt", sagte Gutzmer.

Prof. Dieter Grebe, Geschäftsführer von AVL List, gab in seiner Keynote einen Ausblick auf die Rolle des Verbrennungsmotors im weltweiten Energiemix. Der nahezu lineare Anstieg von menschengemachten CO_2-Emissionen über die vergangenen 60 Jahre kann seiner Ansicht nach ohne Maßnahmen wie Carbon Capture nicht umgekehrt werden. Dabei müsse dringend auch die globale Bestandsflotte berücksichtigt und ertüchtigt werden. Schon mit dem Einsatz von E20 könne der CO_2-Ausstoß bis 2030 um 10 % gesenkt werden. Wasserstoffträger zum Transport sollten idealerweise am Produktionsstandort Energie zur Herstellung benötigen und nicht am Nutzungsort, um den gesamten Footprint möglichst klein zu halten. Um die Energiewirtschaft darzustellen, benötige man molekulare Energiespeicher, wodurch automatisch Methanol und synthetische Kraftstoffe vorhanden seien. Durch die CO_2-Gesetzgebung und langfristig betrachtet sind BEVs und FCEVs laut Grebe der richtige Weg, für die Bestandsflotte sei der Anteil biogener Kraftstoffe jedoch schnell zu erhöhen. Weltweit gesehen sei der eingeschlagene europäische Technologiepfad mit einer alleinigen Konzentration auf BEVs sehr isoliert. Japan zeige sich hier als Antipode mit einer klar technologieoffenen Bewertung aller möglichen Antriebstechnologien.

Japan setzt auf systemische Lösungen

Takahiro Nagai, Deputy Director der NEDO in Japan, nahm den Ball direkt auf und präsentierte den, in seinen Worten, realistischen japanischen Weg hin zu einer CO_2-neutralen Gesellschaft. Die NEDO dient hier als Beratungsorganisation für die Regierung. Diese setze auf eine starke Diversifizierung der Maßnahmen, darunter auch den Aufbau einer Wasserstoffinfrastruktur. Bis 2028 soll zudem eine Produktion für synthetische Kraftstoffe aufgebaut werden. Ziel ist es, die CO_2-Emissionen von HEV und HD-Dieselantrieben um 55 % zu senken. Emissionen von Verbrennungsmotoren müssten mit technischem Fortschritt und synthetischen Kraftstoffen gesenkt und gleichzeitig der Durchbruch bei elektrischen Fahrzeugen erreicht werden.

Dr. Monika Griefahn, Vorstandsvorsitzende der eFuel Alliance wies darauf hin, dass E-Mobilität ein wichtiger Faktor, aber nicht überall einsetzbar sei. Die weltweite

Gesamtflotte von 1,3 bis 1,4 Milliarden Fahrzeugen müsse klimafreundlicher betrieben werden: „Jeder Liter E-Fuel ist hier ein direkter Beitrag zum Klimaschutz." Dies müsse schnell umgesetzt werden, um nennenswerte Skaleneffekte zu erzielen. Die Gesetzgebung sei der Knackpunkt, da nach wie vor eine Tank-to-Wheel-Diskussion geführt werde. „Wir hoffen, dass mit der neuen Kommission die Diskussion um eine Cradle-to-Cradle-Betrachtung sofort wieder aufgenommen wird."

Zum Abschluss der ersten Session beschrieb Thorsten Herdan, CEO von HIF EMEA, die Veränderung des Geschäftsmodells einer Anlage zur Produktion von synthetischen Kraftstoffen im Gegensatz zu den gegenwärtigen Modellen. Der Return-of-Invest liege dabei im Bereich von 15 Jahren. Haru-Oni sei als Pilotanlage auch trotz der enormen Investitionssummen von 80 Mio. EUR wichtig, um Erfahrungen zu sammeln und Fehler im Aufbau zu erkennen, bevor hochskaliert werde. Herdan geht davon aus, dass es zusätzlich zu Carbon Capture eine CO_2-Infrastruktur geben müsse. HIF baut aktuell weitere Anlagen in Uruguay, den USA und Chile. Methanol als Plattformkraftstoff könne verschickt und am Nutzungsort raffiniert werden. „Alle Produkte daraus sind erneuerbar", so Herdan, „das Endprodukt sollte aber vor Ort hergestellt werden, um die Liefersicherheit aus verschiedenen Regionen zu gewährleisten."

Dr. Alexander Heintzel
Marc Ziegler

Der dogmatische Weg zur E-Mobilität verlagert das Problem

Wie kann der Ausstoß von CO_2 im Verkehrs- und Energiebereich reduziert werden? Worauf es jetzt ankommt, war Thema der Podiumsdiskussion auf dem 11. Internationalen Motorenkongress.

„Ohne Verbrennungsmotor kein Klimaschutz!" Der bereits durch den provokanten Titel der Podiumsdiskussion gesetzten These, dass effektiver Klimaschutz nur unter Einbeziehung von Verbrennungsmotoren und nicht durch deren Verbot möglich ist, stimmten prinzipiell alle Diskutanten zu.

„Der Vorstoß in Europa, dogmatisch in Richtung Elektromobilität zu gehen, verlagert das CO_2-Problem nur", sagte etwa Michael Fleiss, CEO von Aurobay. Denn die notwendige Energie müsse so oder so bereitgestellt werden, und bei einer Knappheit regenerativ erzeugter Energie geschehe dies eben über fossile Energieträger. Daher kann und darf die Klimafreundlichkeit von Fahrzeugen nicht einfach am Auspuff gemessen werden. „Wichtig ist, dass wir defossilisieren nicht dekarbonisieren", stellte Professor Dr. Christian Beidl heraus, der die Diskussion zusammen mit ATZ/MTZ-Chefredakteur Dr. Alexander Heintzel leitete.

E-Mobilität muss organisch wachsen

Nach Ansicht von Takahiro Nagai, Stellvertretender Direktor der New Energy and Industrial Technology Development Organization (NEDO), ist ein abrupter Umstieg von Verbrennungsmotoren auf elektrische Antriebe schon deshalb nicht zielführend, da sie nicht alle Anwendungen abdecken können. Daher müssen seiner Ansicht nach beide Technologien parallel und evolutionär weiterentwickelt werden. Zumal batterieelektrische Fahrzeuge derzeit aufgrund des Strommixes und der energieintensiven Produktion der Batterien noch keinen Vorteil bringen, stellt Michael Fleiss fest: „Der CO_2-Rucksack ist derzeit größer als die Einsparungen in der Nutzungsphase."

Laut Dr. Markus Schwaderlapp, Leiter Forschung & Entwicklung bei der Deutz AG, wird die Herausforderung der Zwischenspeicherung von Energie von der Politik unterschätzt. Die Annahme, dass die Nutzung elektrischer Energie immer effizienter ist, sei falsch, wenn man die Notwendigkeit der Zwischenspeicherung und die effizientere Produktion regenerativer Energie in anderen Teilen der Welt berücksichtigt. Zudem sind die Batterien laut Dr. Monika Griefahn, Vorstandsvorsitzende der eFuel Alliance, aktuell noch nicht auf eine Wiederverwertbarkeit und Recycling ausgelegt. Ein zentraler Aspekt sei es, die derzeit etwa 1,3 Milliarden Bestandsfahrzeuge weltweit miteinzubeziehen.

Das eine tun, ohne das andere zu vernachlässigen

Die Verkehrs- und Energiewende muss sich also auf zwei Säulen stützen: Elektromobilität und regenerativ erzeugte molekulare Energieträger, so Professor Beidl. Wie kann die Entwicklung beschleunigt werden? Hier hilft laut Griefahn ein Blick nach Japan. Dort stehe das Ziel, CO_2 zu senken, über allem, und alles, was dazu beiträgt, ist gut.

Es ist also ein pragmatisches Vorgehen notwendig. „Durch eine einseitige Entscheidung macht sich Europa zum Museum.“ Das betreffe übrigens nicht nur die Automobilindustrie, sondern auch den Maschinenbau. „Wir müssen beides tun, das ist keine Entweder-oder-Entscheidung.“ Der gängige Weg ist also, „dass beide Technologien antreten und entweder eine gewinnt oder beide parallel bestehen bleiben“, ergänzt Fleiss.

„Es wundert nicht, dass derzeit viele gute Entwicklungen aus China kommen und nicht aus Europa“, sagt Benjamin Krieger, Generalsekretär des Europäischen Verbands der Automobilzulieferer (CLEPA). Es fehle den europäischen Akteuren wegen des angedachten Verbots von Verbrennungsmotoren an Perspektive. Daher fordert er eine Zielvorgabe ohne Technikvorgabe. Allerdings müsse man sich auch mit Blick auf die Wettbewerbsfähigkeit selbst hinterfragen und Lösungen bringen, die die Kunden auch wirklich brauchen.

Molekulare Speicher dringend benötigt

Die Notwendigkeit regenerativ erzeugter Kraftstoffe sieht auch Dr. Holger Becker, Mitglied des Deutschen Bundestags und Mitglied im Parlamentskreis „Regenerative Kraftstoffe“. „Wir brauchen natürlich molekulare Speicher, das ist unbestritten, da sich die E-Mobilität für manche Anwendungen nicht eignet.“ Allerdings sei die Frage, in welchem Bereich sie zum Einsatz kommen sollen. Er glaube nicht, dass im Pkw-Bereich die Entwicklung Richtung Elektrifizierung aufzuhalten ist. Zudem bestehe bei politischen Akteuren die Befürchtung, dass die Elektrifizierung im Verkehrsbereich unter einem breiten Einsatz regenerativer Kraftstoffe leide, weil dann wieder eine Diffusion von Investitionen stattfinden könnte, die man sich nicht leisten könne.

Diese Befürchtung allerdings ist in den Augen der übrigen Teilnehmer unbegründet. Die Investitionen in die Elektrifizierung sind längst getätigt, und die Unternehmen sind darauf angewiesen, die neuen Produkte in der Zukunft auch an den Kunden zu bringen.

Die Essenz ist also, „dass wir letztlich alle verfügbaren nachhaltigen Energieträger brauchen, um Verkehr und Energieerzeugung künftig klimafreundlicher darzustellen“, fasst Dr. Alexander Heintzel zusammen. „Dazu müssen alle Beteiligten an einem Strang ziehen, denn nur ein systemisches Vorgehen im gesellschaftlichen Konsens wird letztlich zum Erfolg führen. Wenn wir nicht in Kreisläufen denken – Cradle-to-Cradle – und die Politik das nicht versteht, sind wir zum Scheitern verurteilt, und das wäre jammerschade.“

Thomas Schneider

Conference Report MoKo 24

Combustion Engines for Climate Protection

The 11th edition of the International Engine Congress takes place in Baden-Baden on February 27 and 28, 2024. With 280 participants on site and a further 50 who have joined online.

In his welcoming address, Scientific Director of the Congress and Editor-in-Charge Prof. Peter Gutzmer once again emphasized the urgent need for technology-open development and referred to the ongoing negotiations at European level. Gutzmer emphasized the global relevance of the combustion engine and its role in effective climate protection. He also pointed out that consumers are not prepared to follow the path towards fully electric vehicles. "No technology has ever prevailed due to coercion," said Gutzmer.

In his keynote speech, Prof. Dieter Grebe, Managing Director of AVL List, gave an outlook on the role of the combustion engine in the global energy mix. In his opinion, the almost linear increase in man-made CO_2 emissions over the past 60 years cannot be reversed without measures such as carbon capture. The global fleet of existing vehicles also urgently needs to be considered and upgraded. The use of E20 alone could reduce CO_2 emissions by 10% by 2030. Hydrogen carriers for transportation should ideally require energy for production at the production site and not at the place of use in order to keep the overall footprint as small as possible. Molecular energy storage systems are needed to represent the energy industry, which means that methanol and synthetic fuels are automatically available. Due to CO_2 legislation and in the long term, BEVs and FCEVs are the right way to go, according to Grebe, but the proportion of biogenic fuels in the existing fleet needs to be increased quickly. From a global perspective, the European technology path taken with a sole focus on BEVs is very isolated. Japan is proving to be the antipode here with a clearly technology-open assessment of all possible drive technologies.

Japan Relies on Systemic Solutions

Takahiro Nagai, Deputy Director of NEDO in Japan, picked up the ball directly and presented what he described as Japan's realistic path towards a carbon-neutral society. NEDO serves as an advisory organization for the government. It is focusing on a strong diversification of measures, including the development of a hydrogen infrastructure. Production of synthetic fuels is also to be established by 2028. The aim is to reduce CO_2 emissions from HEV and HD diesel drives by 55 %. Emissions from combustion engines must be reduced with technical progress and synthetic fuels, while at the same time achieving a breakthrough in electric vehicles.

Dr. Monika Griefahn, CEO of the eFuel Alliance, pointed out that e-mobility is an important factor, but cannot be used everywhere. The total global fleet of 1.3 to 1.4 billion vehicles must be operated in a more climate-friendly way: "Every liter of e-fuel is a direct contribution to climate protection." This must be implemented quickly in order to achieve significant economies of scale. Legislation is the sticking point, as the tank-to-wheel debate is still ongoing. "We hope that the new Commission will immediately reopen the discussion on a cradle-to-cradle approach."

At the end of the first session, Thorsten Herdan, CEO of HIF EMEA, described the change in the business model of a synthetic fuel production plant compared to the current models. The return on investment is in the region of 15 years. Haru-Oni is also important as a pilot plant, despite the enormous investment of 80 million euros, in order to gain experience and identify errors in the construction before scaling up. Herdan assumes that there will have to be a CO_2 infrastructure in addition to carbon capture. HIF is currently building further plants in Uruguay, the USA and Chile. Methanol as a platform fuel can be shipped and refined at the point of use. "All products from this are renewable," says Herdan, "but the end product should be produced locally to ensure security of supply from different regions."

Dr. Alexander Heintzel
Marc Ziegler

The Dogmatic Path to E-mobility Shifts the Problem

How can CO_2 emissions be reduced in the transport and energy sectors? What is important now was the topic of the panel discussion at the 11th International Engine Congress.

"No climate protection without the combustion engine!" In principle, all of the panellists agreed with the provocative title of the panel discussion, which stated that effective climate protection is only possible by including combustion engines and not by banning them.

"The push in Europe to move dogmatically towards electromobility is only shifting the CO_2 problem," said Michael Fleiss, CEO of Aurobay, for example. This is because the necessary energy has to be provided one way or another, and if there is a shortage of renewable energy, this is done using fossil fuels. Therefore, the climate friendliness of vehicles cannot and must not simply be measured by the exhaust pipe. "It is important that we defossilize, not decarbonize," emphasized Professor Dr. Christian Beidl, who chaired the discussion together with Editor-in-Chief ATZ | MTZ Dr. Alexander Heintzel.

E-mobility Needs to Grow Organicall

According to Takahiro Nagai, Deputy Director of the New Energy and Industrial Technology Development Organization (NEDO), an abrupt switch from combustion engines to electric drives is not expedient, as they cannot cover all applications. In his opinion, both technologies must therefore be developed in parallel and in an evolutionary manner. Especially as battery electric vehicles are not yet advantageous due to the electricity mix and the energy-intensive production of the batteries, Michael Fleiss notes: "The CO_2 rucksack is currently greater than the savings in the usage phase."

According to Dr. Markus Schwaderlapp, Head of Research & Development at Deutz AG, the challenge of intermediate energy storage is underestimated by politicians. The assumption that the use of electrical energy is always more efficient is wrong when you consider the need for intermediate storage and the more efficient production of renewable energy in other parts of the world. Furthermore, according to Dr. Monika Griefahn, CEO of the eFuel Alliance, batteries are not yet designed for reusability and recycling. A key aspect is to include the approximately 1.3 billion existing vehicles worldwide

Doing One Thing Without Neglecting the Other

The transport and energy transition must therefore be based on two pillars: Electromobility and renewably generated molecular energy sources, according to Professor Beidl. How can development be accelerated? According to Griefahn, a look at Japan can help. There, the goal of reducing CO_2 is paramount, and anything that contributes to this is good. A pragmatic approach is therefore necessary. "By making a unilateral decision, Europe is turning itself into a museum." Incidentally, this applies not only to the automotive industry, but also to mechanical engineering. "We have to do both, it's not an either-or decision." The usual path is therefore "that both technologies compete and either one wins or both continue to exist in parallel," adds Fleiss.

"It is not surprising that many good developments are currently coming from China and not from Europe," says Benjamin Krieger, Secretary General of the European Association of Automotive Suppliers (CLEPA). The European players lack perspective due to the planned ban on combustion engines. He is therefore calling for a target without a technology specification. However, we must also question ourselves in terms of competitiveness and provide solutions that customers really need.

Molecular Memory Urgently Needed

Dr. Holger Becker, Member of the German Bundestag and member of the parliamentary group "Regenerative Fuels", also sees the need for regeneratively produced fuels. "Of course we need molecular storage systems, that is undisputed, as e-mobility is not suitable for some applications." However, the question is in which area they should be used. He does not believe that the trend towards electrification in the passenger car sector can be stopped. There is also a fear among political players that electrification in the transport sector will suffer from the widespread use of renewable fuels, as this could lead to a diffusion of investments that cannot be afforded.

However, this fear is unfounded in the eyes of the other participants. The investments in electrification have long since been made and the companies are dependent on bringing the new products to customers in the future.

The essence is therefore "that we ultimately need all available sustainable energy sources in order to make transport and energy generation more climate-friendly in the future," summarizes Alexander Heintzel. "To achieve this, everyone involved must pull together, because only a systemic approach based on social consensus will ultimately lead to success. If we don't think in cycles – cradle-to-cradle – and politicians don't understand this, we are doomed to failure, and that would be a shame."

Thomas Schneider

Inhaltsverzeichnis

Green Innovation Fund of NEDO toward a Carbon-neutral Future

Takahiro Nagai(✉)

National Research and Development Agency New Energy and Industrial Technology Development Organization (NEDO), Kawasaki City Kanagawa, Japan
nagaitkh01@nedo.go.jp

1 About NEDO

New Energy and Industrial Technology Development Organization (NEDO) is a national research and development agency in Japan that creates innovation by promoting technological development necessary for realization of a sustainable society. NEDO acts as an innovation accelerator to contribute to the resolution of social issues by developing and demonstrating high-risk innovative technologies having practical application.

1.1 NEDO's Missions

NEDO's missions are addressing energy and global environmental problems and enhancing industrial technology. NEDO actively undertakes the development of new energy and energy conservation technologies. It also conducts research to verify technical results. Through these efforts, NEDO promotes greater utilization of new energy and improved energy conservation. NEDO also contributes to a stable energy supply and the resolution of global environmental problems by promoting the demonstration of new energy, energy conservation, and environmental technologies abroad based on the knowledge obtained from domestic projects.

With the aim of raising the level of industrial technology, NEDO pursues research and development of advanced new technology. Drawing on its considerable management expertise, NEDO carries out projects to explore future technology seeds as well as mid- to long-term projects that form the basis of industrial development. It also supports research related to practical application.

1.2 Positioning of NEDO

Contribute to the resolution of social issues, NEDO formulates technology strategies and project plans and, as part of its project management, establishes project implementation frameworks by combining the capabilities of industry, academia, and

A. Heintzel (Hrsg.): Int.Engine.Congr. 2024, Proceedings, S. 1–11, 2025.
https://doi.org/10.1007/978-3-658-47966-4_1

government. NEDO also promotes technology development by carrying out, evaluating, and allocating funding to promising projects to accelerate the practical application of project results (Fig. 1).

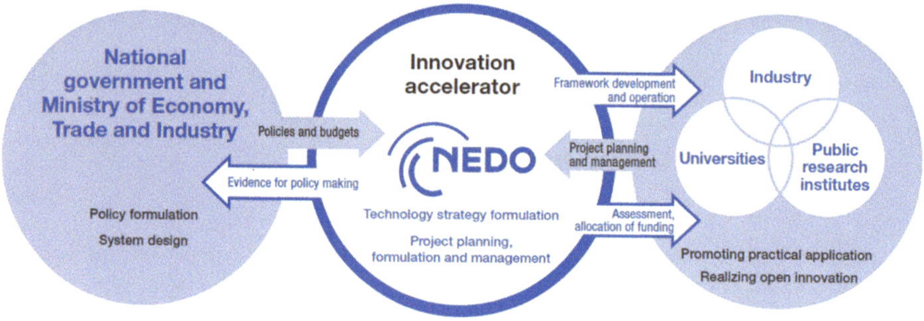

Fig. 1. Positioning of NEDO

1.3 Background Information of NEDO

- Foundation
  Originally established on October 1, 1980, reorganized as an incorporated administrative agency on October 1, 2003.
- Foundation Purpose
  The purpose of NEDO is to enhance industrial technology and promote commercialization by comprehensively performing functions such as promoting research and development carried out using skills from the private sector.
  Promoting research and development carried out by the private sector about technology for non-fossil energies, combustible natural gas, and coal.
  Promoting the technology required for the rational use of energy and technology in mining and industry; and promoting the utilization of such technology in cooperation with the international community.
- Details of Major Operations
  Operations relating to research and development management of national projects and practical application promotion activities.
- Minister in Charge
  Minister of Economy, Trade and Industry (METI)
- Personnel 1,464 (as of April 1, 2023)
- Budget Approximately 1.14 billion US dollars (initial budget for FY 2023)
  In addition to the above, the following programs will be funded and conducted as publicly solicited research and development projects.
  - Moonshot Research and Development Program 196 million US dollars.
  - Research and Development Project of the Enhanced Infrastructures for Post-5G Information and Communication Systems 5.97 billion US dollars.
  - Green Innovation Fund Projects 17.2 billion US dollars.

- Program for Developing Important Economic Security Technologies 1.87 billion US dollars.
- Programs for Specified Semiconductor Production-Related Development 8.02 billion US dollars.
- Support Project to Secure a Stable Supply of Specified Important Goods 6.23 billion US dollars.
- Deep-Tech Startups Support Program 752 million US dollars.

2 Japan's Strategy toward Carbon Neutrality by 2050

In October 2020, Japan declared that it aims to achieve carbon neutrality by 2050, with the goal of reducing overall greenhouse gas emissions to zero by this year. Carbon neutrality by 2050 cannot be realized through ordinary efforts. It is necessary to significantly accelerate efforts toward structural changes in the energy and industrial sectors, undertake bold investment for innovation. For this reason, the Ministry of Economy, Trade and Industry (METI), in collaboration with other ministries and agencies, formulated the "Green Growth Strategy through Achieving Carbon Neutrality in 2050."

- 2015 Adoption of the "Paris Agreement"
- 2019 Cabinet Decision on "Long-Term Strategy as a Growth Strategy Based on the Paris Agreement" Formulation of "Carbon Recycling Technology Roadmap" Formulation of a technology roadmap for carbon recycling that uses CO_2 as fuel and raw materials.
- 2020 Jan. Formulation of the "Revolutionary Environmental Innovation Strategy" Oct. "Realization of a carbon-neutral, decarbonized society in 2050" declared by Prime Minister Suga
- 2020 Dec. Formulation of the Green Growth Strategy for Carbon Neutrality in 2050 An action plan that clearly states current issues and future efforts for each of the fourteen important areas. "Green Innovation Fund" established at NEDO.
- 2021 Apr. "Green Innovation Fund" project started.

3 Green Innovation Fund Projects

The strategy specifies fourteen promising areas expected to demonstrate future growth and provides action plans for each of these areas from the viewpoints of both industrial and energy policies. The strategy reflects Japan's determination to pursue ambitious goals while highlighting to the greatest extent possible realistic pathways toward their achievement (Figs. 2 und 3).

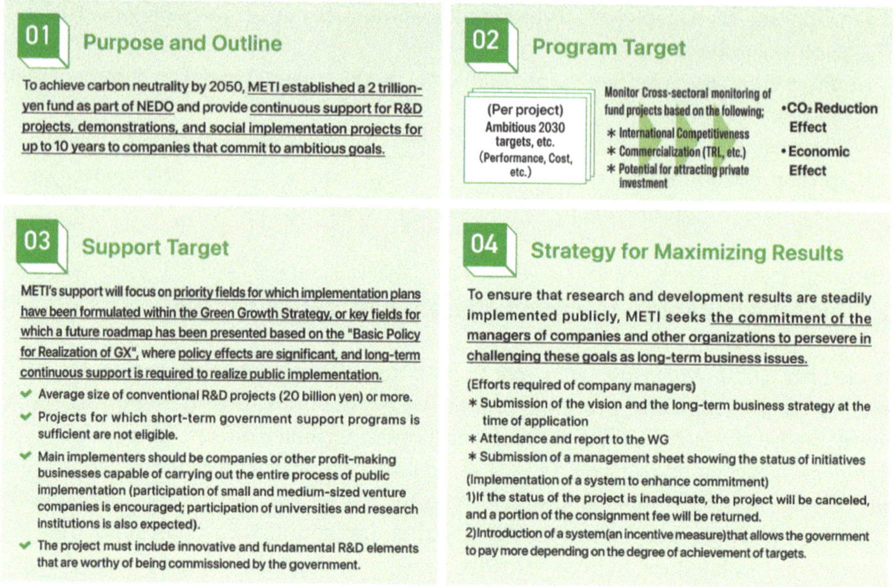

Fig. 2. Basic policies for green innovation fund

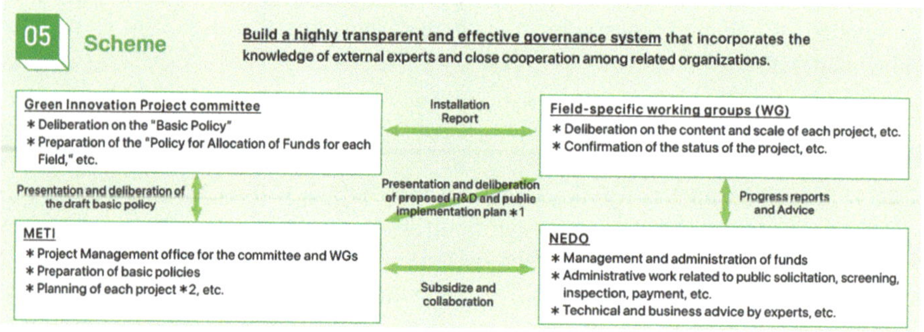

Fig. 3. Scheme for green innovation fund

The Green Innovation Fund Projects, carried out based on ambitious goals shared by the public and private sectors, aims to provide continuous support to companies and other organizations committed to addressing such goals as part of their business activities. Such support, available for up to a maximum of ten years, ranges from research and development (R&D) activities and demonstrations to social implementation of project outcomes (Fig. 4).

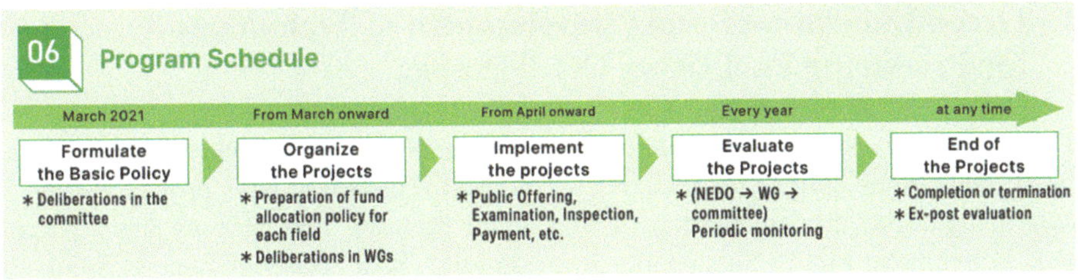

Fig. 4. Program schedule of green innovation fund

Moreover, by leveraging government funding, this program is intended to serve as an impetus encouraging private sector investment in R&D activities and infrastructure development, and attract global ESG-related funding, estimated to total thirty-five trillion dollars, to Japan. Through such efforts, Japan aims to realize carbon neutrality by 2050.

Moreover, to support the forward-looking efforts of companies aiming to realize these goals, many kinds of policy measures are being mobilized. As part of this effort, a two trillion yen* fund has been established at NEDO in March 2021 to launch the Green Innovation Fund Projects (Fig. 5).

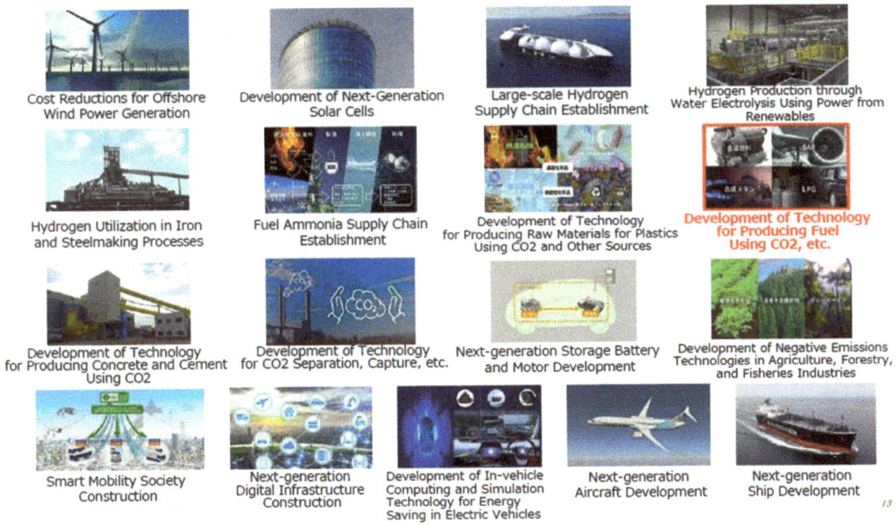

Fig. 5. Green innovation fund projects

4 Green Innovation Fund "Development of Technology for Producing Fuel Using CO_2" Project

It is necessary to promote the development of technology for carbon recycling Fuel as one of the various options for realizing a decarbonized society. This project will work toward the social implementation of two liquid fuel and two gaseous fuels. Liquid fuel; 1) Synthetic fuel 2) Sustainable Aviation Fuel (SAF) Gaseous fuel; 3) Synthetic methane 4) Green LPG.

4.1 Positioning of Synthetic Fuel in Carbon Neutral Policy

- 2019 Cabinet Decision on "Long-Term Strategy as a Growth Strategy Based on the Paris Agreement" Formulation of "Carbon Recycling Technology Roadmap"
- 2020 Jan. Formulation of the "Revolutionary Environmental Innovation Strategy"
- 2020 Oct. "Realization of a carbon-neutral, decarbonized society in 2050" declared by Prime Minister
- 2020 Dec. Formulation of the Green Growth Strategy for Carbon Neutrality in 2050. An action plan that clearly states current issues and future efforts for each of the fourteen key areas. Synthetic fuel are specified for the first time in the position of carbon neutralization of fuels in the automotive and storage battery industries.
- 2021 Apr. Synthetic fuel study group interim report by METI (Fig. 6)

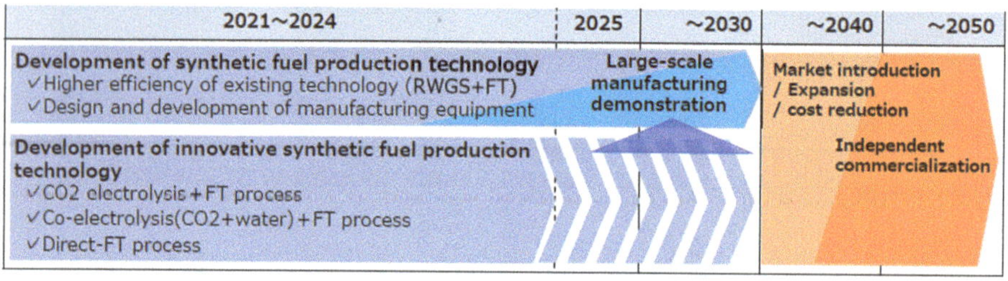

Fig. 6. Carbon Neutralization of Fuel

- Consider expanding support for the current Green Innovation fund project (highly efficient large-scale FT synthesis process).
- Considering support for capital investment by business operators who attempt early supply using existing technologies, and for demonstrations aimed at establishing business models.
- Promoting cooperation with other countries and the development of information platforms.

4.2 Overview of Synthetic Fuel

Synthetic fuel production (conventional technology). Material: H2 (produced from renewable energy-derived power) and CO_2. Conversion to fuel: FT reaction (Conversion to raw material syngas: RWGS). New technology attracting attention. Conversion to syngas: Electrolysis using renewable energy-derived power. Co-electrolysis (CO_2 + water), CO_2 electrolysis. Conversion to fuel: Direct-FT reaction (Fig. 7).

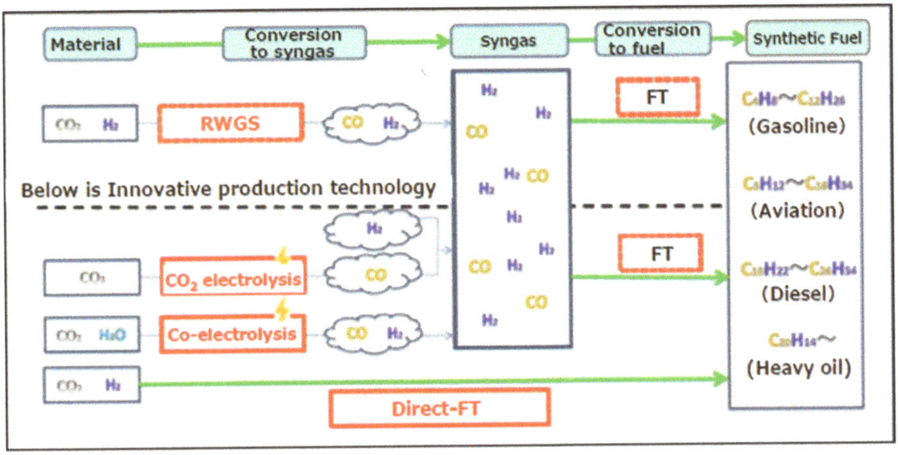

Fig. 7. Synthetic fuel production methods

5 Green Innovation Fund of Synthetic fuel Production and Utilization at NEDO

5.1 R&D Activities of Green Innovation Fund project

Promoting the development of integrated production process technologies to produce Synthetic fuel from CO_2 and hydrogen as part of the following projects. One is to improve production yield and mass production of Synthetic fuel. New technologies will be developed to expand the utilization of Synthetic fuel on a large scale based on conventional technologies. Other project is to promote comprehensive technology development to reduce CO_2 emissions from internal combustion engine vehicles.

5.2 Green Innovation Fund Projects

1. Synthetic fuel production technology
   Development of a high-efficiency production process for liquid fuels from CO_2 with synthetic reaction by ENEOS Corporation (Fig. 8).

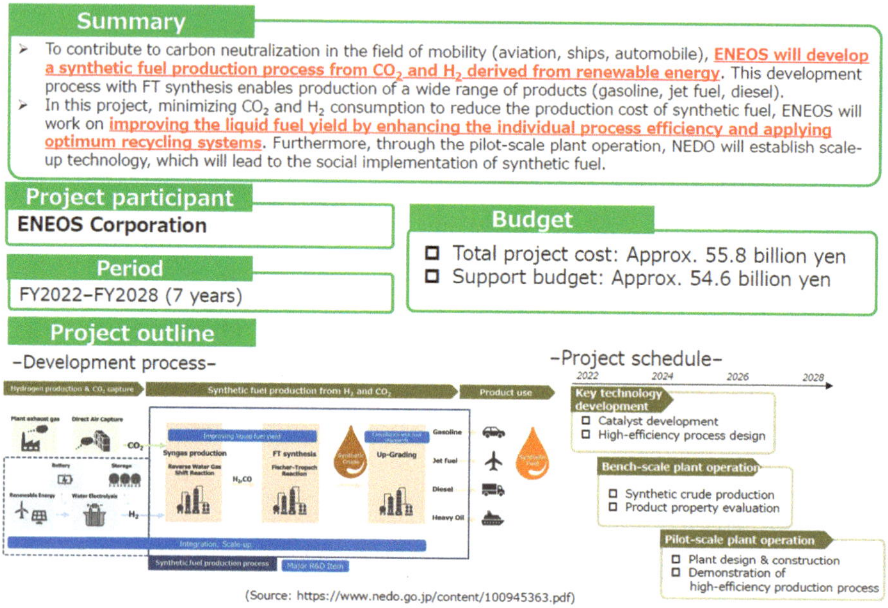

Fig. 8. Synthetic fuel production project of green innovation funds

2. Synthetic fuel utilization technology
   Development of technology to increase the efficiency of synthetic fuel utilization and solve the reciprocal phenomenon in passenger cars and heavy-duty vehicles by Research Association of Automotive Internal Combustion Engines (AICE) (Figs. 9 und 10).

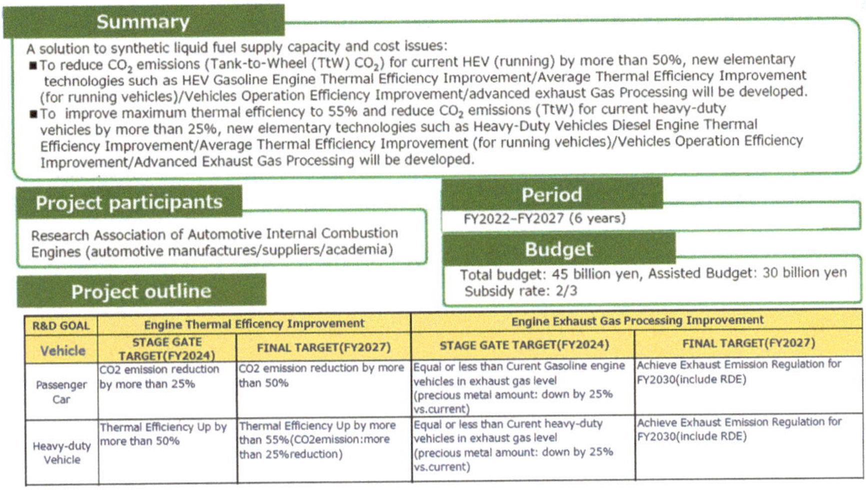
Summary

A solution to synthetic liquid fuel supply capacity and cost issues:

- To reduce CO_2 emissions (Tank-to-Wheel (TtW) CO_2) for current HEV (running) by more than 50%, new elementary technologies such as HEV Gasoline Engine Thermal Efficiency Improvement/Average Thermal Efficiency Improvement (for running vehicles)/Vehicles Operation Efficiency Improvement/advanced exhaust Gas Processing will be developed.
- To improve maximum thermal efficiency to 55% and reduce CO_2 emissions (TtW) for current heavy-duty vehicles by more than 25%, new elementary technologies such as Heavy-Duty Vehicles Diesel Engine Thermal Efficiency Improvement/Average Thermal Efficiency Improvement (for running vehicles)/Vehicles Operation Efficiency Improvement/Advanced Exhaust Gas Processing will be developed.

Project participants

Research Association of Automotive Internal Combustion Engines (automotive manufactures/suppliers/academia)

Period

FY2022–FY2027 (6 years)

Budget

Total budget: 45 billion yen, Assisted Budget: 30 billion yen
Subsidy rate: 2/3

Project outline

R&D GOAL	Engine Thermal Efficency Improvement		Engine Exhaust Gas Processing Improvement	
Vehicle	STAGE GATE TARGET(FY2024)	FINAL TARGET(FY2027)	STAGE GATE TARGET(FY2024)	FINAL TARGET(FY2027)
Passenger Car	CO2 emission reduction by more than 25%	CO2 emission reduction by more than 50%	Equal or less than Curent Gasoline engine vehicles in exhaust gas level (precious metal amount: down by 25% vs.current)	Achieve Exhaust Emission Regulation for FY2030(include RDE)
Heavy-duty Vehicle	Thermal Efficiency Up by more than 50%	Thermal Efficiency Up by more than 55%(CO2emission:more than 25%reduction)	Equal or less than Current heavy-duty vehicles in exhaust gas level (precious metal amount: down by 25% vs.current)	Achieve Exhaust Emission Regulation for FY2030(include RDE)

Fig. 9. Synthetic fuel utilization project of green innovation funds

Fig. 10. About research Association of Automotive Internal Combustion Engines (AICE)

6 Next Step and Conclusion

6.1 Roadmap of Production and Utilization Promotion of Synthetic Fuel

In case of the introduction of Synthetic fuel, the compatibility between fuel quality and engines performance needs to be confirmed and verified. Synthetic fuel is more environmentally valued in comparison with conventional fossil fuels, but on the other hand, a business scheme and model of supply-chain need to be established because it can be predicted that the amount of Synthetic fuel supply will be small, and the price will also be higher at the beginning of the launch of Synthetic fuel (Fig. 11).

Fig. 11. ENEOS and TOYOTA joint performance event on May 28, 2023

Information regarding demonstration tests of engines with Synthetic fuel, the study of synthetic fuel business schemes and models, and social demonstration need to be shared (Fig. 12).

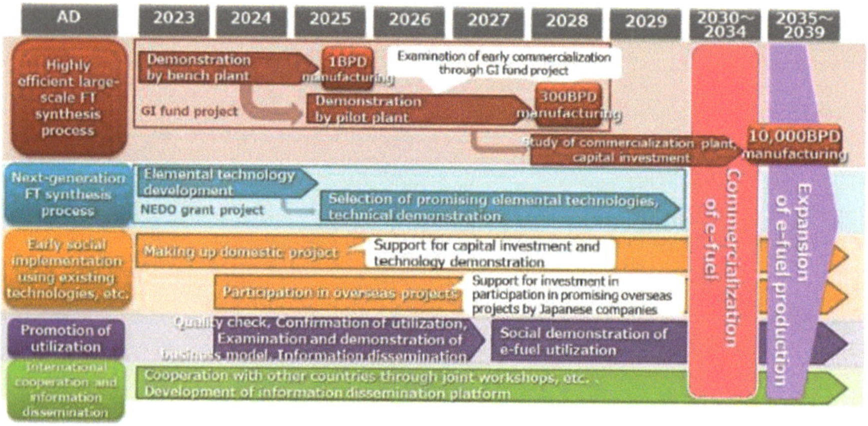

Fig. 12. Policy trends: Introduction roadmap of Synthetic fuel

6.2 Conclusion

Policy Trends in Japan towards Carbon Neutrality in 2050 is after declaring "Realization of a carbon-neutral and decarbonized society by 2050", the "Green Growth Strategy for Carbon Neutrality in 2050 has been formulated, and various measures have been implemented to realize it.

In the Green Innovation Fund Project, which is one of the measures, technology development that contributes to future carbon neutrality, such as synthetic fuels, is underway.

NEDO aims to realize a carbon-neutral and sustainable society by promoting necessary technological development and creating innovation through the Green Innovation Fund Projects.

References

1. NEDO Publications.: Profile of NEDO 2023 (2023).
2. NEDO Publications.: Green Innovation Fund Projects (2023).
3. METI Homepage, https://www.enecho.meti.go.jp/en/category/special/article/detail_168.html, last accessed 2023/12/21.
4. ENEOS Homepage, https://www.eneos.co.jp/newsrelease/upload_pdf/20230529_01_01_1080097.pdf, last accessed 2023/12/21.

Renewable Electricity in the Tank—Fact or Fiction?

Thorsten Herdan(✉) und Dorothea Nold

HIF EMEA GmbH, Berlin, Germany
{thorsten.herdan,dorothea.nold}@hifglobal.com

1 HIF's Project Development Activities Across the Globe

1.1 From Power Point to Power Plant

When you see numbers floating around in the eFuels market, litres or tonnes that will be available at a certain cost in 2030, be careful. So far, they're all power points. To date, no industrial-scale eGasoline project has been signed. Most project developments worldwide are at the pre-feasibility or feasibility stage. And as long as no customer is willing to participate further in this endeavour, we can all go home and take a deep nap.

Well, maybe not. We—that is my company HIF and myself—want to make it happen. Together with Porsche as offtaker, and other partners we have been able to build the first ever eGasoline production with the energy stemming purely from wind.

The volumes produced in this demonstrator are just a drop in the ocean, one might say. And rightfully so. This can just be the start. This project is a showcase to demonstrate that what we are talking about is real. And more importantly, it helps us as a company to learn and grow every day. The learnings we have from operating our demonstration plant flow directly into our global project development activities.

Our HIF Project Development Activities Span the globe. We want to be first and fastest to market. We aim to be the world's largest eFuel supplier. Next to our demonstration facility in the South of Chile, Magallanes, we have leased land that can host up to 20 GW of wind power. The first phase currently in development in Chile, with around 280 MWof wind power, that could produce up to 75 million litres per year. Assuming no efficiency improvements and no new technology developments, we could theoretically produce up to 4 billion litres of e-gasoline per year with 20 GW of wind power.

A. Heintzel (Hrsg.): Int.Engine.Congr. 2024, Proceedings, S. 12–18, 2025.
https://doi.org/10.1007/978-3-658-47966-4_2

Let us compare that number of eFuels needed annually in agriculture and special vehicles, that cannot be easily electrified. The BENIVER study, commissioned by the German Ministry for Energy and Climate and published in December 2023[1], found that a total of around 4 billion litres of fuel are currently used annually for agriculture and special vehicles[2], that cannot easily be electrified. This represents a substantial 6% of Germany's transport sector's yearly fuel consumption, which amounted to around 70 billion liters of fuel in 2019.

Our Most Advanced Project Is in USA, Texas. With the Inflation Reduction Act, the US is showing how to encourage the development of a new industry. We aim to produce 1.4 million tonnes of eMethanol before 2030 with our first US project. This fuel is targeted at the marine sector, where for the first time mandatory GHG reduction targets are being developed or are already in place, such as in the EU. A second US phase is already in development, targeting eSAF (Sustainable Aviation Fuel) for the aviation sector.

In Australia, we are in the engineering phase (pre-FEED) of our first industrial scale project. Our target market is Japan, which has limited electrification options and has taken a fresh look at eFuels to defossilize its transport sector. In our second and third phase, we will be able to produce either eMethanol or eGasoline, depending on customer demand.

In Uruguay, we successfully won a public tender for sustainable biogenic CO_2 as one main feedstock of eFuels. We are currently planning to go in phases to produce approximately 300 million liters per year. Again, the endproduct will depend on the customer. Either it can be used as shipping fuel, or further processed to eGasoline or eSAF.

In the Middle East and Africa, we are currently evaluating different project opportunities for an industrial scale project. Europe would be the natural offtaker market, due to its vicinity.

1.2 Key Ingredients for Competitive eFuels

As we have global capacities of eMethanol at different places, our strategy involves to be present at every place on earth that has the main ingredients for competitive eFuels: cheap electricity with abundance of wind, sun and land, sustainable CO_2, and low capital costs.

Cheap Renewable Electricity Coupled with land form the basis for competitive eFuels. Self explanatory, Europe itself, apart from a few areas in northern Europe, is not ideal as a production location. As an industrial powerhouse, blessed with skilled

[1] BEniVer, Roadmap für strombasierte Kraftstoffe, 2023, https://dx.doi.org/10.5281/zenodo.10208039.

[2] BEniVer defines "special vehicles" as vehicles for various niche applications such as the German Armed Forces, rescue vehicles, construction machinery and mining vehicles as well as other sectors, in particular agriculture, scheduled and long-distance coaches and rail transport.

people but little wind and sun, it needs to use renewable electricity directly and as efficiently as possible. In addition, there is not much land available to accommodate the hundreds of thousands of wind turbines needed to achieve the required output. In Chile, because of the constant wind, our output is over three times as much as it would be for the same production facility if it were located in Germany.

Availability of Sustainable CO_2 Sources Is a prerequisite for eFuels. Our objective is to test the Direct-Air-Capture technology in our demonstration plant and from there, upscale it in our first projects until it is fully commercially available. This year, we will install a proof of concept of a Direct Air Capture Unit developed by the Porsche AG at our demonstration plant Haru Oni in the South of Chile. We are confident that we will be able to use Direct Air Capture to competitive prices within this decade. Until then, we will focus on sustainable biogenic CO_2 sources and recycled industrial CO_2.

The Price of Capital Is one of the most important and often underlooked factors influencing eFuel prices and production locations. eFuels are capital intensive to produce. The energy is free, it just needs to be harnessed from the sun and wind. So all the investment has to be made upfront. This is a departure from the conventional fuel market dynamic. Unlike the traditional pay-as-you-go model, the eFuel landscape requires a long-term commitment in the form of off-take agreements of 15 years or more. This is a 180degree turnaround for the conventional fuel market.

Recognising the central role of capital in the eFuel equation is essential to understanding the dynamics of the industry. It also highlights the need for stakeholders to adopt a forward-thinking mindset, characterised by sustained commitment and strategic financial planning, in order to usher in the market development of this sustainable energy source.

2 Key Challenges: It's Political and it is You, our Clients

2.1 Indirect Electrification as a Game Changer

As the world seeks sustainable alternatives to traditional fossil fuels, indirect electrification emerges as a pivotal strategy with far-reaching implications for the automotive sector, and many others. Indirect electrification addresses the limitations of direct electrification, particularly in hard-to abate sectors and sectors where high energy density or specific energy requirements are critical, such as maritime and aviation sector, but also agriculture and construction machinery, emergency response, the chemical sector and the military.

Indirect electrification allows us and many others to keep the combustion process going without adding CO_2 to the atmosphere. In Chile, we use to say "we are bottling the wind".

The Technology is there We use electricity to produce of eFuels. With our Haru Oni demonstration project, HIF is providing living proof that it can be done. So what is missing? Why do we not see projects like mushrooms coming out of the ground?

2.2 Political Barriers to eFuels

The European Union aims to become a world leader in the production and use of green hydrogen and eFuels such as eMethanol, eGasoline or Sustainable Aviation Fuel. In 2023, it has agreed an ambitious target of 42.5% renewable energy in all sectors by 2030 and has explicitly stated that a significant part of this will have to be met by imports of green hydrogen and its derivatives. With specific regulations for shipping and aviation, and the extension of the EU ETS to these sectors, it will slowly force these industries to reduce their greenhouse gas emissions over time. This will lead to a switch to low-carbon fuels such as eFuels in the medium and long term.

The success of this regulatory story will depend on the availability and competitiveness of these eFuels in 2030. Importating these eFuels into Europe from competitive production locations such as Chile or the MENA region is a necessity to complement limited European production. However, the European Union (EU) faces a significant challenge in sourcing significant volumes of efuels, which are critical to meeting climate change targets, with the regulatory landscape enacted in 2023.,

The Certification of a "green" eFuels Or RFNBO in EU terminology, faces unseen obstacles in third countries within the confines of Brussels bureaucracy. Unknown to many future customers of eFuels, the EU Commission has defined the production criteria for eFuels in the EU and globally.[3] Any fuel produced globally must meet these principles. The challenge, and the main stumbling block, is that these production criteria are rooted in the concepts and ideas of the EU energy market.

The main regulatory barriers to the production and certification of"green" eFuels in third countries dedicated to the EU market are as follows:

Electricity Subsides will make your EFuel as bad as a Fossil based Fuel —you loose your green attribute. The consequence is dire. All green hydrogen and eFuels produced in the USA will not be certified as "green" by the EU. The heavily subsidized US eFuels will go to other markets. This regulatory icing on the cake has its roots in the European energy and electricity market. In Germany, for example, renewable electricity is subsidized by the state as a means of planning and controlling the further uptake of renewable electricity. It is used as a tool to control the amount of renewable electricity needed to meet the EU's renewable energy targets. However, it simply does not make sense to transfer this EU approach to other markets around the world. Fuels from renewable electricity are green—whether the electricity was subsidized or not. To ensure that this electricity is not "double counted", the concept of rules of origin can be used. These ensure that the "renewability" of the electricity is attribute to only one product—either the electricity or the eFuel.

Recycled Carbon Emission from industrial Point Sources cannot be used Outside of the EU The EU regulation is twisted in such a way, that no project finance is possible with recycled industrial CO_2, including non-abatement emissions from ce-

[3] Delegated Acts on Articles 27 and 28 of the Renewable Energy Directive 2018/2001 (RED II).

ment. The EU requires a national "effective carbon pricing mechanism" for imports of RFNBO with industrial CO_2, the conformity of which must be confirmed by the EU Commission. However, the assessment criteria for "effectiveness" cannot be derived with legal certainty, and such a pricing mechanism comparable to the EU ETS does not exist anywhere else in the world. More importantly, the EU will re-certify this "effectiveness" every year the plant is in operation.

This means that the certification of your product essentially depends on whether the EU Commission likes the national carbon pricing policy of any nation producing green fuels. Under these conditions, no investment will take place. Neither a bank nor a customer would take the risk, that in the production location the national government needs to uphold an effective carbon pricing mechanism over the lifetime of the plant.

The result is a powerful import barrier, particularly for all products, in the MENA region, where there are few sustainable biogenic sources.

Sustainable biogenic CO_2 Sources are currently the only option where regulatory clarity is strong enough to allow project financing outside the EU. However, even in this case there are many barriers and challenges associated with certifying the sustainability of your biomass source. The EU does not accept internationally recognized sustainability certificates and requires additional audits under EU certification schemes. There are also other conditions attached to the use of biogenic CO_2 sources that need to be further clarified by the EU Commission. Otherwise, project developers will wait or look to other markets.

What is the EU's Motivation Behind this regulatory set-up, one might wonder. Several factors contribute to this regulatory conundrum. First, climate policy considerations that emphasize strict definitions of recycled CO_2 have influenced these regulations. These considerations are valid, reducing CO_2 emissions is a priority for all of us. Therefore, the use of recycled industrial point sources should be clearly limited. However, an indirect total ban, as is the case today, does not help the climate. It blocks the testing and ramp-up of sustainable alternatives such as the Direct Air Capture technology and results in more fossil emissions from the burning of conventional fuels.

Second, these regulations have been drafted in relation to EU laws and regulations. No impact assessment has been carried out to determine whether the requirements laid down are appropriate for other energy market environments such as the US, Latin America or Africa and the MENA countries.

And third, misguided industrial policy interests to protect European producers from foreign competition may have played a role. From a European perspective, we may have protected the wrong industry. Our economy and our way of life depend on cheap energy. If our airlines, shipowners, logistics companies, agricultural sector and consumers have to cope with overpriced eFuels—what can we gain from a handful of industrial scale eFuels projects in the EU?

The result is a prohibitive regulatory framework that denies a "green certificate" to eFuels produced in third countries entering the European single market. The unintended consequence is that the attractiveness of the EU as an eFuels market cannot contribute to the rapid expansion of the global eFuels economy.

In the face of these challenges, it is imperative that the current regulatory framework be adapted as soon as possible. Putting a band-aid on the regulatory wound is not enough. The regulatory wound needs to be operated on. Aligning climate policy with the broader goal of promoting a global sustainable eFuels economy is essential to overcome these barriers. In addition, a nuanced approach to industrial policy that meets the needs of European technology providers as exporters and European eFuels customers as importers is crucial.

2.3 You, our Clients

The imperative for change within the automotive industry extends beyond mere sales aspirations for BEVs. The responsibility lies also on the shoulders of automotive companies to actively engage in mitigating climate change. The profit-centric approach often adopted by automotive companies is themselves a barrier to the uptake of eFuels.

eFuels are more expensive than conventional fuels. The first litres of a new technology are always the most expensive. If this market is to take off, there must be first movers willing to bear the higher price and able to bring the first projects to market. The car industry should be one of them.

Car manufacturers are compelled to address the pressing concerns associated with existing vehicle fleets they have brought into the market. Rather than shifting blame onto others, the industry must take ownership of its role in climate protection, marked by public scandals and a growing demand for sustainable alternatives.

By prioritizing sustainability over immediate financial gains, the public claims for taking climate responsibility could become reality. Public adherence to climate goals must extend beyond new sales of net-zero" battery electric and address the existing fleet's contribution to air pollution and climate change.

We propose several key actions that industry leaders can undertake:

Create Demand and Drive Market Momentum Industry players should take the lead in creating demand for sustainable fuels by securing long-term procurement contracts for the first wave of large-scale industrial projects. This includes exploring options such as the use of eFuels for internal fuel needs to stimulate market growth.

It is the automotive industry itself that can get eFuels to market for their own demand:

- First factory fill
- Own fleet, i.e. company cars
- Logistics fleet
- Development and Testing
- Racing Series
- Press Events

The impact of these examples on the take-up of the first e-fuels in the EU should not be underestimated. Consider the production volume of vehicles in the year 2023 in Germany, approximately 4 million cars. If all these vehicles were initially fueled with eFuels before leaving the factory, it would amount to 40 million liters per year (assuming a 20-liter initial fill-up with 50% eFuels blend). This example illustrates the substantial impact that adopting eFuels in the automotive industry could have on annual fuel consumption. This volume of 40 million liters annually would be the production volume of over half of our first industrial scale eFuels project in Chile. This shows that these volumes can enable the first eFuels projects to come to market and be a kick-start to the whole industry.

Collaboration for Industry-Wide Impact Organizations like VDA and ACEA must unite to collectively champion the cause of indirect electrification. All automotive manufacturers, regardless of their individual successes, share a collective responsibility to foster change and reduce their climate impact.

For a genuine commitment to addressing climate change, automotive companies must initiate a comprehensive transformation of their existing fleets, actively reducing the carbon footprint associated with vehicles already in circulation.

The VDA could act as a catalyst and coordinator for the industry, i.e. to create sufficient voluntary demand for initial projects that could have a real impact on the CO_2 emissions of the existing fleet that this industry has brought into the world of transport.

Now is the time by the automotive industry to move beyond rhetoric and embrace tangible actions.

The New V8 Powertrains in the Porsche Cayenne

Thomas Wasserbäch(✉)

Dr.-Ing. h.c. F. Porsche AG, Weissach, Deutschland
thomas.wasserbaech@porsche.de

1 Introduction

In 2002, Porsche implemented the vision of an off-road vehicle with sporty driving characteristics and outstanding off-road capabilities combined with everyday practicality for the first time in the newly created Cayenne model series—the sports car in the SUV segment was born.

In the more than 20-year success story of the Cayenne, the vehicle and the powertrain have been further developed and continuously improved. Always with the target of offering the customer a unique driving experience with increased performance and efficiency while implementing the globally applicable emission regulations. As in the first generation of the Cayenne, a V8 engine is the core of the powertrain in the top range drivetrain in the new generation. While the first Cayenne in 2002 was powered by a 4.5-litre V8 engine with 250 kW as a naturally aspirated engine and 331 kW as a turbocharged variant, the new electrified V8 powertrain now has a total system output of 554 kW and 950 Nm, making it the most powerful Cayenne of all time. In addition, the 4.0-litre twin-turbo V8 engine engine is used as a conventional powertrain in the Cayenne S and Cayenne S Coupé models. The powertrain portfolio of the new Cayenne is rounded off by the revised turbocharged 3.0-litre V6 turbo engine, which provides an output of 260 kW as a conventional powertrain and a system output of up to 382 kW in the hybridised powertrains (Fig. 1).

A. Heintzel (Hrsg.): Int.Engine.Congr. 2024, Proceedings, S. 19–35, 2025.
https://doi.org/10.1007/978-3-658-47966-4_3

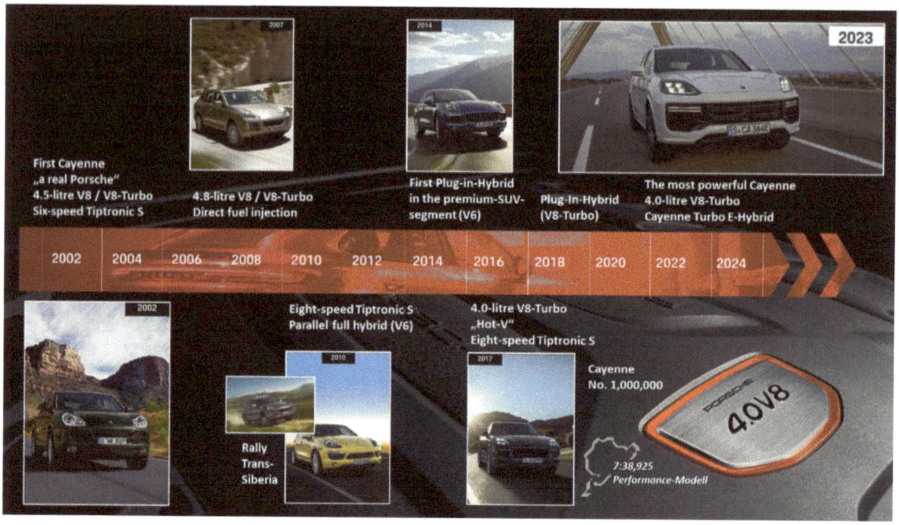

Fig. 1. Milestones in the powertrain development of the Porsche Cayenne

2 The Powertrain of the New Porsche Cayenne

2.1 Overview

Even in the first generation of the Porsche Cayenne, the new developed 4.5-litre V8 engine, which was used as a naturally aspirated and turbocharged engine in combination with a six-speed Tiptronic S transmission and all-wheel drive, was at the centre of the powertrain. This powertrain has been consistently developed over three generations of the vehicle. In doing so, it was necessary to ensure the implementation of far-reaching goals, such as excellent off-road capabilities while increasing sportiness or maintaining everyday usability and comfort requirements with extensive measures to increase the efficiency of the powertrain. And, of course, for worldwide use (Fig. 2).

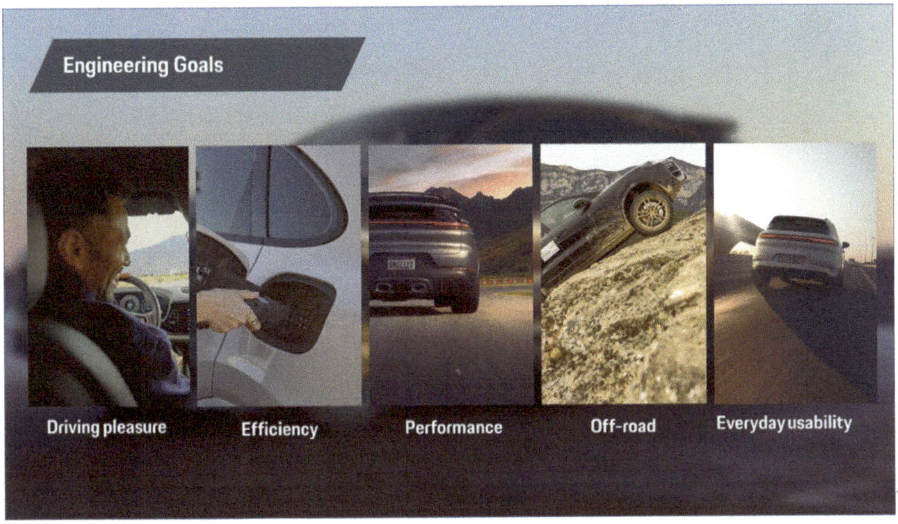

Fig. 2. Engineering goals in the powertrain development of the Cayenne

The result is the new powertrain of the Cayenne with

- an extensively reworked 4.0-litre twin-turbo V8 engine
- an enhanced eight-speed Tiptronic S transmission
- a new electric motor with a newly developed high-voltage battery
- an optimised all-wheel drive system (Fig. 3)

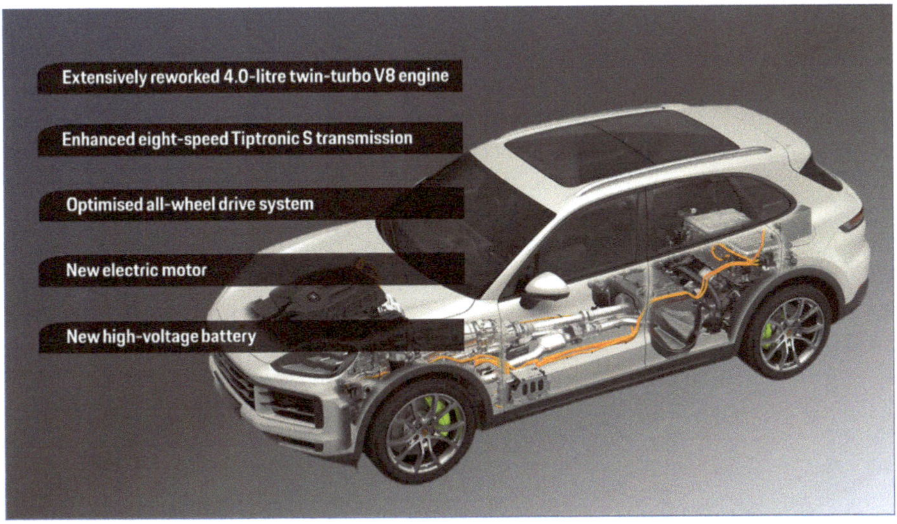

Fig. 3. The new V8 powertrain and its components

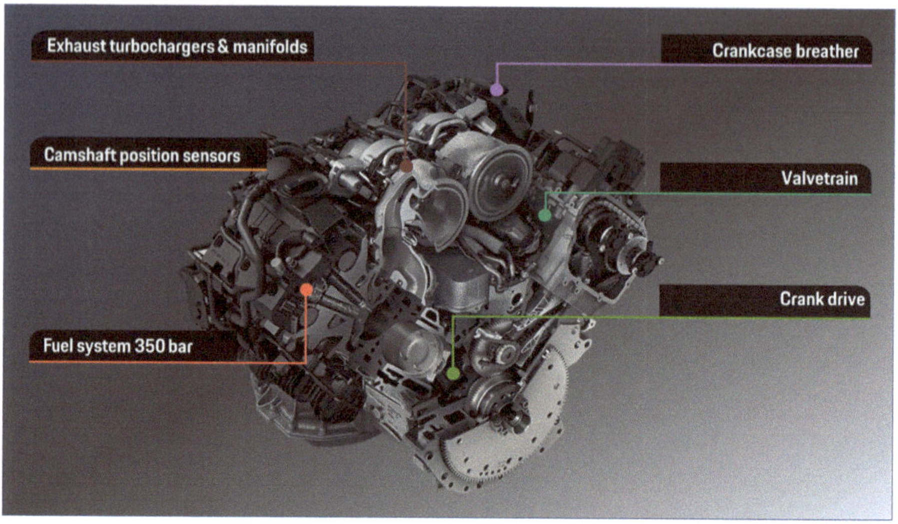

Fig. 4. Technical focus of the new 4.0-litre twin-turbo V8 engine

2.2 V8 Twin-turbo Engine

Engine Concept The new turbocharged V8 engine is an extensively reworked engine, based on the engine introduced in 2016 and already used successfully in several model series. The main engineering goal of the further development was to meet the worldwide stricter becoming emission regulations and those that are expected in the future, without compromises in the typical Porsche driving performance requirements. Also, the focus was on further efficiency optimisation and an adapted design for hybridized powertrains.

The basic engine architecture of the predecessor with the arrangement of the exhaust manifolds and turbochargers inside the cylinder V and its compact design provided an excellent foundation for the further development of the engine. The key design points with displacement, stroke/bore ratio, valve angle/diameter and the typical Porsche firing sequence could be retained. The basic combustion chamber design with the central injector position and the design of the inlet duct routing for a tumble-assisted combustion process with intensive charge movement also proved to be expedient and were retained. Thanks to the extended electrification of the drive trains, the mechanical vacuum pump could be omitted and a weight optimisation of 4.5 kilograms could be implemented compared to the predecessor engine (Fig. 4).

Significant changes were made to the system components relevant for the implementation of the thermodynamic concept

- Base engine components/crank drive
- Valvetrain and charge exchange control
- Camshaft position determination
- Turbocharging system
- Engine Ventilation

in order to meet the development targets set in terms of operational behaviour, performance and emission reduction. In addition, further optimisations for future legal and emission requirements have been integrated into the basic engine architecture, such as the possibility for a secondary air system. The main motor parameters are shown in Table 1.

Table 1 Engine parameters

Designation	Unit	Value
Engine design		V-petrol engine, 90° bank angle
Number of cylinders		8
Number of valves/cylinders		4
Ignition sequence		1-3-7-2-6-5-4-8
Cubic capacity	[cm^3]	3.996
Stroke/Bore	[mm]	86/86
Cylinder spacing	[mm]	93
Main Bearing/Connecting Rod Bearing Diameter	[mm]	65/56
Compression ratio	[ε]	9,7
Valve lift inlet large/small	[mm]	10,5/6
Valve lift outlet cyl. 1, 2, 6, 8/3, 4, 5, 7	[mm]	9/10
Inlet valve disc diameter	[mm]	32
Exhaust valve disc diameter	[mm]	28
Inlet Valve Angle	[°]	23,6
Exhaust valve angle	[°]	25,2
Camshaft bearing diameter inlet/outlet	[mm]	23/28
Rated power	[kW]	441
Nominal Torque	[Nm]	800
Maximum speed	[$^{min-1}$]	6.800

Base Engine/Crank Drive The new engine also continues to use the familiar square stroke/bore ratio with a bore diameter of 86 mm. The iron coating applied in atmospheric plasma spraying process, which has already been successfully used in the predecessor unit and has proven itself in worldwide use, is used as the cylinder raceway. In order to improve the internal efficiency, especially in the power range, increased mechanical requirements were defined for the base engine with regard to peak combustion pressure and pressure rise gradients as part of the combustion process optimisation. Those were mainly achieved by design optimisation measures on the connecting rod and piston. The further developed crank drive is now characterised by a peak pressure load capacity (average value) of 140 bar with maximum peak pressures of up to 167 bar (3σ). For this purpose, the forged connecting rods were significantly revised in order to implement the higher loads with a minimum material expenditure of +0.5%.

The pistons, which are now used as pistons with cooling channels, have undergone a complete redesign. These new pistons are characterized by a special ring cavity in the area of the ring package/fire bar, which is supplied with oil via the piston spray nozzles. This makes it possible to cool the fire bar and the piston crown more effectively. The achievement of an even temperature distribution during operation was achieved with the help of extensive simulation and test series. The piston is manufactured using the casting process (Fig. 5).

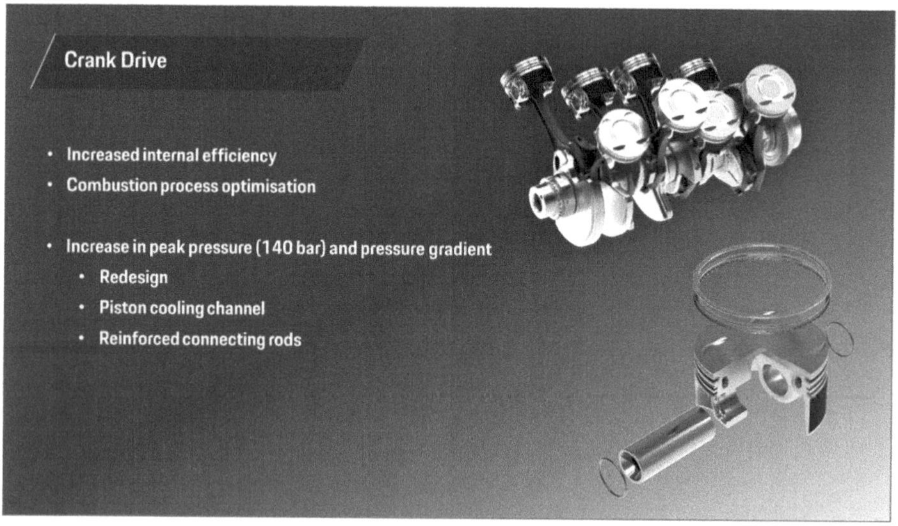

Fig. 5. Piston with optimised temperature distribution

Control and Valvetrain/Charge Exchange Control The camshafts are driven by a chain drive with intermediate shaft in order to maintain the compact design of the unit. The previously used cylinder deactivation was replaced by an intake valve lift switch on all cylinders. Map-controlled or depending on dynamic load requirements, the system switches between a partially load-optimised stroke curve with reduced valve opening time (six millimeter valve lift) and a fill-oriented intake cam profile (10.5 millimeter valve lift) for the higher load range. This variability is achieved with the sliding cam system familiar from the predecessor engine. Compared to the cylinder shutdown system, this measure opens up a significantly larger map range in which

an improvement in fuel consumption is possible through reduced throttle losses. This is particularly advantageous with regard to the operating strategy and the operating collective in hybrid engine operation, since cylinder shutdown has lower consumption potentials due to its limitation to the low load range of the combustion engine map. In addition, the implementation of the intake valve lift switch has further potential for reducing emissions, especially during cold starts and in the catalytic converter heating phase. Here, particulate emissions could be reduced by up to 80% compared to full-engine operation with a large valve lift (which is active in the cylinder shutdown concept during these phases).
The exhaust camshaft was implemented as a classic built camshaft. Exhaust cam profiles with different valve opening times are used on cylinder groups 1, 2, 6, 8 and 3, 4, 5, 7. This measure is due to the switch from the twin-scroll turbocharging with ignition-sequence-oriented, flood-separated exhaust manifolds of the predecessor engine to the mono-scroll turbines with single-flow manifolds (see turbocharging system). The "exhaust opens" timings, which vary from one cylinder to another, counteract the unfavourable transverse influences resulting from the ignition distances of individual cylinders during the charge exchange. This concept was also used in the 4.5-litre V8 engine developed for the first generation of the Cayenne.

Camshaft Position determination The new Porsche V8 engine uses the intake and exhaust camshaft adjustment with the compact vane adjusters of the predecessor engine. The target of the further development was to determine the real opening and closing times of the charge exchange valves more precisely. For this purpose, the position detection of the intake and exhaust camshafts was completely revised and the previous system was converted to a newly developed magnetoresistive phase sensing.

For this magnetoresistive phase sensing, a ring-shaped permanent magnet is placed in the camshaft ends, which forms a directed magnetic field between the poles. With the help of two Hall sensors offset by 90 degrees, the magnetic field strength is measured and the absolute angle of the magnetic field and thus the angular position of the camshaft is calculated.

Several advantages can be realized through the magnetoresistive phase determination. Thanks to the permanent magnet, it is possible to teach the camshaft sensors in the production process and to minimize the deviations in the timing setting process. In addition, with this measuring principle, the position of the magnetic field and thus the position of the camshaft can be determined even without a relative movement, and the position of the camshaft is already available during engine stop phases. The system is also insensitive to axial and radial tolerances, which can be compensated for like the effects of temperature. In addition, the accuracy and resolution of the position detection has been significantly improved with the new magnetoresistive phase determination compared to the usual encoder wheel/Hall sensor detection. The phase position of all four camshafts is made available to the engine control system with a sampling rate of six degrees crank angle with an accuracy of +/−0.5 degrees crank angle, thus offering advantages in calculating the current cylinder filling and camshaft position control, especially in dynamic engine operation (Fig. 6).

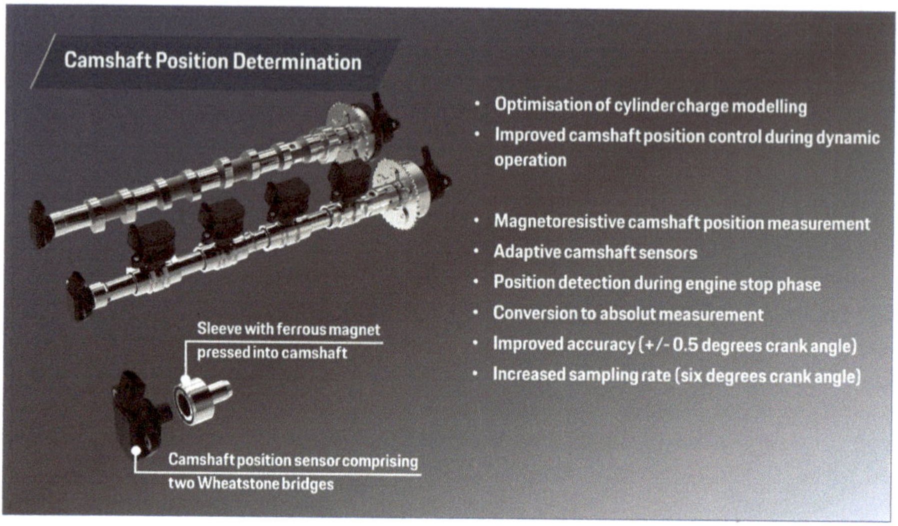

Fig. 6. Magnetoresistive camshaft sensing

Turbocharging System With the conversion of the turbocharging system from a twin-scroll to a mono-scroll turbine, both the requirements and objectives in terms of emissions reduction as well as the performance potential of the engine were taken into account.

For example, the single-flow manifold and turbine of the mono-scroll exhaust gas turbocharger (ATL) are characterized by a smaller exhaust gas surface area and a lower thermal mass, which has a positive effect on reduced heat loss of the exhaust gas, especially after cold start in catalytic converter operation. Furthermore, the early merging of the exhaust gas of all cylinders promotes the exhaust/air mixing and exothermic reaction of an optional secondary air system already in the manifold. Concept studies have shown that the flood separation of the twin-scroll system in conjunction with a secondary air system cannot achieve the full potential. In this context, it should also be noted that the coupling of the exhaust ducts via a common secondary air gallery entails a significant reduction in the low-end torque (LET) achievable by a twin-scroll concept, as the gas-dynamic separation of the cylinders, which have an unfavourable influence on the exhaust gas side charge exchange, is reduced by the secondary air system.

The advantages of a mono-scroll concept are also offered by greater degrees of freedom in the design and thermomechanical properties of the thermally highly stressed components exhaust manifold and ATL turbine. The absence of flood separations results in less structural complexity with reduced material accumulations and wall thickness jumps, while at the same time the flow capacity of the gas paths is positively influenced.

With the help of a simulation-based optimisation of the wall thicknesses and geometries, a permanent temperature resistance of 1020 degrees could be realized. Compared to the twin-scroll system of the predecessor engine, this made it possible to apply a limit temperature that was about 50 degrees higher (Fig. 7).

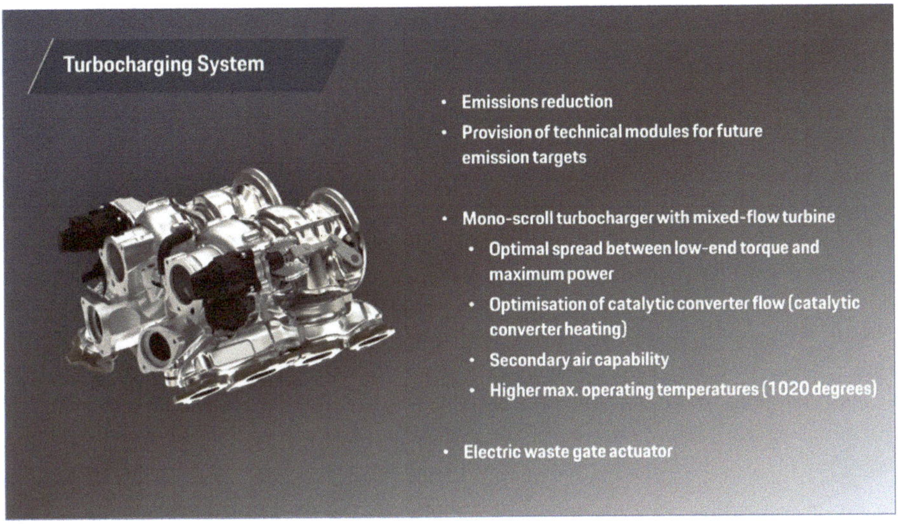

Fig. 7. Charger unit with electric waste gate actuator

Electric wastegate actuators are used for boost pressure control, which offer a higher control quality and speed than pneumatic actuation of the bypass valves. In addition, the wastegate flaps can be completely opened at engine start and during the catalytic converter heating phase in order to direct a large part of the hot exhaust gas mass flow directly to the catalytic converter face, bypassing the turbine, thereby reducing heat loss to the turbine housing and turbine wheel (Fig. 8).

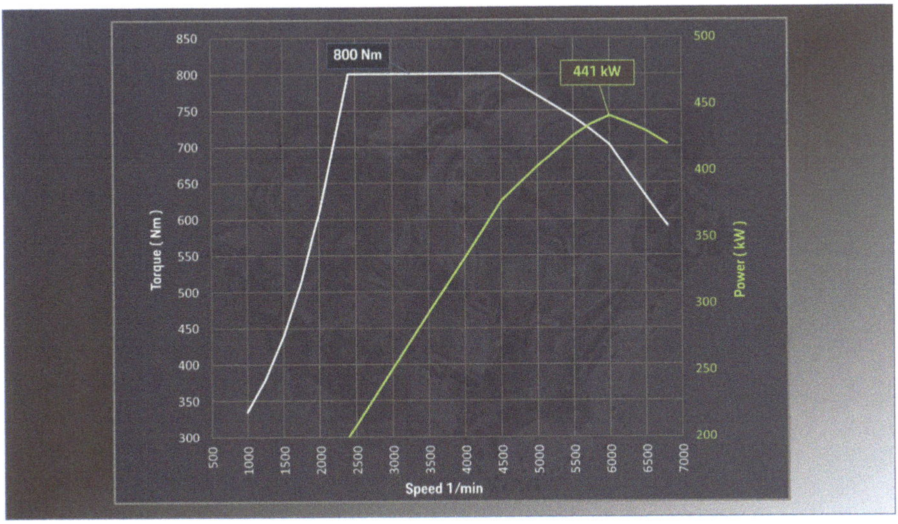

Fig. 8. V8 twin-turbo engine full load curve

Fuel System The system pressure of the fuel injection system has been increased from 250 bar to 350 bar in conjunction with the use of a new generation of high-pressure injectors and high-pressure pumps. On the one hand, the injection pressure increase improves the fuel atomisation quality of the injector and the mixture formation, and on the other hand, the injector dimensioning, which is adapted to the pressure level, enables an extended throughput spread between the minimum quantity injections and the full-load quantity requirement. The improved electric injector properties compared to the previous generation (minimum injection time, minimum pause time) have been consistently used to implement optimised multiple injection strategies depending on the operating point. The spray pattern of the centrally positioned, solenoid-driven multi-hole valve (seven spray holes) was developed with the help of CFD simulations, optical investigations and engine test bench measurements, with particular attention paid to the improvement of wall wetting effects. As a result, a further improvement in particulate emissions and fuel input into the engine oil could be achieved, which is particularly relevant in hybrid engine operation with frequent cold starts.

Engine Ventilation As with the previous unit, the crankcase ventilation system is a bank-symmetrical system, which essentially consists of two cylinder bank selective oil separators and takes over the function of coarse/fine oil separation, pressure control in the crank compartment, the switching of partial and full load path and the recirculation of the separated oil (Fig. 9).

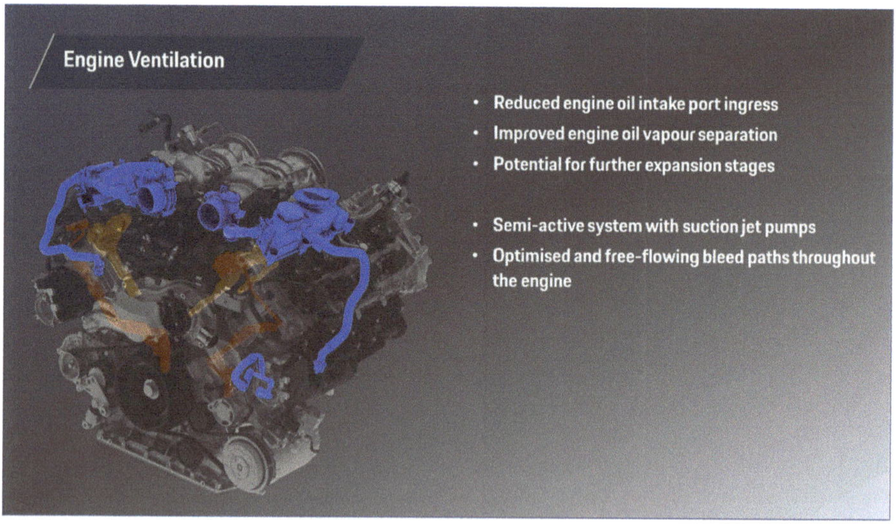

Fig. 9. Engine ventilation with integrated suction jet pump

The blow-by gases are fed into the valve covers via integrated channels in the crankcase and cylinder heads. In the further development of the unit, the ventilation paths were once again heavily revised and dethrottled. As a result of the optimisations, the crude oil input into the oil separators could be reduced by up to 90%, depending on the operating point. In the oil separators, the gases are further cleaned of oil by means of targeted deflections and a fleece. The system is supported by an additional suction jet pump per cylinder bank, which is active in the map area of the full-load path and thus allows higher pressure losses/separation rates in the oil separator. This semi-active overall system shows a similar performance to active systems (e.g. disc separators) in measurements and achieves very high separation rates, which ensures good separation efficiencies even under marginal operating conditions.

2.3 Transmission, All-wheel Transfer Case

Objective In addition to the new V8 engine, the entire components of the powertrain have been further developed in order to consistently implement the goals of performance, sustainability, efficiency and suitability for everyday use, and to offer the customer an even greater powertrain experience both in everyday life and under extreme conditions on the race track. The further development was aimed at both the technical and the functional properties.

Automatic Converter For the new Cayenne, the eight-speed Tiptronic S transmission has been optimised to offer a significant increase in performance in conjunction with the V8 engine. The focus was on improving the gearshift sequences and the increased driving dynamics that could be experienced as a result. For example, in the pull-up shifts, the speed gradients in the Sport/Sport Plus drive mode have been increased by up to 15% compared to the predecessor application, and the thrust-reverse shifts have been shortened by up to 30%, especially in the response time experienced by the customer to tip commands via the shift paddles. The result is faster gear changes for an even sportier driving experience (Fig. 10).

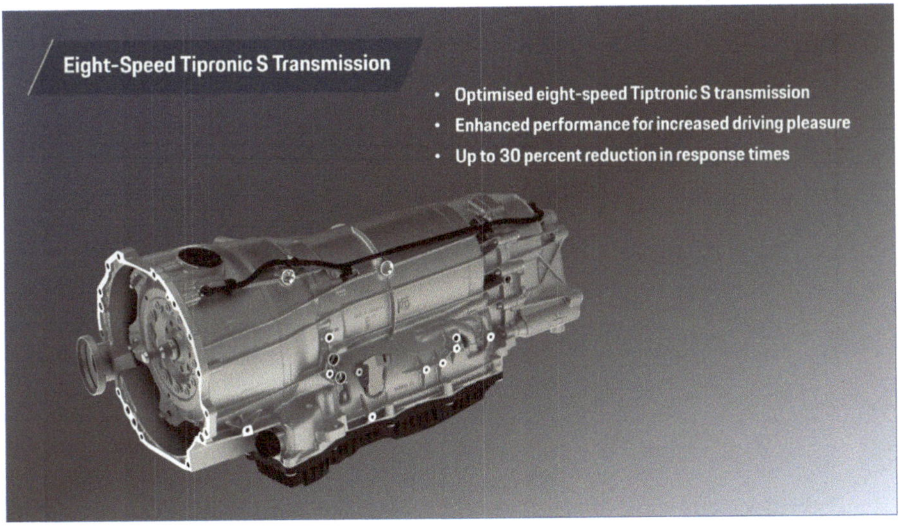

Fig. 10. Eight-speed Tiptronic S transmission

In addition to the optimised switching sequences, the switching strategy has been significantly optimised. On the one hand, the use of predictive route data for "forward-looking" gear selection has been evolved. On the other hand, the recognition of the driver's dynamic driving style in the area of curve detection has been implemented beyond the general driver type recognition, so that the optimum gear is engaged when accelerating out of corners. This enhances the dynamic driving experience without negatively affecting the confidence in everyday use.

All-wheel Transfer Case HangOn In the Cayenne's new powertrains, the torque is distributed via a HangOn transmission. For this purpose, the front axle is "attached" by means of a controlled multi-plate clutch (HangOn), while the rear axle is rigidly driven. The amount of drive torque to the front axle is controlled by the contact force of the multi-plate clutch. The multi-plate clutch is actuated by a ball ramp system and an electric actuator with integrated control unit. The clutch is operated in continuous slippage and has a capacity of 1200 Nm. The evolutionary stage of the HangOn system for the new Cayenne represents a new benchmark in terms of power density. With a total weight of just 14 kilograms including 0.6 liters of oil volume, a performance level of the fully variable all-wheel drive is made possible in conjunction with a new active water cooling system, which further increases traction capability and agility and ensures high availability of the all-wheel drive system even at high power levels to be transmitted (e.g. racetrack operation) (Fig. 11).

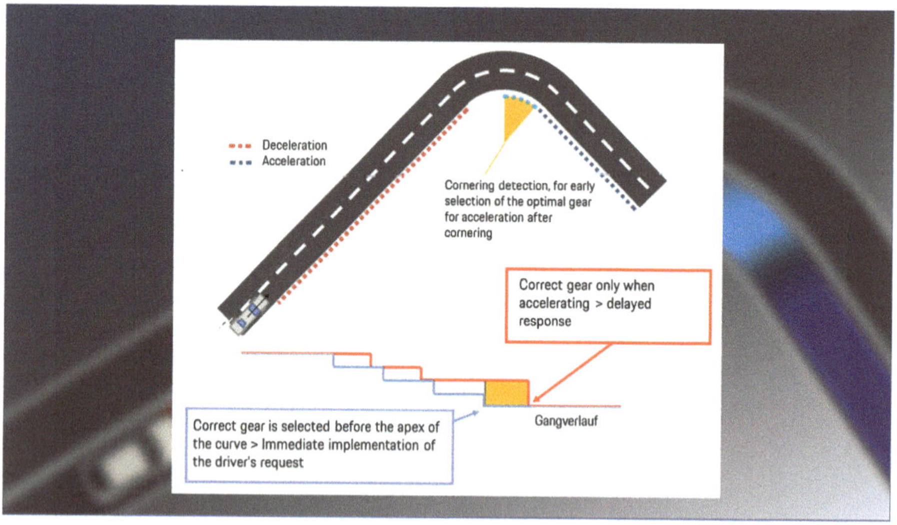

Fig. 11. Optimised gear selection in dynamic cornering

2.4 Hybrid System

Objective With the new Cayenne, Porsche has upgraded hybrid technology in several ways, significantly increasing electric range and performance in particular.

Electric Motor At the heart of the hybrid system is a new electric motor. One of the main goals was to increase the electrical power in a space-neutral manner, which was achieved by changing the magnetic and electrical sheet material in conjunction with an optimised winding variant of the coils. As a result, the output of the electric motor could be increased by 30 kW to 130 kW. At the same time, the maximum torque has been increased by 60 Nm to 460 Nm. In addition, the new electric motor can convert 30% more deceleration power into electrical energy in generator mode and store it in the traction battery: the new Cayenne E-Hybrids now recuperates with an output of up to 88 kW (predecessor 62 kW) and up to a speed of two km/h (predecessor 14 km/h). In addition, the efficiency of the electric motor in the customer-relevant load range of 200–400 Nm has been improved by a further two percent, thus making a significant contribution to the longer range of the plug-in-hybrid powertrain. The compact electric engine is integrated into the revised eight-speed Tiptronic S transmission (Fig. 12).

Fig. 12. Components of the hybrid system

Pulse Inverter In addition to the electric motor and the high-voltage battery (HV battery), the pulse inverter, which forms the interface between the HV battery and the electric motor and controls it, had a central role in the further development of the hybrid system. Among other things, it is characterized by a variable switching frequency and different modulation methods, which are optimizsd depending on the current operating point. Thanks to the adapted control, it was possible to improve the efficiency of the pulse inverter and thus increase the performance of the electric motor by up to ten percent. In order to counteract possible acoustic effects that can occur due to the adjusted clock frequencies, an artificial noise is generated by targeted control, which leads to a harmonious overall acoustic.

In the hybrid applications, the powertrain is supplemented by a new high-voltage battery, which now provides 25.9 kWh (gross), an additional capacity of eight kWh compared to its predecessor.

3 Consumption, Emissions, E-performance

With the target of building the sportiest SUV on the market, the Cayenne has focused on the system performance of the powertrain since its market launch. It is not only the consistent further development of pure combustion engines that is noteworthy, but also the hybrid technology that has been introduced over the years and extensively revised.

The first V8 variants of the Cayenne, with their 250 kW and 331 kW, were relatively tame by today's standards. From the Cayenne S to the Cayenne Turbo E-Hybrid with GT package, the current V8 variants cover a system output range between 349 kW and 544 kW. Maximum torque ranges from 600 Nm in the Cayenne S to 950 Nm in the Cayenne Turbo E-Hybrid. This corresponds to a power increase of 44 kW compared to the previous Cayenne Turbo S E-Hybrid (Fig. 13).

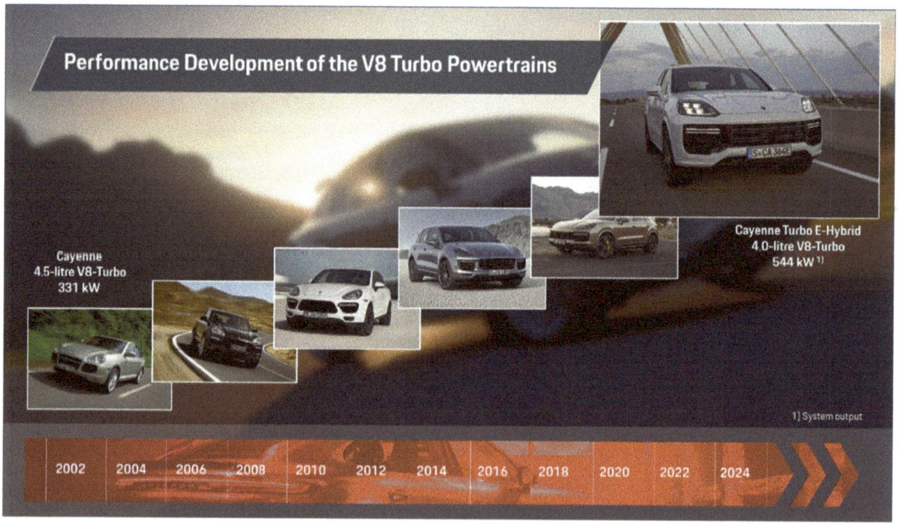

Fig. 13. Performance development of the V8 powertrains in the Cayenne since market launch

The figures of the top-of-the-range Cayenne Turbo E-Hybrid with GT package point out the Cayenne's claim to be the sportiest SUV on the market. The power-to-weight ratio is just 3.3 kilograms/hp. The 2495 kilogram (kerb weight) SUV easily sprints from 0–100 km/h in just 3.6 seconds. This figure is 0.4 seconds faster than the current 911 Carrera 4 (992) and only 0.1 seconds slower than the 911 Carrera S (992) with additional sport-chrono package. The top speed of the Cayenne models with V8 engines is between 273 km/h and 305 km/h. Compared to the first Cayenne Turbo from 2002, which managed the sprint from 0–100 km/h in a range of 5.6 seconds, there is an impressive increase in drive performance of the V8 powertrains over the individual vehicle generations.

In addition to these impressive performance figures, the Cayenne with its V8 powertrains is also a comfortable touring vehicle and feels absolutely at home, both off-road and in city traffic. In off-road mode, the driver has the option of choosing between different driving modes, the characteristics of which are adapted to different tyre friction coefficients (e.g. gravel, mud or snow). Both the pedal characteristics and the shift program of the eight-speed Tiptronic S transmission are tuned in such a way that the Cayenne can master all tasks with flying colours in conjunction with the Porsche Torque-Vectoring Plus (PTV Plus) (Fig. 14).

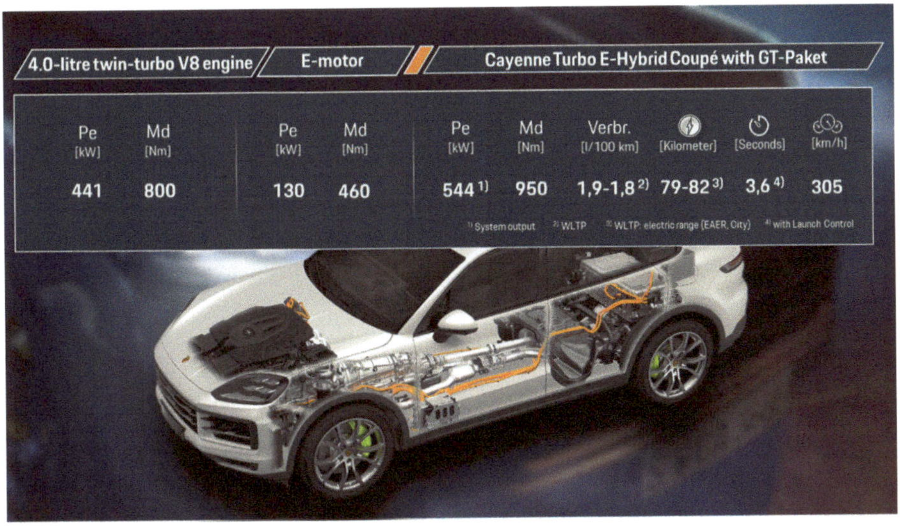

Fig. 14. The new V8 powertrain in the Cayenne

But the Cayenne also wants to be at the forefront in terms of environmental performance and sustainability. As a result, both the electric range and the charging power have been significantly increased. The Cayenne Turbo E-Hybrid, with its 130 kW electric motor, now manages to drive up to 82 km (EAER city) purely electrically. This corresponds to a combined consumption according to WLTP of only 1.9 l/100km. Charging the 25.9 kWh high-voltage battery with the eleven kW on-board charger requires a charging time of less than 2 hours and forty minutes at a suitable wall charger or charging station. Another highlight of the plug-in-hybrid powertrain is the operating strategy, which, in hybrid car mode in conjunction with activated route guidance, using environmental and navigation data, ensures that the share of electric driving within built-up areas is maximized and thus emissions in urban areas are reduced to a minimum.

4 Summary and Outlook

Whether on the circuit, on an unpaved road or in normal everyday life. Thanks to the extensive revision of the powertrains, the new Cayenne Turbo E-Hybrid now has the most powerful V8 powertrain with increased efficiency and extended electric range, which, together with the V6 powertrains, offers the driver a unique driving experience in all life situations (Fig. 15).

Fig. 15. The new V8 powertrains in the Cayenne—summary

But also for future worldwide applications with new emission standards or for use in other model series and high-performance offerings, this powertrain represents an excellent basis and an important component in the Porsche trio of all-electric drives, plug-in hybrids and highly emotional combustion engines (Fig. 16).

Fig. 16. The Porsche trio

Acknowledgment. Many thanks to the team of authors for creating the article (in alphabetical order):

B. Eng Nico Bieber, Dipl.-Ing. Thomas Breitinger, Dipl.-Ing. Nico Brill, Dr. Ing. Thomas Casper, Dipl.-Ing. Ole Hufendiek, Dipl.-Ing. Sven Kessler, Dipl.-Ing. Andreas Kramer, Dipl.-Ing. Matthias Krämer, Dipl.-Ing. Hans-Werner Lanz, Dr.-Ing. Wolfgang Liesen, Dipl.-Ing. Bohdan Serafin, Dr.-Ing. Frank Maier, Dipl.-Ing. Thomas Wasserbäch and all other participants involved.

Reference

1. Dipl.-Ing. Jörg Kerner, Dr.-Ing. Thomas Günther, Dipl.-Ing. Martin Werner, Dr.-Ing. Alexander Kronich, Dipl.-Ing. Matthias Krämer, Dipl.-Ing. Andreas Kramer, Dipl.-Ing. Christoph Pleuß, Dipl.-Ing Sven König, Dr.-Ing. Wolfgang Liesen, Dr.-Ing. Frank Maier: The new V8 turbo engine from Porsche. 37th International Vienna Motor Symposium

Optimization of Combustion System, Emission Concept and Hybrid Operating Strategy for an LCV powered by H_2 Engine

Jan Niklas Geiler[1(✉)], Klaus Moritz Springer[2], Peter Bloch[1], and Markus Kirzinger[3]

[1] Power Solutions, Robert Bosch GmbH, Stuttgart, Germany
{jan.geiler,peter.bloch}@de.bosch.com
[2] Ford-Werke GmbH, Cologne, Germany
kspring1@ford.com
[3] Forschungsinstitut für Kraftfahrwesen und Fahrzeugmotoren Stuttgart (FKFS), Stuttgart, Germany
markus.kirzinger@fkfs.de

1 Introduction

Within the joint project "H_2 ICE Democar", funded by the German Federal Ministry for Economic Affairs and Climate Action, Ford, Bosch, MAHLE, Umicore, Institute of Automotive Engineering (IFS, University of Stuttgart), Research Institute for Automotive Engineering and Powertrain Systems Stuttgart (FKFS), Chair of Thermodynamics of Mobile Energy Conversion Systems (tme), DHL, PUREM by Eberspächer and Shell are identifying key requirements for a serial hybrid light commercial vehicle (LCV) powertrain basing on a H_2 engine. The carbon free fuel is not only capable of CO_2 free operation and long driving range with fast refueling. In combination with a lean burn combustion system, H_2 is a key for zero-impact exhaust emissions and allows for highest thermal efficiency. When it comes to the powertrain operation strategy, the serial hybrid technology approach unlocks optimization potential but also enables all use cases that are closely linked to electric driving capability.

Figure 1 outlines the "H_2 ICE Democar" concept. The technical solution will be verified by two demonstrator vehicles for logistics application (transporter) and passenger transportation (people mover). The small transporter version aims at daily delivery logistic applications. Current issues with H_2 infrastructure can be overcome by fleet operator hubs with centralized H_2 filling stations and hence, H_2 LCV can be prone to H_2 automotive scaling. Complementary to Fuel Cell Electric Vehicles (FCEV), it must be verified that H_2 engine powered vehicles offer viable solutions in the field of H_2 economy and hence can contribute to increase the demand for H_2 infrastructure.

A. Heintzel (Hrsg.): Int.Engine.Congr. 2024, Proceedings, S. 36–50, 2025.
https://doi.org/10.1007/978-3-658-47966-4_4

The "H2 ICE Democar" project aims to identify essential building blocks for a H_2 engine propelled serial hybrid LCV powertrain. Understanding the interaction of the lean H_2 engine with a H_2 dedicated aftertreatment system, the hybrid operating strategy and new software functions allow for the holistic assessment and optimization of the H_2 engine based serial hybrid powertrain.

The key elements of the H_2 engine demonstrator vehicle with lowest tailpipe emissions and high thermal engine efficiency are: H_2 engine (Sect. 2), H_2 specific exhaust gas treatment (Sect. 3), H_2 specific operation strategy (Sect. 4) and the H_2 supply system integrated into the base vehicle (Sect. 5).

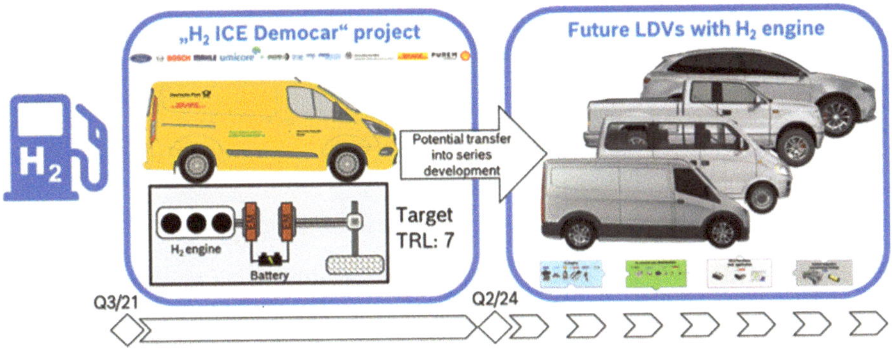

Fig. 1. Project scope of the "H_2 ICE Democar"

2 Hydrogen Engine

The H_2 engine is derived from the well-known 1.0 l Ford EcoBoost three-cylinder gasoline engine [1]. Table 1 summarizes the main technical features and specifications. Figure 2 shows the system layout of the H_2 engine concept including all sensors and the exhaust gas treatment system, see Sect. 3 for more details. The sensors and actors are based on standard automotive technology. However, some must be adapted for the measuring range, e.g., boost pressure sensor. In addition, the fuel system and injection equipment as well as the exhaust system are completely revised. The temperature and nitrogen oxide (NO_x) sensors as well as the urea dosing modules in the exhaust system are controlled by the Dosing Control Unit (DCU). The hydrogen fuel system (discussed in Sect. 5) including various software functions for a reliable and safe vehicle operation is controlled by the Tank Control Unit (TCU) in close interaction with the ECU.

Table 1. Technical data of 1.0 l EcoBoost H_2 engine concept

Engine Type	In-line three-cylinder
Displacement/l	0.999
Fuel	Compressed gaseous H_2
Bore x Stroke/mm x mm	71.9 × 82.0
Compression Ratio	10.0
Fuel System	H_2 direct injector, centrally mounted
Ignition System	Prototype H_2 spark plug, modified ignition coil
Cylinder Head	Aluminum
Cylinder Block	Cast iron
Crankshaft	Steel
Connecting Rod	Steel
Piston	With cooling channel
Turbocharger	Single stage VGT
Rated Power/kW	60

H_2 Engine, Cylinder Head and Base Engine Special attention has been paid to increase the robustness against high combustion peak pressure. The engine crank train has been reinforced with a steel crank shaft and steel conrods, enhanced bearings and an enhanced piston with a ring carrier for high wear resistance in the first ring groove and a salt core cooling gallery for efficient piston cooling. Furthermore, MAHLE updated the intake and exhaust valves and valve seat material with H_2 resistive steel that also allows for "dry" operation. For minimization of oil consumption, the cylinder honing has been optimized and the tribological system has been enhanced with a revised liner honing and matching piston rings [2, 3]. Minimizing the oil input into the combustion chamber is essential, as oil is the only remaining source of engine-out unburnt hydrocarbons (HC), carbon monoxide (CO) and carbon dioxide (CO_2) emissions in an H_2 engine. In addition, oil droplets or oil coal particles are a potential origin of preignition. To balance the project effort, no major modification of the cylinder head has been planned (e.g., adapt charge motion by intake port modification).

Boosting System For H_2 engines, the resulting engine-out NO_x emissions are a major challenge that must be considered in the H_2 engine design concept. A homogeneous lean combustion based on hydrogen is characterized by high dilution tolerance due to the high laminar flame speed of H_2. The lean combustion concept enables a high thermal efficiency in combination with lowest engine-out exhaust gas emissions. Both high air demand and low exhaust gas temperatures come along with lean engine operation and hence special focus must be put on the boosting system for the delivery of the required air mass. In result, the mass production engine wastegate turbocharger has been replaced by an all-new single stage VGT turbocharger that has been matched to the H_2 engine concept (see [4] for details regarding design parameters).

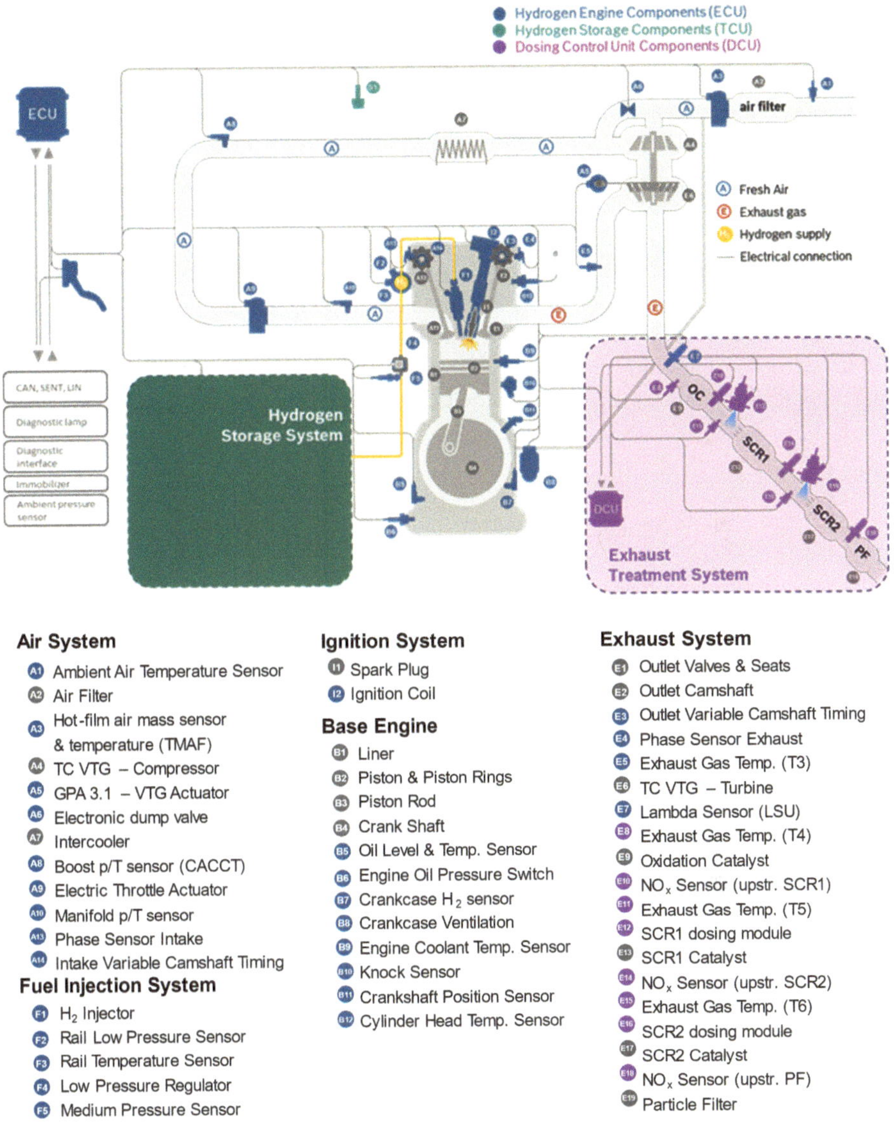

Fig. 2. 1.0 l EcoBoost H_2 engine system layout

Crankcase Ventilation As known from conventional combustion engine, also for a H_2 engine, a so-called blow-by gas occurs due to leakage flows past the piston rings into the crankcase. In addition to gaseous components including unburnt hydrogen, it also contains water and oil droplets and ligaments. An active crankcase ventilation system, the MAHLE high pressure impactor (HPI), has been adapted to the engine in series flow design downstream of the ventilation system elements. This system should remove the liquid components efficiently and lower H_2 concentration in the blow-by flow before it is fed into the air intake.

Figure 3 shows the measured H_2 concentration in the crankcase gas atmosphere without (Fig. 3(a)) and with (Fig. 3(b)) active crankcase ventilation. Without active ventilation, the H_2 concentration in the crankcase atmosphere increases with engine load. Hence blow-by only worst-case measurements are shown in Fig. 3(a), which show H_2 concentrations up to 8.48 vol.-%.

With the active ventilation system, a significant H_2 concentration reduction can be achieved. In Fig. 3(b), the threshold limit is set to a concentration of 3 % with further potential for even lower concentration – however, it costs additional parasitic system losses and has an effect on oil separator efficiency due to higher gas flow.

Furthermore, it should be mentioned that the scavenging air mass supplied to the intake manifold must be considered in the ECU charge determination.

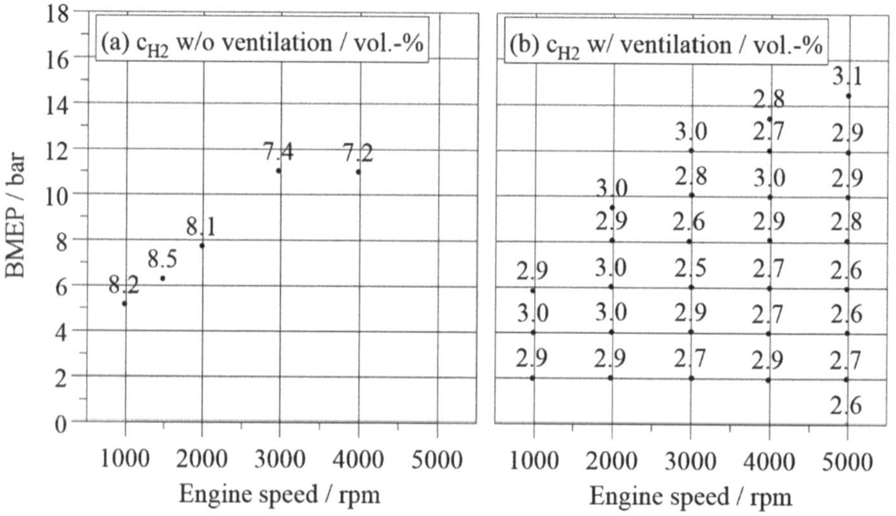

Fig. 3. H_2 concentration (c_{H2}) measurement (a) w/, (b) w/o active crank case ventilation

The adapted system is shown in Fig. 4. The MAHLE HPI creates a controllable vacuum downstream the crankcase that draws ambient air via an air filter into the crankcase and conveys the diluted crankcase gases via an oil separator into the air intake system and hence into the combustion process.

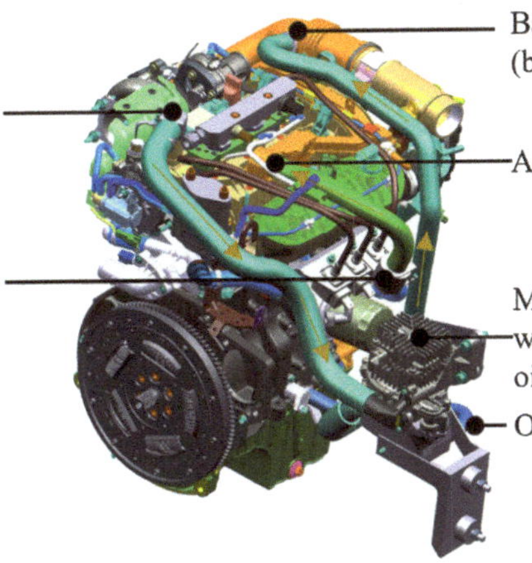

Fig. 4. Applied MAHLE HPI active crank case ventilation system

Ignition System A "cold" prototype H_2 spark plug as well as modified ignition coils optimized for H_2 operation have been introduced by Bosch.

Fuel Injection System Bosch provided dedicated low pressure H_2 direct injection prototype injectors. This injector is characterized by an outward opening nozzle with high flow cross section and hence delivers a high flow rate when operated with low density gaseous H_2 [5]. The selected injector variant has a static H_2 flow rate of 5.8 g/s at 25 barA, 85 °C and is capable to inject the required H_2 mass in short injection windows (< 60° CA, see Fig. 5(e)). The maximal operating pressure has been set to 21 bar (see Fig. 5(f)) to utilize the fuel tank efficiently and maximize the driving range. For combustion system optimization, the injectors can also be adapted with tailored Jet Guiding Caps, which enable jet targeting optimization and thus enhanced mixture formation as discussed in [4, 6]. In this paper, the Bosch H_2 DI injector is mounted in central position with recess (negative protrusion of −3 mm) to form a focused jet [4, 7]. In this way, early H_2 contact with hot surfaces at the spark plug and exhaust valves can be avoided and the risk of preignition in this engine configuration is mitigated. At the same time, H_2 backflow into the intake port can be minimized for open valve injection. A pressure and a temperature sensor were integrated into the fuel rail to account for density changes resulting in better fuel metering. The rail pressure is a function of engine speed and load and can be controlled by a pressure regulator.

H_2 Engine Test Results Are shown Fig. 5. The operating point at 5000 rpm, full load, corresponds to the power of ~ 60 kW. The peak power is constraint by the peak in-cylinder pressure (140 bar) and the minimum air mass which is required for the lean combustion limit that determines the level of engine-out NO_x emissions.

The engine can be operated with a lean relative air/fuel ratio ($\lambda \geq 2.0$, see Fig. 5(a)) over the entire engine map. High efficiency (Fig. 5(c)) across a wide engine map area enables degrees of freedom to optimize the serial hybrid vehicle operation strategy for low engine-out NO_x emissions without effect on the powertrain thermal efficiency.

To maximize time for mixture formation, early start of injection (SOI) timings are advantageous at low loads up to BMEP = 4 bar. For engine speed ≤ 3000 rpm, the injection windows are shifted towards late injection timing after intake valve closing to enhance trapping efficiency in the high load area. At higher engine speed, SOI during intake valve opening phase is beneficial in terms of NO_x emissions (n.b. COV and efficiency remain almost constant), see [6].

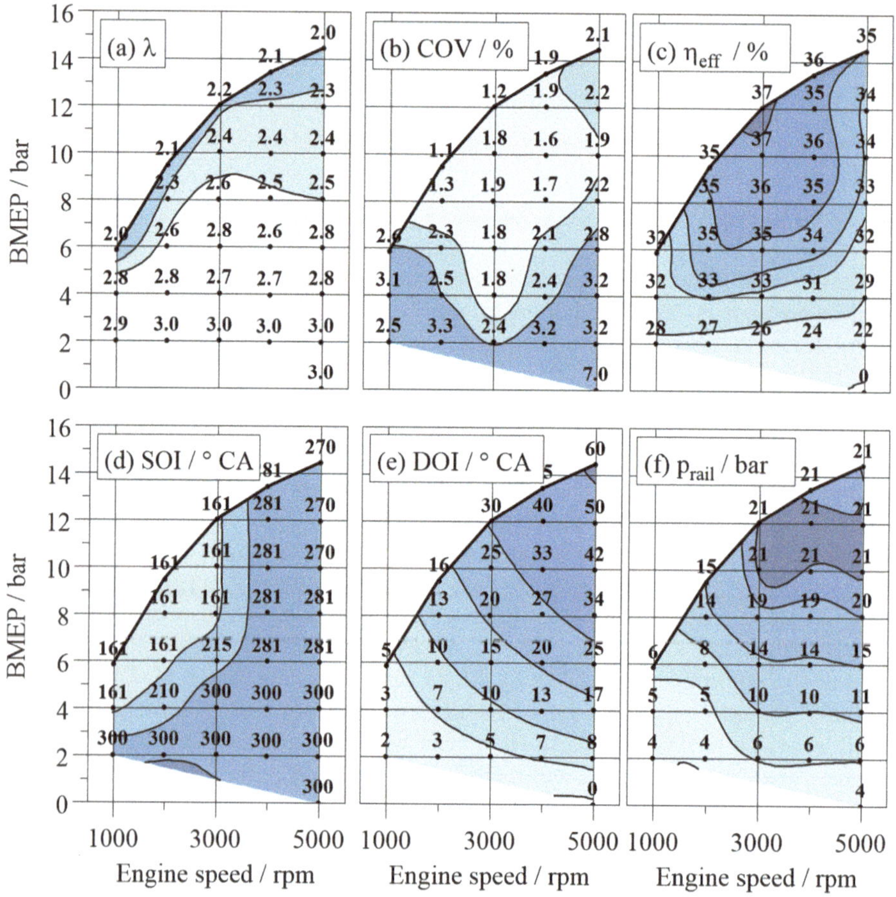

Fig. 5. H_2 base engine test results: **a** exhaust λ, **b** COV, **c** effective efficiency, **d** SOI, **e** DOI, **f** rail pressure p_{rail}

As shown in Fig. 5(e, f), required H_2 mass can be injected in less than 60° CA (DOI: Duration of Injection) at low rail pressures between 4 bar (low load) and 21 bar (rated power). To ensure independence of injected H_2 mass from in-cylinder backpressure, the ratio of injection pressure to in-cylinder pressure was set above the critical pressure ratio (~ 2) during the complete injection window.

Figure 6 shows the engine-out emissions of HC, CO and nitrogen oxides (NO_x) normalized to the emissions of the fully optimized series gasoline engine. The very low figures illustrate that there are negligible HC (< 0.6 %) or CO (< 0.2 %) emission contributions from engine oil. This is important since engine oil is the only source of engine-out HC or CO emissions. The residual amounts of engine-out HC and CO can be reliably oxidized by the oxidation catalyst (OC, see Sect. 3).

In summary, even at the most unfavorable engine operation conditions (e.g. max BMEP at 1000 rpm) the H_2 engine concept with not fully optimized operation settings is characterized by 6.5 times lower NO_x engine-out emissions when compared to the fully optimized gasoline engine. At rated power, even 10 times lower NO_x engine-out emissions were measured. As reported in [6], Jet Guiding Caps can lead to additional NO_x benefits. Only slightly lower full load torque targets easily reduce engine-out NO_x significantly. This outlines the importance to consider a full load lean λ limit when designing the engine combustion system in combination with the boosting concept.

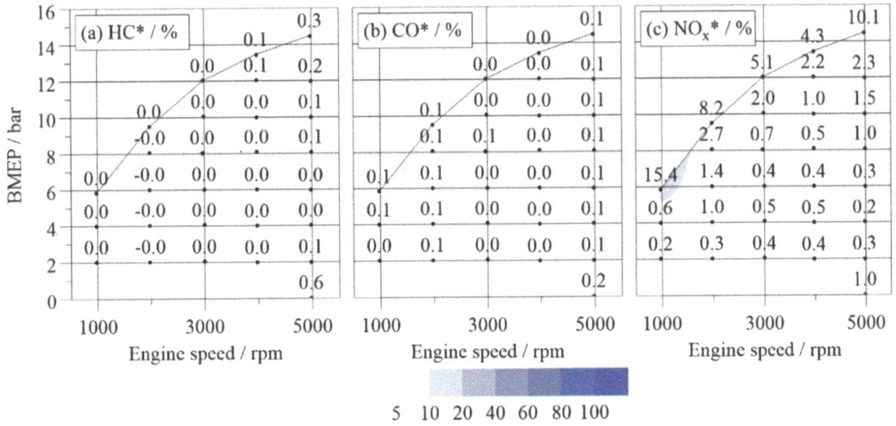

Fig. 6. **a** HC*, **b** CO*, **c** NO_x* engine-out emissions normalized to the values of the gasoline engine in %

3 Hydrogen Engine exhaust Gas Treatment

Due to the carbon-free fuel, hydrocarbons (HC), carbon monoxide (CO) and particulate matter (PM) are on very low level. As demonstrated in Fig. 6 the main engine-out pollutant emissions are NO_x emissions. Lean steady-state operating conditions result in extremely low NO_x formation in the combustion process. Additionally, emission increases during highly transient engine operation can be limited by the serial hybrid operation strategy (see Sect. 4).

The engine results presented in Sect. 2 demonstrate the trade-off between highest efficiency (operation at higher loads) and lowest engine-out NO_x emissions (leaner operation at lower load). Hence, for a lean H_2 engine an efficient EGT system is key to combine lowest tailpipe emissions with highest thermal engine efficiency at full load operation.

The EGT system that has been developed in close cooperation with the project partners Umicore and Purem by Eberspächer is shown in Fig. 7. It consists of five catalysts and hence must be regarded as "demonstrator vehicle solution". This system has been designed to investigate as much as possible interactions of the selected catalyst technologies in the context of a lean H_2 engine powertrain. Besides the opportunity to get a deeper understanding, it is obviously challenging to install five catalysts with reasonable volume into an existing vehicle package.

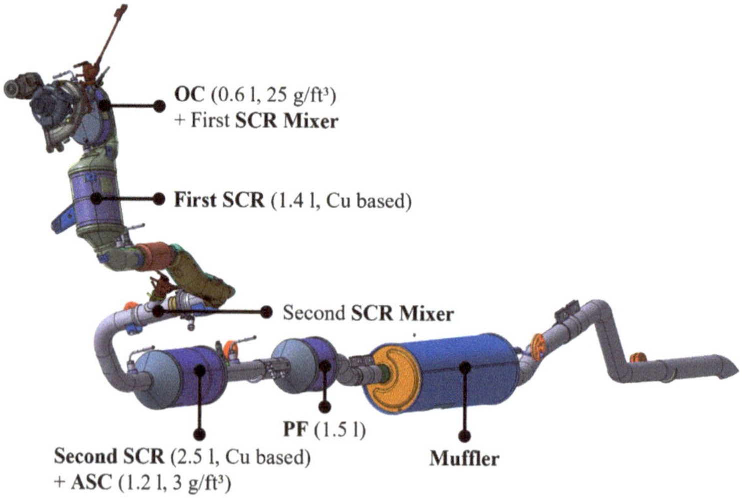

Fig. 7. H_2 EGT system: close-coupled part with OC and first SCR and underbody part with second SCR and particulate filter

The democar EGT system consists of the following components:

- **Oxidation catalyst (OC)** with a volume of 0.6 l and 25 g/ft3 platinum loading: because of incomplete combustion, unburnt H_2, HC and CO can be detected in the exhaust gas. The OC oxidates these residuals in the lean regime. The exothermic H_2 conversion by the OC can be exploited to light off downstream components to ensure fast conversion. The latter combined with an additional late H_2 injection in the expansion stroke can help to improve catalyst light-off especially at cold start conditions. In addition, unburnt H_2 can influence SCR performance in a negative way [8] and hence needs be oxidized. Furthermore, oxidizing NO to NO_2 (NO_2/NO_x-ratio) leads to a more efficient NO_x conversion in the downstream SCR.
- **Two selective catalytic reduction (SCR) catalysts** on Cu-basis with 1.4 l (first SCR) and 2.5 l (second SCR) volume: the available vehicle package space for

a close-coupled SCR catalyst with large volume system is limited. Hence, a split (close-coupled and an additional underbody catalyst) SCR catalyst system design has been selected to achieve the required catalyst volume. Due to the still existing limitations of H_2 SCR systems [9], the EGT system is based on industry standard AdBlue-SCR technology. Homogeneous mixing of urea with the exhaust gas is of major importance for an efficient NO_x conversion in the SCR.

- **Ammonia slip catalyst (ASC)** with a volume of 1.2 l and 3 g/ft3 platinum: to prevent tailpipe NH_3 emission slip in case of NH_3desorption.
- **Uncoated particulate filter (PF)** with a volume of 1.5 l: since limited experience regarding H_2 engine wear over lifetime is available, the possible increase of engine oil combustion cannot be predicted yet. To address a possible increase of particulate emissions due to engine oil consumption over vehicle mileage, a PF was integrated into the EGT system. The demonstrator vehicle tests in the next project phase "vehicle calibration" will be used for further evaluation.

4 Hydrogen Specific Operation Strategy

A serial plug-in hybrid electric powertrain as shown in Fig. 8 is characterized by having no mechanical link between the H_2 engine and the wheels. Vehicle traction is built up by an electrical machine (M) that is mechanically linked with the wheels and sources it's power either from the HV battery and/or the secondary e-machine (G-Generator) which is mechanically coupled to the H_2 engine (H_2E). This allows for an H_2 engine speed and load control that is independent from the vehicle speed and driver torque demand. Therefore, power is not directly deployed by the driver but by the controller of the electrical traction and storage system and the traction battery state of charge is a parameter of major importance.

The driver torque demand is processed in the vehicle control unit which controls the e-machine torque and manages the battery state of charge (SOC) and generator power. The power request to the engine is translated into an engine operating point, which is defined by engine speed and engine torque. An algorithm based on a "minimization of a cost function" calculates H_2 engine speed and torque accordingly, depending on various parameters (e.g., SOC, temperature, H_2 engine efficiency, emissions etc.).

The minimization of a cost function enables a multi-criteria based optimization of various parameters according to their weighted "price". This weighted price is based on the "importance" of a set of the considered parameters and aims – in case of the H_2 powertrain – to realize the requested power at lowest possible cost.

The main optimization parameters for the H_2 engine are the effective engine efficiency (η_e) and engine-out NO_xemissions. In a pre-processing step of the powertrain calibration process the engine mapping results are associated with "characteristic curves" with various weightings. The "characteristic curve" definition results from the minimization of the cost function. Figure 9 shows characteristic curves with different weightings, which are reflecting the importance of the optimization parameters in the cost function. The price minimization algorithm processes the best weighting curve (lowest price) for iso-lines of power. In this way, the best associated operation strategy and the underlying engine operation parameters can be determined.

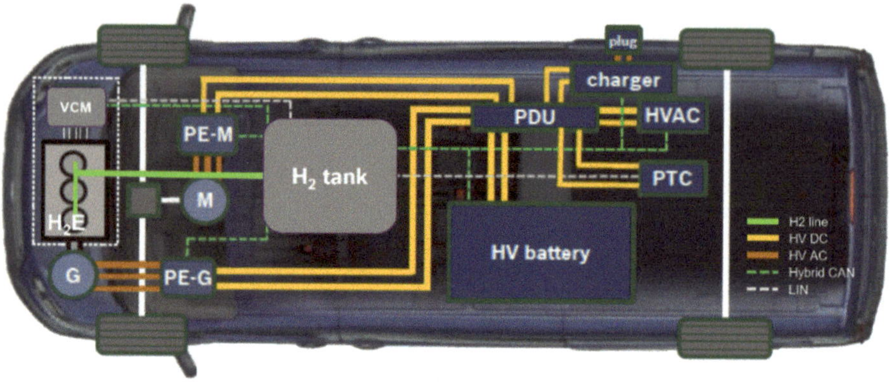

Fig. 8. Serial plug-in hybrid vehicle H_2 engine powertrain architecture (Main components: H_2E: H_2 Engine, G: Generator, M: traction e-machine)

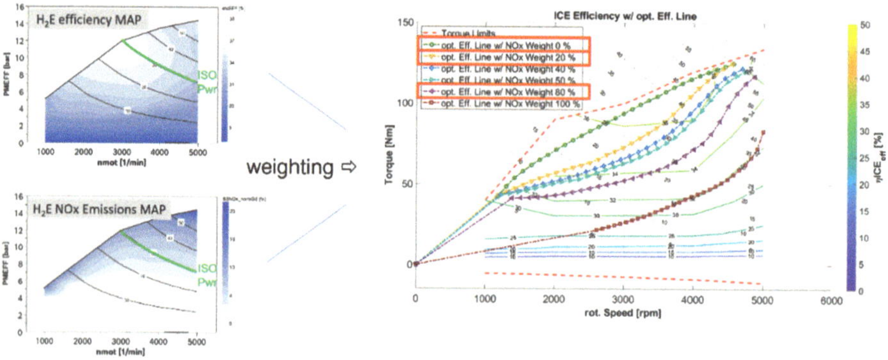

Fig. 9. Determination of engine operating points from stationary mappings by applying weightings on optimization parameters

Exemplarily, for a WLTC cycle driven in charge sustaining mode, the resulting operating points are shown in Fig. 10 for three different weightings. A pure focus on engine efficiency (weighting η_E/NO_x emissions = 100/0) leads to full load operation at 3000 rpm where highest efficiency is achieved. Due to limitations in the available air mass, operation below $\lambda < 2.5$ is necessary to achieve the full load target. In consequence, engine-out NO_x emissions (compare Figs. 5 and 6) increase. With this constraint, a power demand of 30 kW would be realized by an engine operation point at 3000 rpm/95 Nm.

Shifting the weighting towards lower engine-out NO_x emissions (weighting η_E/NO_x emissions = 80/20, resp. 20/80) reduces the torque level and further increases the engine speed demand. Accordingly, the power demand of 30 kW can be realized by an engine operation point at 3500 rpm/82 Nm (weighting: 80/20), respectively 4000 rpm/72 Nm (weighting: 20/80).

As demonstrated, the serial hybrid powertrain architecture enables additional degrees of freedom for a lean H_2 engine. By varying the weighting between H_2 engine efficiency and engine-out NO_x emissions, it is possible to minimize engine-out NO_x emissions at only slightly reduced efficiency. Details regarding the vehicle operation strategy are discussed in [10].

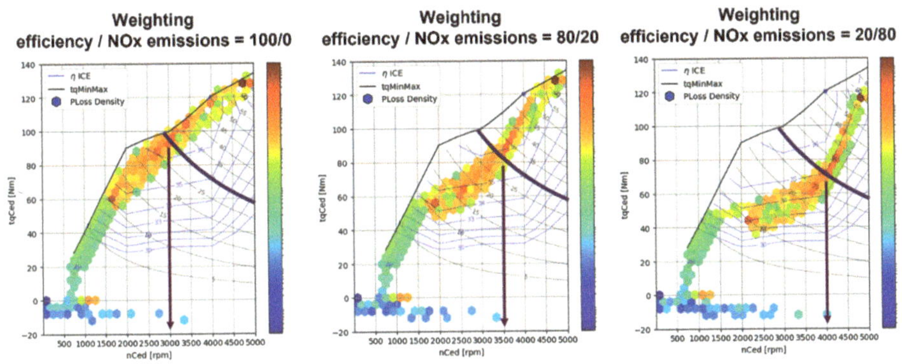

Fig. 10. H_2 engine operating points in WLTP cycle (charge sustaining mode) with different weightings of optimization parameters η_E and engine-out NO_x emissions

5 Hydrogen System

Figure 11 shows the realized hydrogen supply system. Since the base vehicle has not been designed for H_2 application, the vehicle package does not offer sufficient package space for the high-pressure tank installation. To balance the project effort, the H_2 storage has been installed on the vehicle roof. In this way, 4.5 kg H_2 can be stored at 700 bar. The system is structured in three pressure ranges:

- **"high-pressure"** with a pressure level up to 700 bar;
- **"medium-pressure"** with a constant pressure of 30 bar set by a mechanical pressure regulator;
- **"low-pressure"** with the target rail pressure set by an electric pressure regulator according to the engine load between 3 and 21 bar (see Fig. 5(f)).

Four hydrogen concentration sensors monitor potential H_2 leakage and hence ensure the vehicle safety as key elements. In case of detected H_2 leakage, the high pressure H_2 supply can be cut off by closing the on-tank valve (OTV) and additionally the mid- and low pressure H_2 system will be closed by system isolation valves (SIV). Pressure release valves (PRV) ensure that the maximum pressure in the respective area is not exceeded even in the event of a system fault. With an IR data interface located at the fuel receptacle, the H_2 system has been designed in a way that the demonstration vehicles can be refueled at standard hydrogen refueling stations.

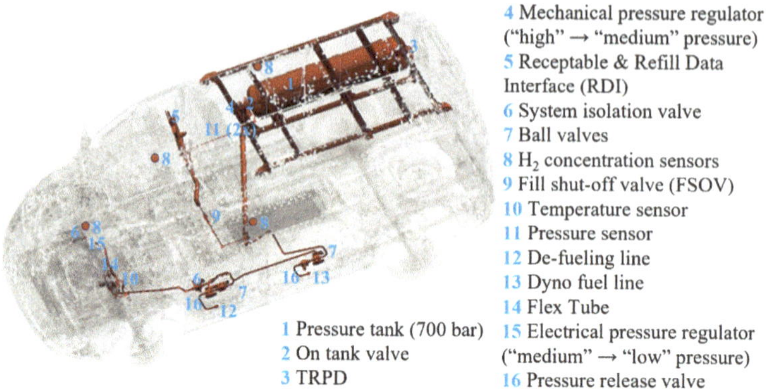

Fig. 11. Realization of the H_2 system in the "H_2 ICE Democar"

6 Summary and Outlook

6.1 Summary

This paper provides an overview of the main building blocks to realize a H_2 engine powered hybrid demonstrator vehicle.

A lean combustion system with $\lambda \geq 2.0$ in the entire engine map combined with a low-pressure direct injection was realized. The engine test results discussed in Sect. 2 confirm the potential of lean engine operation, which enables low engine-out NO_x emissions and high powertrain efficiency. Compared to the fully optimized gasoline engine, the lean H_2 concept engine has extremely low engine-out NO_x emissions provided that the engine is operated not richer than $\lambda = 2.0$.

The H_2 EGT system (Sect. 3) has also been outlined in this paper. The main challenge for the design process was the limited available package space in the engine compartment. Hence, the selected catalyst volumes represent a compromise between functionality and package space especially for the close-coupled OC and SCR catalyst.

Section 4 discusses the H_2 specific operation strategy, which is based on a "cost function". The key project objectives "high thermal efficiency" and "low engine-out NO_x emissions" are permanently optimized for every vehicle operation state.

A successful integration of the H_2 storage system into the base vehicle is a prerequisite to investigate the potential of the technical solution in the demonstrator vehicle.

In summary, the presented results outline a high efficiency hydrogen powertrain concept that minimizes engine out emissions by its lean combustion system. It is demonstrated that a serial hybrid architecture enables a H_2 engine to run under optimal conditions in every vehicle operation mode.

6.2 Outlook

Future objectives of the project include vehicle calibration. Also, the optimization of the serial hybrid operating strategy will be investigated, utilizing degrees of freedom offered by a lean H_2 engine.

Fig. 12. People mover version of the "H_2 ICE Democar", exhibited at the event "Wasserstoffmotor LIVE" at IFKM, KIT Karlsruhe (June 29/30th 2023)

Figure 12 shows the already fully established people mover version of the demonstrator vehicle. After completion of the vehicle calibration, performance of the "H_2 ICE Democar" concept will be evaluated on the chassis roller dyno and on the test track in driving cycles relevant for respective vehicle use cases.

Acknowledgement. The authors would like to thank the Federal Ministry for Economic Affairs and Climate Action, TÜV Rheinland and all project partners for their support and contributions to this work.

Funding

Finanziert von der Europäischen Union NextGenerationEU	Gefördert durch: Bundesministerium für Wirtschaft und Klimaschutz aufgrund eines Beschlusses des Deutschen Bundestages	The work reported in this publication has been undertaken as part of the "H_2 ICE Democar" project. This project has received funding from the program "Neue Fahrzeug- und Systemtechnologien" of the Federal Ministry for Economic Affairs and Climate Action on the basis of a decision by the German Bundestag under grant agreement no. 19I21019A

References

1. R. Friedfeldt, T. Zenner, R. Ernst and A. Fraser, "Three-cylinder Gasoline Engine with Direct Injection," MTZ worldwide, vol. 73, pp. 4–11, 2012.
2. H.-P. Böhm, S. Duda, C. Maisch, U. Morawitz and D. Schulz, "Tribological challenges of the cylinder bore in H2 ICEs," in 10th International Engine Congress, Baden-Baden, 2023.
3. C. Trabold, A. Puck, T. Gallas, P. Grabner und K. Gschiel, „Influence of hydrogen combustion on engine components in full load operation," in 10th International Engine Congress, Baden-Baden, 2023.
4. J. Geiler, M. Springer, L. Lorenz, M. Blomberg, J. Achenbach und P. Bloch, „H2 Engine Hybrid Powertrain for Future Light Commercial Vehicles," in 10th International Engine Congress, Baden-Baden, 2023.
5. S. Paltrinieri, M. Olcuire, V. Calia, F. Mortellaro, M. Medda, F. Gullino, K. Stapf, J. Geiler, P. Jochmann, M. Böe, M. Lippisch and C. Wundling, "Experimental and Numerical Investigation of Hydrogen Injection and its Preliminary Impact on High Performance Engines Development," SAE Technical Paper 2023-01-0402, 2023.
6. J. Geiler, T. Rabe, E. Schünemann, M. Springer, M. Blomberg und M. Kirzinger, „Light Commercial Vehicle with a H2 Engine Hybrid Powertrain," in 32nd Aachen Colloquium Sustainable Mobility, Aachen, 2023.
7. D. Lejsek, D. Seboldt, J. Geiler, P. Leick, M. Frank and K. Stapf, "Mixture Formation and Combustion in a Passenger Car Engine with Low Pressure H2 Direct Injection," in 19th Symposium "Sustainable Mobility, Transport and Power Generation", Graz, 2023.
8. E. Schutting, S. Roiser, H. Eichlseder and S. Lux, "Hydrogen Engine Exhaust Aftertreatment," in 43rd International Vienna Motor Symposium, Wien, 2022.
9. E. Esser, S. Kureti, L. Heckemüller, A. Todt, P. Eilts, T. Morawietz, A. Friedrich, W. Waiblinger, S. Hosseiny und F. Bunar, „Low-Temperature NOx Reduction by H2 in Diesel Engine Exhaust," SAE Technical Paper 2022-01-538, 2022.
10. P. Bloch, W. Mäurer, J. Geiler, E. Schünemann und M. Kirzinger, „Light Commercial Vehicle with a H2 Engine Hybrid Powertrain," in Powertrain with Renewable Energy Carriers, Stuttgart, 2023.

PHev towards Zero EmissioNs & Ultimate ICE Efficiency: the PHOENICE Project

Toni Tahouh[1], Mathieu André[1], Federico Millo[2], Luciano Rolando[2(✉)], Giuseppe Castellano[2], Andrea Bottega[2], Francesco Bocchieri[3], Alberto Migliorero[3], and Rosario Loiodice[3]

[1] IFP Energies Nouvelles I. Carnot IFPEN TE, Rueil-Malmaison Cedex, France
{toni.tatouh,mathieu.andre}@ifpen.fr
[2] Politecnico di Torino, Torino, Italy
{federico.millo,luciano.rolando, giuseppe.castellano,andrea.bottega}@polito.it
[3] FEV Italy, Via Paolo Borsellino, Torino, Italy
{bocchieri,migliorero,loiodice}@fev.com

1 Introduction

Nowadays the most effective path to tackle the urgent demand for a sustainable mobility is the fast deployment of a wide portfolio of technical solutions with a low carbon footprint [1]. As a matter of fact, single technologies could be very effective for a specific application, but they could hardly represent a silver bullet for the whole transport sector. In such a framework, the development of a new generation of Internal Combustion Engines (ICEs) capable to achieve efficiency levels beyond 45% can still represent a valuable method to curb the production of Green House Gas (GHG), especially when combined with powertrain hybridization [2].

A possible approach to achieve such an ambitious efficiency target relies on the combined use of high compression ratios and of ultra-lean combustion processes. Indeed, lean mixtures can mitigate both thermal and exhaust energy losses and could enhance the combustion phasing thanks to the reduced risk of knock due to their lower reactivity [3, 4]. An additional strategy enabling a further increase of the compression ratio could be the Exhaust Gas Recirculation (EGR) since it can further reduce the mixture reactivity [5, 6]. Furthermore, EGR could be applied during part load operation to minimize the pumping losses [7] through engine de-throttling. Similar benefits can also be achieved by integrating a Variable Valve Actuation (VVA) system. Indeed, the capability to control the valve lift profile could not only avoid the use of a throttle valve at part load but also enable a reduction in knock likelihood through the Millerization of the engine cycle [8, 9].

On the other side, the abovementioned technologies, particularly the Millerization of the cycle, typically require the development of advanced turbocharging configurations to achieve the desired levels of boost. Among the available solutions, electrified

A. Heintzel (Hrsg.): Int.Engine.Congr. 2024, Proceedings, S. 51–64, 2025.
https://doi.org/10.1007/978-3-658-47966-4_5

turbochargers represent one of the most promising to solve this task ensuring, at the same time, faster transient responses and partial recovery of waste enthalpy of the exhaust gases [10]. This recuperated energy could supply the electric ancillaries of the vehicle, yielding to an additional fuel consumption reduction up to 2–3% on the Worldwide Harmonized Light Vehicles Test Cycle (WLTC) [11].

In such a context, the PHOENICE (PHev towards zerO EmissioNs & ultimate ICE efficiency) H2020 European project aims to develop a plug-in hybrid powertrain integrated into a C-class Sport Utility Vehicle (SUV) striving for a 10% improvement in fuel economy compared to the baseline vehicle currently available in the European market and being compliant with the upcoming EU7 regulation [12]. The achievement of these challenging targets mainly relies on an innovative engine concept which is designed to attain an indicating peak efficiency of 47% and a Brake Thermal Efficiency (BTE) beyond 42% in a wide region of its operating map.

Therefore, this paper aims at proving the fuel economy potential of the PHOENICE engine prototype through an extensive experimental campaign performed on a fully dynamic test rig. First, the calibration of the main engine operating parameters performed in steady state conditions will be discussed proving the capabilities to achieve the target efficiency levels. Particular attention will be devoted to the optimization of the EGR levels and of the Air/Fuel Ratio (AFR) to maximize their benefits at both high and low loads. Afterwards, the digital twin of the engine, developed in a previous stage of the project, will be exploited to highlight the contribution of the new technologies integrated in the PHOENICE engine.

The paper is organized as follow: in Sect. 2 the key features of the engine prototype will be presented, while Sect. 3 describes the experimental campaign performed at the dynamic test rig. Sect. 4 assesses the benefits of the Dual Dilution combustion on a set of representative operating conditions previously identified through 0D/1D simulations. Sect. 5 highlights, through numerical simulations, the contribution of the new engine technology to the efficiency improvements. Finally, Sect. 6 wraps up the main findings of these studies and provides some hints on the future activities of the project.

2 The PHOENICE Engine Concept

The PHOENICE engine is based on a state-of-the-art 4-cylinder 1.3 L turbocharged Direct Injection Spark Ignition (DISI) engine [13] featuring an all-aluminum engine structure, a high stroke-to-bore ratio, a compact 4-valve combustion chamber with side fuel injectors, a MultiAir variable valve actuation system [14] and a cylinder head with integrated exhaust manifold. As it can be seen from Fig. 1, one of the main upgrades carried out during the project deals with the combustion system which was improved to ensure a good combustion stability in highly diluted conditions and to achieve 47% of peak gross indicated efficiency.

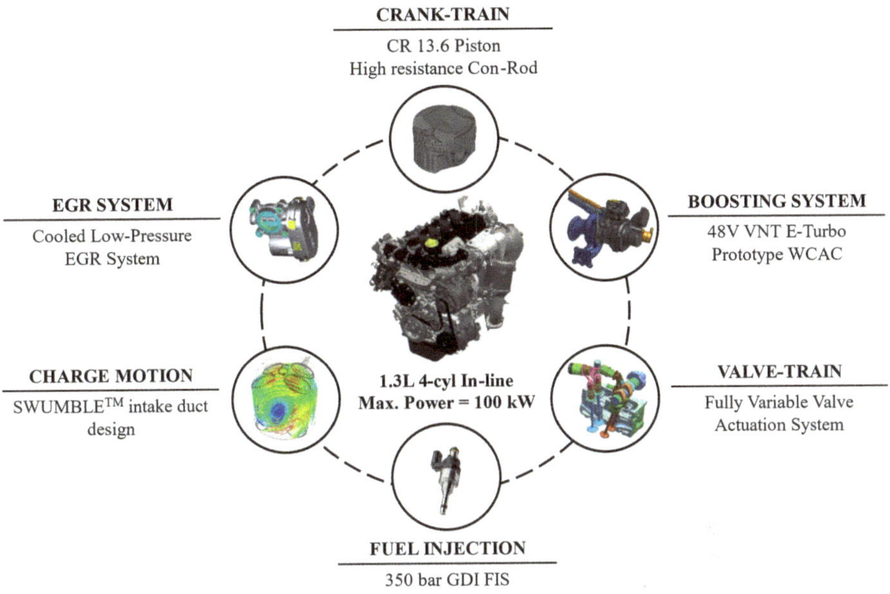

Fig. 1. Overview of the PHOENICE engine features

As a matter of fact, the PHOENICE concept adopts a Dual Dilution Combustion Approach (DDCA) which synergistically blends lean mixture with cooled low pressure EGR. To support the DDCA, the intake ducts and ports geometries have been redesigned to implement the Swumble™, a newly developed charge motion concept capable of consistently improving turbulence intensity by combining swirl and tumble in-cylinder motions with valuable effects on flame propagation [15–18].

Moreover, the Compression Ratio (CR) of the original engine was increased up to 13.6 through a new piston design and combined with aggressive Miller strategies obtained through the MultiAir System to limit the knock tendency.

On the boosting system side, a 48V electrified turbocharger embedding a Variable Nozzle Turbine (VNT) can be found. Its specifications were optimized in a previous work of the authors [18] to meet the requirements of highly diluted mixtures and aggressive Miller cycle. Finally, a 350 bar Fuel Injection System (FIS) has been implemented. The combination of all these technologies leads to the engine specifications summarized in Table 1.

Table 1. PHOENICE engine specifications

Specification	Value
N. of Cylinders	4 in-line
Displacement	1332 cm^3
Bore x Stroke	Ø 70 mm × 86.5 mm
Stroke/Bore	1.24
Compression Ratio	13.6:1
Number of Valves	16
VVA System	MultiAir III (intake only)
Fuel Injection System	350 bar GDI
Turbocharging	48V VNT E-Turbo
Rated Power (target)	100 kW @ 4500 RPM
Rated Torque (target)	218 Nm @ 3500 RPM

To guarantee EU7-compliant tailpipe emission levels, a high-performance Exhaust After-Treatment System (EATS) is required. Moreover, the highly lean engine operating conditions makes the use of a Selective Catalytic Reduction (SCR) mandatory for the conversion of NOx.

The EATS designed for the PHOENICE project is schematized in Fig. 2. This system is made up of two sections: a close-coupled section and an underfloor section.

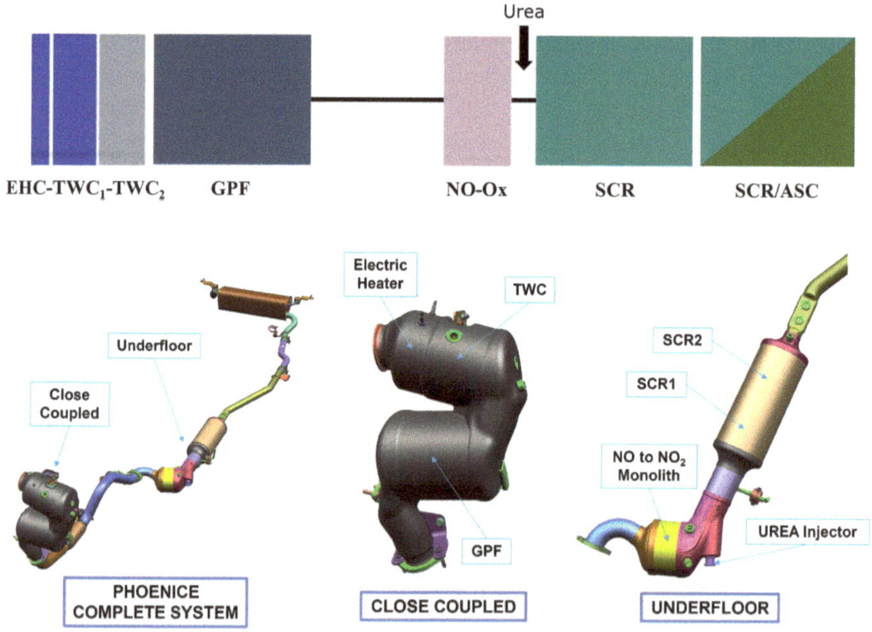

Fig. 2. PHOENICE EATS layout

The close-coupled section comprises an Electrically Heated Catalyst (EHC) and two Three-Way Catalyst (TWC) devices to control CO, HC and NOx gaseous emissions under cold-start and stoichiometric ($\lambda = 1$) conditions. The EHC features a coated metallic substrate which can be electrically heated during a cold-start to significantly shorten the TWC light-off time. If the vehicle can move on electrical power alone, then this procedure might be carried out even before the engine is started. Moreover, a coated Gasoline Particulate Filter (GPF) is used to trap particulate matter. To meet the demanding PN emissions limits prescribed by EU7, an innovative coating technology was used to deliver a step gain in filtration efficiency over the current EU6d-level technologies with an equivalent penalty in backpressure [17].

The underfloor section includes an SCR device to convert NOx under lean ($\lambda > 1$) conditions. Its conversion efficiency at low temperature is further enhanced by an upstream oxidation catalyst (NO-Ox) which converts a portion of the NO into NO_2. Finally, an Ammonia-Slip Catalyst (ASC) is integrated downstream to abate NH_3 slip which may otherwise result from urea overdosing or as a by-product of the reactions in the upstream catalyst devices under stoichiometric or rich ($\lambda < 1$) operating conditions.

3 Experimental Campaign at Engine Test Bench

An extensive experimental campaign was carried out on the test bench to optimize, under steady-state conditions, the calibration of the main PHOENICE engine actuations. In particular, injection pressures and phasing, Intake Valves Opening (IVO) and Closing (IVC) angles, boost pressures, ignition timings, AFR and EGR rates were taken into account. The activity covered all the relevant real-world operating conditions identified through 0D numerical simulations of Real-World Driving Emissions (RDEs) compliant test cycles (see Fig. 3).

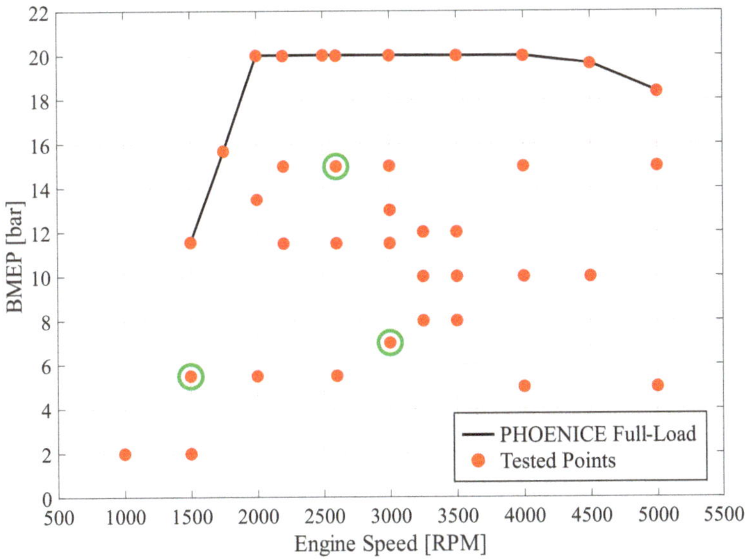

Fig. 3. Engine operating points for steady-state calibration

The experimental campaign was supported by 1D-CFD simulations: a digital twin of the PHOENICE engine previously developed by the authors [18, 19] was exploited to define a preliminary calibration of the valve phasing and to preliminary assess the capabilities of the combustion system [20]. In this work the same virtual test rig was used to identify the contribution of the new PHOENICE engine technologies to the achievement of the final efficiency levels.

4 Dual Dilution Combustion Assessment

This section discusses the effects of both lean A/F ratio and EGR on engine thermal efficiency and engine-out NOx emissions for the three relevant engine operating points highlighted by a green circle in Fig. 3: 1500 RPM and 5.5 bar BMEP, 3000 RPM and 7.0 bar BMEP, 2600 RPM and 15.0 bar BMEP. Next, the efficiency maps resulting from the optimal engine calibration are presented, with an emphasis on the improvements achieved over the reference engine.

4.1 1500 RPM and 5.5 bar BMEP

Figure 4 shows the trends of BTE, Coefficient of Variation (CoV) of Indicated Mean Effective Pressure (IMEP), engine-out Brake Specific NOx (BSNOx) and Spark Timing (ST) as a function of λ and EGR rate for 1500 RPM and 5.5 bar BMEP.

Under this condition, the engine can operate with acceptable levels of CoV up to a λ of 1.43 without EGR or up to a recirculation level of 19% in stoichiometric conditions. Further dilution of the charge cannot be realized since the potential efficiency benefits will be jeopardized by slower combustion speed. This effect is directly visible on the Spark Timing (ST), which must be advanced as the enleanment and/or the EGR rate increase to maintain the optimal combustion phasing.

From the BTE perspective, the increase of the A/F ratio represents the best solution with an improvement of more than 3% at λ=1.43. As a matter of fact, the higher air flow demand results in a wider throttle angle and, in turn, in a reduction in pumping losses. Furthermore, lower the combustion temperature, lower the heat losses.

On the other side, the EGR shows a marginal impact on BTE, but it has a significant benefit on engine-out NOx emissions. As a matter of fact, when the EGR increases, NOx emissions exhibit a sharp decrease with a reduction of more than 80% without appreciably compromising combustion stability.

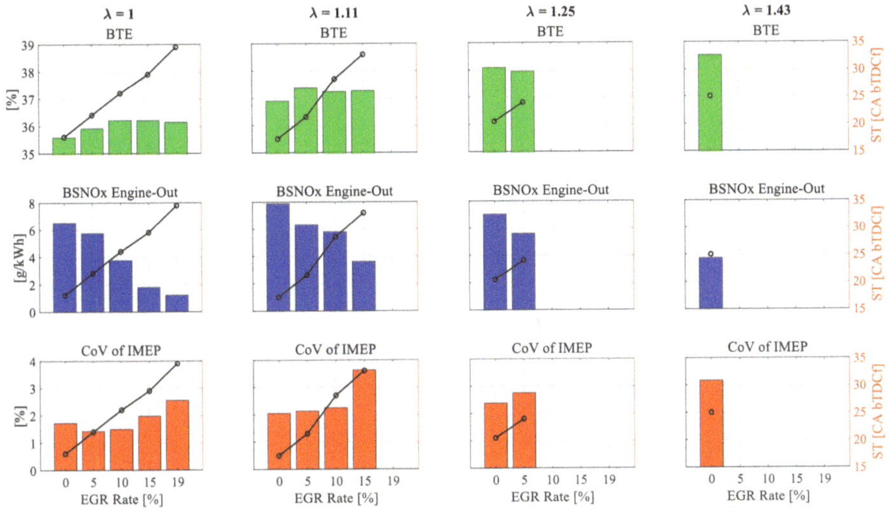

Fig. 4. 1500 RPM and 5.5 bar BMEP: BTE, BSNOx and CoV of IMEP

4.2 3000 RPM and 7.0 bar BMEP

Figure 5, which shows the trends of BTE, CoV of IMEP, engine-out BSNOx and ST as a function of λ and EGR rate for a medium speed and medium load operating point (i.e., 3000 RPM and 7.0 bar BMEP), mainly confirms the behavior highlighted in the previous paragraph. Indeed, lean A/F ratios are more beneficial for the efficiency while the use of EGR produces a higher NOx reduction.

Nevertheless, some differences can still be pointed out. First, greater dilution limits were achieved, reaching λ values up to 1.43 with EGR rate = 10% and EGR rates up to 21.5% in stoichiometric conditions thanks to the increased turbulence level resulting from the higher engine rotational speed which limits the cycle-to-cycle variability.

Secondly, in this case, the efficiency improvements can be mainly related to the reduction of the heat losses. As a matter of fact, at 7 bar BMEP, even in stoichiometric conditions, the throttle valve is almost fully open thanks to the Millerization of the engine cycle. With respect to the stoichiometric operation, the BTE increases by more than 2% when lambda is set to 1.43.

On the other side, EGR has a negligible impact on BTE (less than 1%) while being highly effective on the engine-out NOx emissions. At λ=1, NOx can be cut down by almost 90% at the maximum EGR rate of 21.5% without significantly compromising combustion stability.

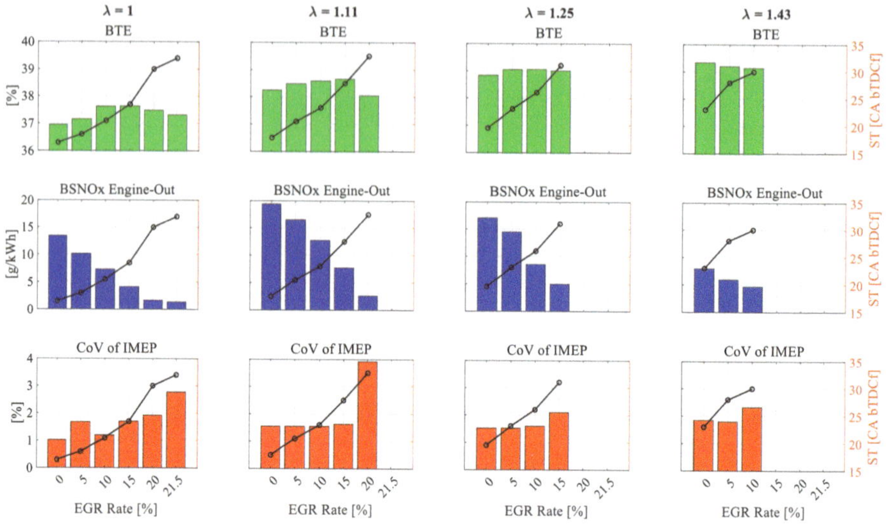

Fig. 5. 3000 RPM and 7.0 bar BMEP: BTE, BSNOx and CoV of IMEP

4.3 2600 RPM and 15.0 bar BMEP

As can be observed from Fig. 6, at high loads the range of available A/F ratios and EGR levels is quite limited with respect to the previous cases. Indeed, beyond these values, an excessive deterioration of the combustion stability was observed.

Due to the increased CR, the engine shows a high knock tendency which requires a very retarded combustion phasing. In this case, the knock mitigation capabilities of the EGR allow achieving BTE levels comparable with a lean A/F but with significantly lower NOx emissions. In stoichiometric conditions and with an EGR rate of 14.3%, the efficiency increase is 1% with a correspondent NOx emissions reduction of 70%.

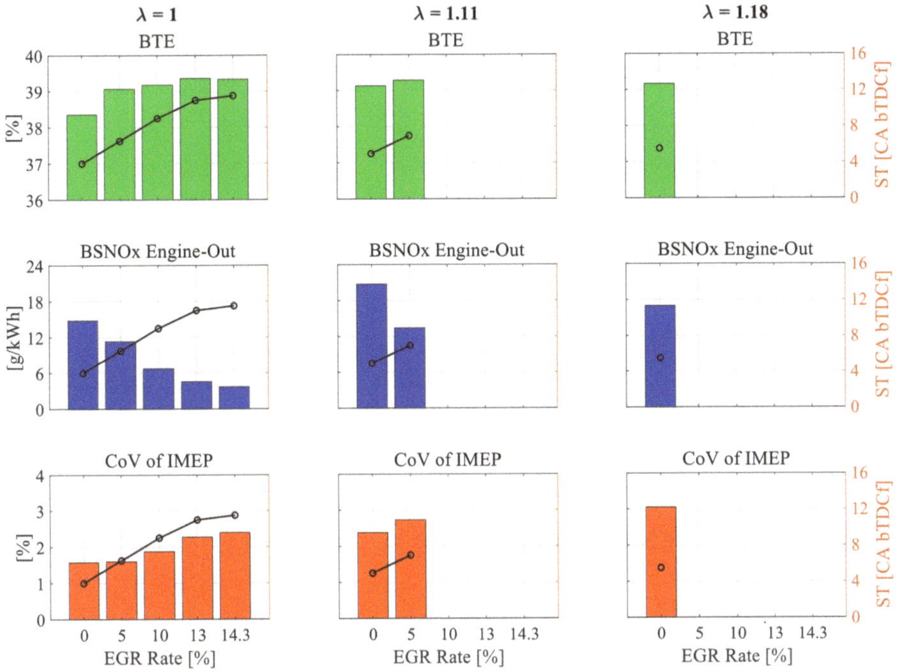

Fig. 6. 2600 RPM and 15.0 bar BMEP: BTE, BSNOx and CoV of IMEP

4.4 Engine Efficiency Assessment

The comprehensive analysis of the trends described in the previous paper section allowed the definition of the optimal engine calibration for the whole test matrix depicted in Fig. 3 and led to the efficiency levels shown in Fig. 7.

As demonstrated by Fig. 7(a), the target of 47% on gross Indicated Thermal Efficiency (ITE) set by the project was achieved at 3500 RPM 10 bar BMEP.

Moreover, Fig. 7(b) shows the relative BTE improvements of the PHOENICE engine compared to the reference engine. For a large portion of the map, these improvements are greater than 10%, proving the fuel economy potential of the PHOENICE engine concept. As will be better discussed in the following section of the paper, most of the contribution derives from the synergic use of the increased CR, of cycle Millerization and of lean operation. However, there are no advantages in the Low-End Torque (LET) region between 1500 and 2000 RPM. Indeed, the surge limit of the compressor limits the charge dilution which therefore cannot be exploited to enhance the combustion phasing thanks to its knock suppression capability.

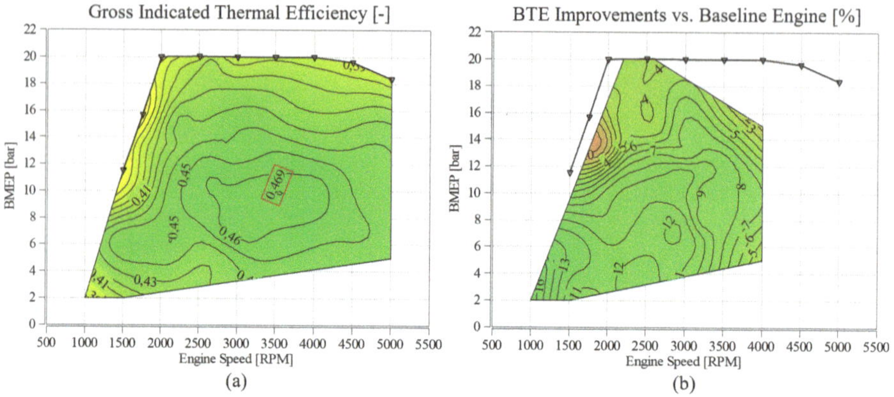

Fig. 7. **a** PHOENICE Gross Indicated Thermal Efficiency and **b** PHOENICE BTE improvements vs. baseline engine

5 Numerical Analysis on the Impact of the New PHOENICE Technologies on Engine Efficiency

Exploiting the digital twin introduced in Sect. 3, 1D-CFD numerical simulations were run to further investigate the benefits of the PHOENICE features on the engine efficiency. The findings of this analysis, which was carried out on a reduced set of meaningful operating points, are presented in Fig. 8 and are expressed in the form of net ITE gains.

The base 1.3L gasoline engine with a CR of 10.5, a waste-gate turbocharger and EU6d EATS, that operates under stoichiometric conditions without EGR, is referred to as the reference configuration.

The first contribution to the efficiency improvement is provided by the CR in combination with the optimized VVA strategy. When the engine speed exceeds 2500 RPM, a gain of more than 2% is recorded. Such a behavior can be motivated by the higher ideal thermodynamic efficiency and by the reduction in pumping losses enabled by the compression ratio and by the VVA system, respectively. However, below 2000 RPM, this efficiency improvement decreases because of the higher knock tendency, which requires the use of retarded spark timings.

As can be noticed, the potential negative effect of the higher backpressure related to the new exhaust system is counterbalanced by the enhanced efficiency of the VNT E-Turbocharger. It is worth mentioning that this analysis did not take electrical assistance into account.

Finally, focusing on EGR and lean mixture, the simulation results confirm what was already highlighted in paragraph 4. The higher air flow rate required by lean mixture and/or EGR results in a substantial reduction of the pumping losses thanks to a wider throttle angle for the low-speed operating points (1000 and 1500 RPM), and of the heat losses for the high-speed points.

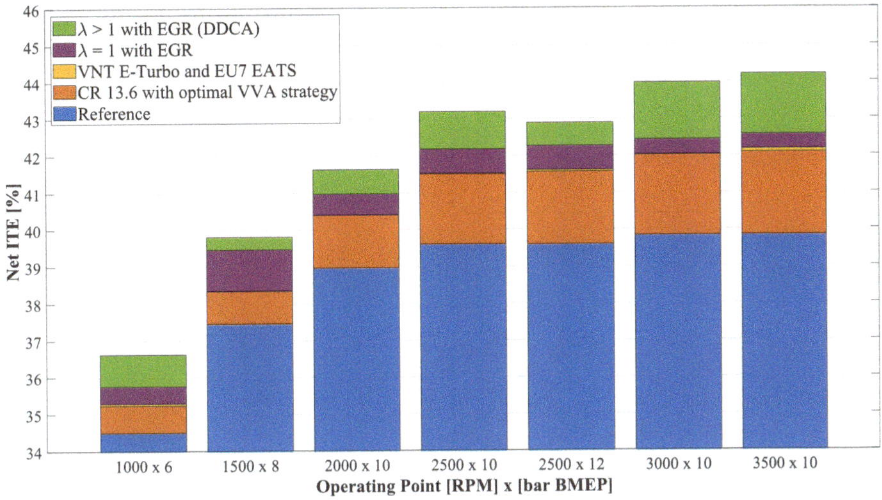

Fig. 8. Numerical results of the impact of the PHOENICE features on Net ITE

6 Conclusions

A new lean-burn spark ignition engine concept was developed based on a state-of-the-art 1.3L gasoline engine in the framework of the PHOENICE H2020 European project. Several innovative technologies were integrated in the engine prototype to support a dual dilution combustion approach and to maximize the synergy with the hybrid architecture of the PHEV demonstrator vehicle.

An extensive experimental campaign on a fully dynamic test rig was performed to optimize the main engine calibration parameters aiming at maximizing the engine thermal efficiency. This calibration effort was carried out on a set of relevant operating conditions previously identified by means of 0D/1D numerical simulations. In particular, the benefits of a dual dilution combustion system, realized through mixture enleanment and EGR dilution, on engine thermal efficiency and engine-out NOx emissions were discussed. Very encouraging results were achieved with a gross indicating efficiency close to the 47% target for the operating points at 3500 rpm × 10 bar BMEP. The comparison of the BTE between the PHOENICE prototype and the reference engine also showed a promising 10% improvements on a wide region of the operating map proving that the project ambitious target to reduce by 10% the fuel consumption of the vehicle on CS mode on WLTC can be achieved.

Afterwards, the digital twin of the engine, developed in a previous stage of the project, was exploited to highlight the contribution of the new technologies integrated in the PHOENICE engine, showing the benefits of the use of an increased CR coupled with an optimal VVA strategy, of the VNT E-Turbo and the EU7 EATS, and at last the impact of lean mixture and EGR dilution.

Future steps of the project will focus on further refining the engine calibration in transient conditions and on the assessment of the engine fuel economy potential in both regulatory driving cycles and real-world driving conditions.

Acknowledgements. This project has received funding form the European Union's Horizon 2020 research and innovation program under Grant Agreement No. 101006841.

Acronyms

AFR: Air/Fuel Ratio
ASC: Ammonia-Slip Catalyst
BTE: Brake Thermal Efficiency
CoV: Coefficient of Variation
CR: Compression Ratio
DDCA: Dual Dilution Combustion Approach
DISI: Direct Injection Spark Ignition
EATS: Exhaust After-Treatment System
EGR: Exhaust Gas Recirculation
EHC: Electrically Heated Catalyst
FIS: Fuel Injection System
GPF: Gasoline Particulate Filter
ICE: Internal Combustion Engines
IMEP: Indicated Mean Effective Pressure
IVO: Intake Valve Opening angle
IVC: Intake Valve Closing angle
ITE: Indicated Thermal Efficiency
LET: Low-End Torque
LIVC: Late Intake Valve Closing
PHOENICE: PHev towards zerO EmissioNs & ultimate ICE efficiency
RDE: Real Driving Emissions
SCR: Selective Catalytic Reduction
ST: Spark Timing
SUV: Sport Utility Vehicle
TWC: Three-Way Catalyst
VNT: Variable Nozzle Turbine
VVA: Variable Valve Actuation
WLTC: Worldwide Harmonized Light Vehicles Test Cycle

References

1. Ulrich Kramer, David Bothe, Christoph Gatzen, André Pfannenschmidt, Carolin Baum, Fabian Schrogl, and Osama Mahmood, "Transformation of Mobility to the GHG-neutral Post-fossil Age," Frankfurth, 2022.
2. Yang, D., Lu, G., Gong, Z., Qiu, A., and Bouaita, A., "Development of 43% Brake Thermal Efficiency Gasoline Engine for BYD DM-i Plug-in Hybrid," SAE Technical Papers, SAE International, 2021, https://doi.org/10.4271/2021-01-1241.
3. Osborne, R., Lane, A., Turner, N., Geddes, J., Atkins, P., Pohorelsky, L., Gidney, J., and Cleeton, J., "A New Generation Lean Gasoline Engine for Premium Vehicle CO2Reduction," SAE Technical Papers, SAE International, 2021, https://doi.org/10.4271/2021-01-0637.
4. Bunce, M., Peters, N., Pothuraju Subramanyam, S.K., Blaxill, H., Gao, J., and Choi, E., "The Impact of Advanced Fuels and Lubricants on Thermal Efficiency in a Highly Dilute Engine," SAE Technical Papers, SAE International, 2021, https://doi.org/10.4271/2021-01-0462.
5. Francqueville, L. and Michel, J.B., "On the Effects of EGR on Spark-Ignited Gasoline Combustion at High Load," SAE Int J Engines 7(4):1808–1823, 2014, https://doi.org/10.4271/2014-01-2628.
6. Kumano, K. and Yamaoka, S., "Analysis of knocking suppression effect of cooled egr in turbo-charged gasoline engine," SAE Technical Papers, SAE International, 2014, https://doi.org/10.4271/2014-01-1217.
7. Potteau, S., Lutz, P., Leroux, S., Moroz, S., and Tomas, E., "Cooled EGR for a Turbo SI Engine to Reduce Knocking and Fuel Consumption," in: SAE, ed., Powertrain & Fluid Systems Conference and Exhibition, SAE, Rosemont, 2007.
8. Millo, F., Luisi, S., Borean, F., and Stroppiana, A., "Numerical and experimental investigation on combustion characteristics of a spark ignition engine with an early intake valve closing load control," Fuel 121:298–310, 2014, https://doi.org/10.1016/j.fuel.2013.12.047.
9. Luisi, S., Doria, V., Stroppiana, A., Millo, F., and Mirzaeian, M., "Experimental Investigation on Early and Late Intake Valve Closures for Knock Mitigation through Miller Cycle in a Downsized Turbocharged Engine," SAE Technical Papers, SAE International, 2015, https://doi.org/10.4271/2015-01-0760.
10. Davies, P., Bontemps, N., Tietze, T., and Faulseit, E.T., "Electric Turbocharging-Key Technology for Hybridized Powertrains," MTZ Worldwide 80:30–37, 2019, https://doi.org/10.1007/s38313-019-0096-y.
11. Cooper, A., Reader, S., Bassett, M., Hall, J., Stansfield, P., Hartland, J., Cao, L., and Taylor, A., "HyPACE-Hybrid Petrol Advanced Combustion Engine," 27th Aachen Colloquium, Aachen: 167–192, 2018.
12. European Commission, "Regulation of the European Parliament and of the Council on Type-approval of Motor Vehicles and Engines and of Systems, Components and Separate Technical Units Intended for such Vehicles, with Respect to Their Emissions and Battery Durability (Euro 7)," Bruxelles, 2022.
13. Marino, C. De, Maiorana, G., Pallotti, P., Quinto, S., and Sacco, D., "The Global Small Engine 3 and 4 Cylinder Turbo: The New FCA's Family of Small High-Tech Gasoline Engines," in: Fortschritt-Berichte VDI, ed., 39. Internationales Wiener Motorensymposium, Fortschritt-Berichte VDI, Wien, 2018, https://doi.org/10.51202/9783186807120-129.

14. Bernard, L., Ferrari, A., Micelli, D., Perotto, A., Rinolfi, R., and Vattaneo, F., "Electro-hydraulic Valve Control with MultiAir Technology," MTZ Worldwide 70:4–10, 2009, https://doi.org/10.1007/BF03226988.
15. Bourhis, G., Laget, O., Kumar, R., and Gautrot, X., "Swumble In-Cylinder Fluid Motion: A Pathway to High Efficiency Gasoline SI Engines," 27th Aachen Colloquium Automobile and Engine Technology, Aachen: 1113–1132, 2018.
16. Gautrot, X., Bardi, M., Leroy, T., Luca, P., Nowak, L., and Reveille, B., "Swumble TM In-Cylinder Fluid Motion for High Efficiency Gasoline SI Engines: development of the second generation," SIA Powertrain and Energy, 2020.
17. Tahtouh, T., Lucignano, G., Bocchieri, F., Castellanos, J.S. et al., "A Methodology to Develop a Sustainable, High Efficiency Hybrid Propulsion System: the PHOENICE Project.", Proceedings of the FISITA 2023 World Congress, Barcelona, 12 – 15 September 2023.
18. Tahtouh, T., Millo, F., Rolando, L., Castellano, G. et al., "A Synergic Use of Innovative Technologies for the Next Generation of High Efficiency Internal Combustion Engines for PHEVs: The PHOENICE Project," SAE Technical Paper 2023–01–0224, 2023, https://doi.org/10.4271/2023-01-0224.
19. Tahtouh, T., Andre, M., Millo, F., Rolando, L. et al., "Development of a Digital Twin to Support the Calibration of a Highly Efficient Spark Ignition Engine," SAE Technical Paper 2023–01–1215, 2023, https://doi.org/10.4271/2023-01-1215.
20. Rolando, L., Millo, F., Castellano, G., Tahtouh, T. et al., "A Numerical Model for the Virtual Calibration of a Highly Efficient Spark Ignition Engine," SAE Technical Paper 2023–32–0059, 2023, https://doi.org/10.4271/2023-32-0059.

On-Board Monitoring for EU7—Evolution vs Revolution

Martin Piffl(✉), P. Goetschl, K. Klumaier, H. Atzler, R. Ruiz-Carrillo, R. Gunja, H. Wancura, and H. Herrmuth

AVL List GmbH, Graz, Austria
{martin.piffl,peter.goetschl,kurt.klumaier, hannes.atzler,ruben.ruiz-carillo,robert.gunja,hannes.wancura, holger.herrmuth}@avl.com

1 Status Quo

1.1 OBM Legislation Outlook

According to the outcome of the EU7 trialogue negotiations, from 18th of December 2023, OBM is introduced by end of 2026 for new type LD applications (main regulation +30 months) and mid of 2028 for new type HD applications (main regulation +48 months). The latest corresponding documents claim OBM limits for NOx and PM (LD) and NOx, NH_3 and PM (HD). The OBM system should monitor emissions, detecting and reporting any potential exceedance of emission limits by a factor of 2.5. This includes, for example, that the OBM system must be capable of communicating the emission information, metadata, and state-of-health information over-the-air and (or) to the vehicle user [2, 3].

1.2 Conventional OBD (EU6/VI-Approach)

With the initial implementation of OBD I in Europe in 2004, diagnostic requirements have been continuously enhanced, particularly in relation to emission-relevant systems or components (hereafter referred to as 'systems').

System- and attribute-specific monitoring functions have been developed over the years to make emission-critical systems detectable. The legislator has thereby permitted for a higher OBD threshold limit to facilitate clear attribution to a critical system attribute through a prototype application for the entire series, while simultaneously having required a defined monitoring frequency through IUPR. In practice, drifting and aging phenomena of systems may be partly compensated by robust calibration and adaptation functionalities [4]. For CI applications the adaption of the urea dosing control is well known, for example. However, if the system aging or drift exceeds a certain degree, the engine control, especially during high-transient operation, can no longer compensate for the negative effect on tailpipe emissions, leading to violations of legal

A. Heintzel (Hrsg.): Int.Engine.Congr. 2024, Proceedings, S. 65–78, 2025.
https://doi.org/10.1007/978-3-658-47966-4_6

emission limits (see Fig. 1). If the result of a monitoring function consequently implies an exceedance of the OBD threshold limit, the critical system attribute is indicated by an error code, and the driver is notified through the engine control light (or "MIL", Malfunction Indicator Light) [5].

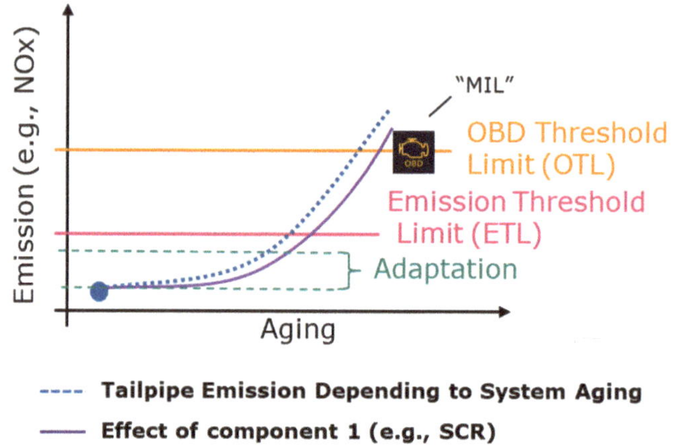

Fig. 1. Aging/Drift Compensation by Adaptation Functions (Simplified View)

1.3 Transition from Component to Emission Monitoring

Robust calibration, adaptation and OBD are established solutions to prevent from emission critical system aging. Still, there are weaknesses of the conventional development approach to be considered. Partial aging or partial drifting of multiple systems simultaneously may result in an overall emission exceedance without beeing sufficiently compensated by adaptation functions or beeing recognized by OBD (see Fig. 2).

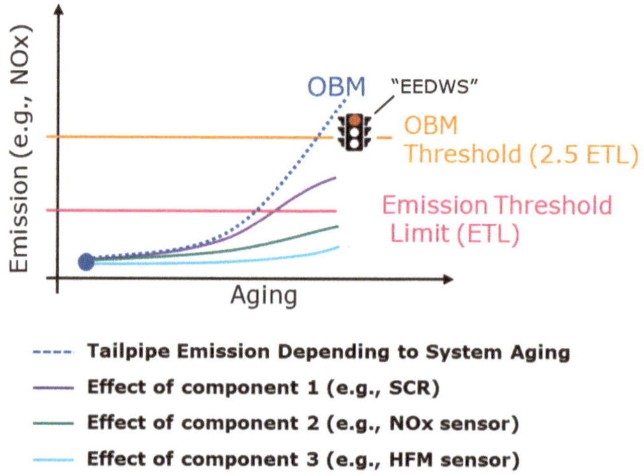

Fig. 2. Emission-Critical Effect of Multiple Partial Degradation (Simplified View)

There are system properties that only become visible under specific operation conditions and, therefore, do hardly meet the legally required monitoring frequency (IUPR). As an example, a turbocharger damage can be considered leading to an over-boost scenario, which is only detectable during full-load operation or in combinations with additional failures, but not under normal use. Last but not least, there is the thread of "intelligent tampering," which involves the deliberate bypassing of monitoring functions by manipulating the engine control unit data or sensor transfer curves, leading to unauthorized but undetected emission exceedances. Latter weaknesses of OBD could be weeded out with the transition to OBM.

The successful transition from emissions-critical system monitoring via OBD to emission monitoring via OBM lies ahead of us. The new requirement comes along with two significant challenges, we want to tackle in the following. These are: If the "EEDWS (Excess emissions driver warning system)-traffic-light" becomes activated due to multiple partial degradated systems, where conventional OBD is "blind", a repair proposal will be missing (see Fig. 2). And secondly, emissions need to be precisely monitored on-board over the entire vehicle's lifetime (i.e., effect of individual system aging is to be properly considered).

1.4 Sensor Concepts for of EU6E Applications and Their Use for OBM

In view of sensor concepts and their availability for the use of OBM, there is a strong differentiation between spark ignited (SI) and compressed ignited (CI) applications to be considered.

While, for SI applications, no direct emission measurements are required to establish the emission control via TWC (Three Way Catalyst) or FWC (Four Way Catalyst), it is widespread practice on CI applications to consider multiple NOx sensors in the control logic. This is typically done to balance the NH_3 Loading on SCR catalysts and to monitor the SCR efficiency (OBD). It is essential for SCR control functions to have reliable replacement values in case of absent sensor information such as during cold start and warm up phases (e.g., dewpoint release), or if sensors are faulty. These replacement values are calculated via models (or so called "virtual sensors"), which have already a high degree of accuracy in CI applications since physical modeling approaches are used for engine out as well as exhaust aftertreatment simulation. Additionally, due to the cross sensitivity of NOx sensors towards NH_3, reliable NH_3 calculations can be performed without a dedicated NH_3 sensor. CI applications are so in a very pleasant state regarding signal availability for the emission calculation for OBM. Here, an evolutionary approach seems to be sufficient to cope with the upcoming OBM requirements.

For SI applications, a revolutionary approach is needed to meet the emphasized challenges of OBM, because there are no NOx sensors available, and the simulation of SI tailpipe emissions is complex.

2 Evolution

The next chapter will discuss the present abilities of SI and CI emission simulation controls functions that can be repurposed and expanded for OBM objectives.

2.1 State of the Art Capabilities for Virtual Emission Sensors

To quantify the OBM requirements related to emission monitoring in situations where sensor information is unavailable, it is crucial to comprehend the accuracy with which emission species can be modeled. The accuracy has a high impact on the sensitivity and precision of the EEDWS activation (OK or NOK System), triggered by calibrated thresholds.

If we look at the capabilities of existing emission models to determine engine-out emissions, which are the emissions before any catalytic post processing of the exhaust gas is done, SI and CI concepts can reach a similar accuracy level. Therefore, map-based empirical approaches and physics-based approaches are imaginable, with "pros and cons" on both sides (see Fig. 3, left image). When it comes to the modelling of tailpipe emissions a stricter distinction between SI and CI application is necessary.

For CI applications the HC, CO and PM emission do not play a major role in lean operation. NOx and NH_3 are usually modeled or measured to a high degree of accuracy to ensure that the SCR system consistently operates at close to 100% overall efficiency once the system is heated up. Due to this, the relevant error on NOx side is minimized and can be understood as negligible. The risk of significant inaccuracies towards the actual tailpipe emissions is less and requires only a minor safety margin on the traffic light setpoint.

On the other hand, SI tailpipe emissions models, as integral part of the ECU, have been no topic until now. Tailpipe emissions formation is mainly driven by cold start engine-out emissions and catalyst light-off behavior. After this phase an accurate downstream lambda control, using binary sensors after the TWC, is the key to keep overall emissions on a low level. This leads to the fact, that ECU-tailpipe emissions models are currently not available. For modelling of SI aftertreatment systems, a correct chemical consideration of all effects is required to ensure a reliable EEDWS traffic light in future vehicles. A simplified approach has the risk that TWC mechanisms are incorrectly predicted, which might lead to high deviations between modelled and real tailpipe emissions, and so to wrong activations of the EEDWS (see Fig. 3, right image).

A holistic new approach for emissions determination of SI is required.

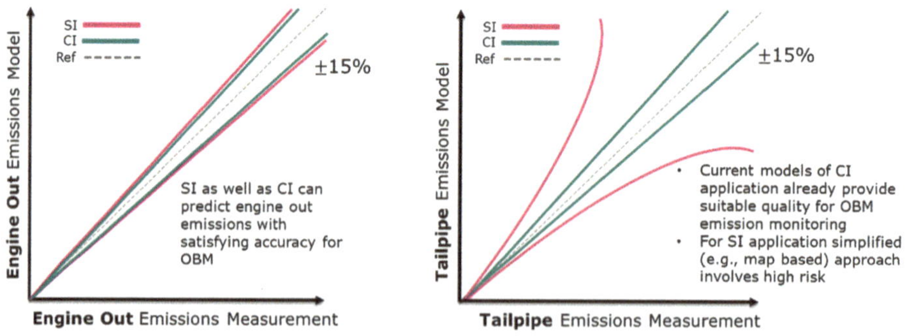

Fig. 3. Emissions modelling—Evolutional Approach (schematic view)

2.2 Refinement of Virtual Emission Sensors for CI Applications

While a satisfactory level of emission modeling has been attained for CI applications under most environmental conditions, models have not been specifically designed to capture emission behavior when the SCR (Selective Catalytic Reduction) is still operating outside its optimal window. This is because its impact on the SCR's balance is low. However, for OBM, there is a need for a continuous signal stream providing reliable emission information. In particular, the existing logic requires enhancement to address cold start conditions and operation at high altitudes. The fundamental concept of the current logic used for emission calculation and modeling of NOx and PM emissions can be retained. The extension of these existing models to simulate NH_3 within the SCR model (AVL kinetic SCR model and Control) and the calculation based on the NOx sensor signal will require special attention to ensure comprehensive signal availability for OBM.

Examining the emission modeling quality illustrated in Fig. 3 and taking into account the straightforward modifications needed for CI applications to achieve the required accuracy of virtual emission sensors for OBM, the evolutionary approach emerges as a viable and effective strategy to meet these new challenges. Nevertheless, the revolutionary approach is a valid development approach too, especially when it comes to the pinpointing needs on emission critical partial degradation.

2.3 Connecting Virtual and Real Emissions Sensors

Regardless of the source of emission information, the OBM system needs to have the ability to accurately provide time-resolved and integrated emission information to the OBD port after every key-cycle. The evaluation of emission information must consider both, technical and legislative boundary conditions. Examples in this context include the readiness of NOx sensors, RDE boundary conditions, or minimum data lengths. Furthermore, OBM system must process emission inputs following the same rules as PEMS, as authorities will compare emissions reported by OBM system with PEMS emissions during the ISC testing [6].

To combine the emission data from virtual and real sensors, currently, AVL is investigating a Kalman Filter-based approach, delivering so far promising results. The evaluation of NOx emissions (1) follows the following formula:

$$NOx_{est} = NOx_{mod} + K1 \cdot (NOx_{sens} - NOx_{mod}) \tag{1}$$

with

$$K1 = \frac{\left(\sigma_{NOx_{mod}}\right)^2}{\left(\sigma_{NOx_{mod}}\right)^2 + \left(\sigma_{NOx_{sens}}\right)^2} \tag{2}$$

Kalman filter ponders the weighting fraction of virtual and real sensor signals based on their quality, by means of the Kalman gain (K1) as described in (2). Kalman gain can adopt values between 0 and 1, weighting between the input data. A higher Kalman-Gain value gives more weight to sensor input, on the contrary, lower values shifts higher priority to the virtual sensor input. The factor K1 is calculated from expected uncertainties from the virtual sensor and their corresponding inputs. On the other hand, for the NOx sensor, the standard deviation is determined from sensor measurements. Self-learning algorithms, like MSA (see Sect. 3.3), might provide an indication of the expected sensor uncertainty, based on the live processing of engine data. Alternatively, uncertainty can be pre-calibrated and corrected with basic information about the state of the system (e.g., ageing effects, manufacturing tolerances).

The Kalman filter continuously updates its formulas (1) and (2) as new data becomes available, effectively balancing the trust in the virtual sensor and the trust in the measurement to provide refined and more accurate calculated emission results. The resulting outcome sets the foundation for time-resolved and ISC-processed distance-specific emissions, to fulfill envisioned OBM requirements. Additionally, the Kalman filter can generate a merged estimation of the emissions uncertainty, providing therefore the necessary means to assess the quality of the emissions values stored in the engine control unit within a predefined interval of confidence. Figure 4 depicts an example based on vehicle data.

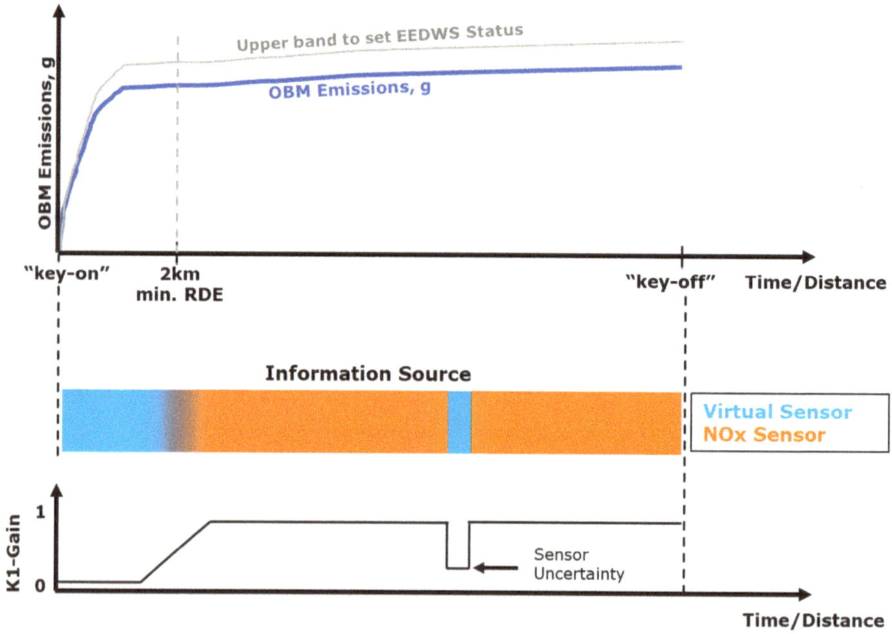

Fig. 4. OBM Kalman filter working principle

2.4 How can Traditional OBD fulfill OBM Requirements?

The OBM system task is to compare recorded emission results with calibrated OBM threshold limits. However, explicitly identifying the root cause of an emission violation is not in the focus of the OBM system. It is important to note that OBD thresholds, as defined in the EU6/VI standards, may not perfectly align with OBM thresholds, and, for the most part, neither do the corresponding monitored KPIs. Despite this disparity, there is an expectation that various OBD functions can be replicated and enhanced specifically for pinpointing issues when violations occur within the OBM system. This adaptability highlights the potential for reusing and further developing OBD functions to address the pinpointing challenge of OBM.

We anticipate evolutionary approaches in monitoring functions aiming to achieve valuable pinpointing results when OBM thresholds are violated. One potential approach involves adapting the thresholds of the replicated OBD functions to OBM thresholds, especially in cases, where monitored KPIs exhibit correlation or even identical characteristics to monitored emissions. As for those the configuration is not mandated by regulations (e.g., IUPR), the ability to detect issues over multiple cycles permits the development of pinpointing functions beyond OBD constraints. This facilitates the acquisition of pinpointing results with a low misdetection rate. A further potential evolutionary pinpointing approach is an on-board supported off-board system for assisting vehicle workshops. The on-board system is an empirical approach for judging the potential root cause for emission exceedance based on development data. The simplified models for determining emission influence in faulty cases are multiplied with the actual sensor/actuator state of health information, providing hints for repair strategies in the workshop.

For future solutions, the selected data will be transmitted Over-The-Air (OTA) from the vehicles to OEM clouds/servers. The stored engine- and aftertreatment models will calculate the actual state of health for all sensors and actuators and predict if, when, and why the emissions will exceed the OBM threshold. These results can be accessed by workshops to propose the best maintenance and repair plan for each vehicle in the fleet.

3 Revolution

3.1 Real-Time Capable Models for Development Tasks

We look back on more of a decade of successful implementation and utilization of real-time capable models in the development process [7] [8]. For specific tasks, the ability to accurately predict tailpipe emissions and to develop system-oriented software components provides a solid basis for the next step, which involves integrating parts of our modeling capabilities into OBM.

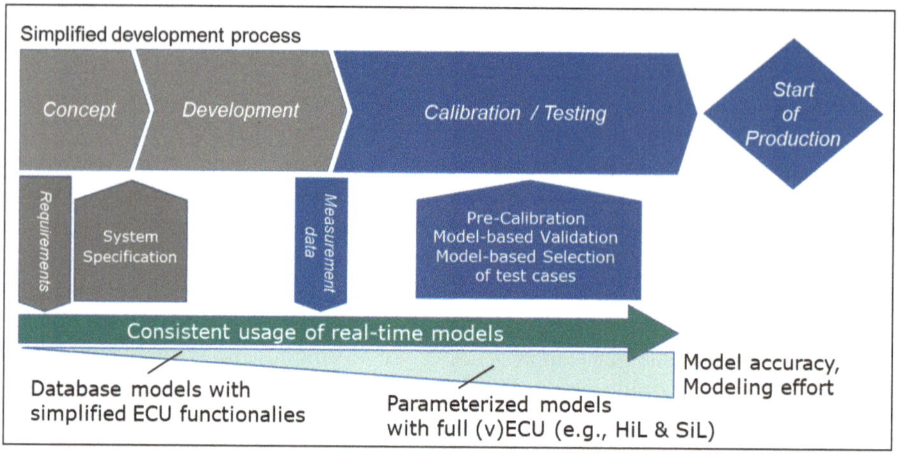

Fig. 5. Simplified Development Process

The simplified development process in Fig. 5 gives an indication of the evolution and the usage of the specific models. At the beginning of the concept phase the simulation environment is set up with all the sub models and interfaces which are required during the development process. The model accuracy is increased stepwise when new data and specifications are available. Our experience collected on the HiL and SiL environments as development tools on the way to SOP serve as profound base for the revolutionary step using intensified virtualization inside the ECU to support OBM fulfillment by virtual sensors.

3.2 Virtual Emissions Sensors for SI Applications

As highlighted in the previous chapters modelling of emissions to fulfill upcoming OBM requirements requires a revolutionary approach for SI applications. Since a first insight in the AVL approach how to model tailpipe emissions for SI applications was given in last year's conference [1] we can focus on respective updates in following section.

AVL favors the approach to model engine out emissions having its origin in the combustion and the aftertreatment system separately for all emission species. Both models are semi-physical to reduce calibration effort on the one hand and being able to extrapolate against hardware deviations (e.g., production tolerances or aging) or even tampering without extensive measurement campaigns during the calibration phase [9]. Our modeling capabilities, as shortly mentioned in Sect. 3.1, presents as effective groundwork for pinpointing potential underlying reasons for an increase in vehicle emissions. It also lays the foundation for future decision-making on how to predict emission species that are presently not detectable by sensors. We proved that an engine and aftertreatment models can be technically operated on a serial ECU. The resulting load was evaluated on a HiL testbed with a state-of-the-art processor is shown in Fig. 6.

CPU Core
- 32-Bit Aurix TC387, Type TC 1.6.2
- 4 Cores, 2 Checker Cores, 300MHz
- Cache per CPU: Program (32kB), Data (16kB)
- SRAM per CPU: PSPR (64kB) DSPR (240kB), DLMU (64kB)
- SRAM global: LMU (128kB), DAM (64kB)
- Flash: Program (10MB), Data (512kB)

	Combustion Model	Aftertreatment Model
Flash	~26.1 kB	~101 kB
RAM	~6.7 kB	~37.6 kB
Execution Time	<1ms	~15ms
Max CPU Load	3%	15%

Fig. 6. ECU Load for SI Emission Models

For the implementation of virtual emission sensors these models are integrated in ECU code in a way that…

- sub model components, which are already existing in the ECU itself—such as air charge determination or exhaust temperature model—are used a potential input.
- real sensor signals serve as an additional input to the models. This helps to increase the vehicle-individual quality and stability of the model (e.g., refer to Sect. 2.3). Possible mismatches between sensor values and model outputs can be revealed too.
- existing OBD function outputs are connected to the models. For instance, the oxygen storage evaluation of catalyst diagnosis is used as an input to the aftertreatment model (e.g., refer to Sect. 2.4).

Regarding the achievable accuracy the following example shall be provided. Data was collected from an ongoing comprehensive vehicle measurement program, which involved vehicle testing in a fault-free condition and with deliberately induced OBD relevant system faults. The virtual emission sensors for NOx, CO, and THC were stimulated using the available ECU data. Tailpipe emission results were compared for various vehicles, test drives and various OBD error-cases. A summary of the results is provided in Figs. 7 (multiple tests for fault free systems) and 8 (tests with induced errors)

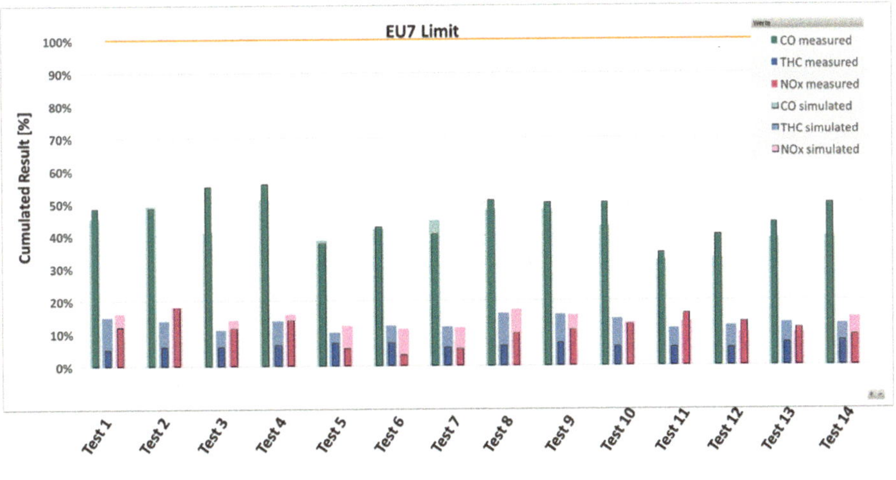

Fig. 7. Vehicle A—*F*ault *F*ree *S*ystem

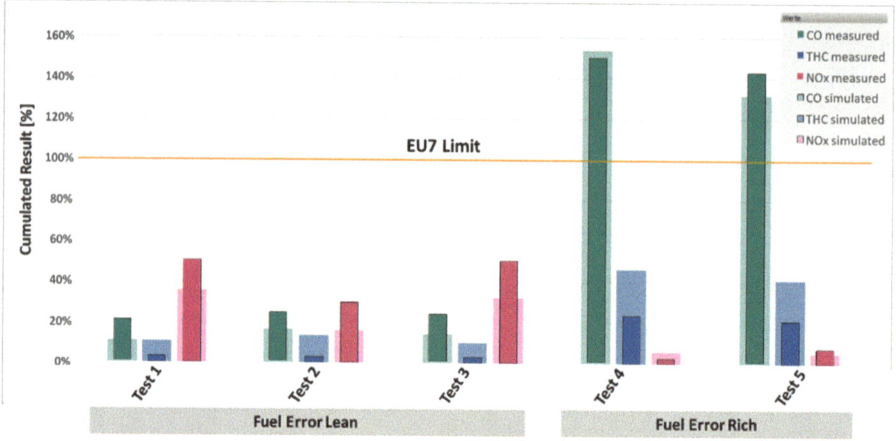

Fig. 8. Vehicle B—WLTC with Induced Fuel System Errors

In both cases, a sufficiently accurate calculation of the virtual emission sensors was achieved.

3.3 Model-based Signal Analysis (MSA), a Revolutionary Approach for Component Pinpointing

The OBM legislation comes along with a significant challenge in addition to the accurate on-board monitoring of emissions. That is developing a working approach capable of robustly and frequently pinpointing on emission critical hardware deviations so that questions as the following can be answered: What should be done in the workshop when there is an OBM exceedance followed by a EEDWS activation, but conventional or evolutionary OBD does not provide any indication of a system fault in the engine? How can it be ensured that the virtual emission sensors age virtually in sync with the real system aging (concept of the "Digital Twin")? We are optimistic to find answers to these questions with our revolutionary "MSA approach".

The concept and calibration of OBD tasks have been largely addressed through expertise in the past. However, with the new legal requirements for OBM, increasing system complexity, and Time-to-Market constraints, the task is becoming more complex and needs to be addressed through data-driven and systematic approaches. In general, the detectability of hardware deviations primarily depends on the vehicle operation. In an exaggerated sense, this should become obvious by the following simple example: Diagnosing more complex components, such as those related to reduced catalytic converter capacity, necessitates specific operational conditions, either highly stationary or transient. In addition to the manner and duration of the vehicle's operation, initial system startup conditions, such as substrate or operating fluid temperatures, learning or adaptation memory, and the state of battery charge, also significantly impact detectability. Considerations may include cold or warm start conditions, which can influence overall tailpipe emissions differently, even with identical vehicle operations. Finally, hardware deviations in other system properties, like sensor drifts or injection system variations, can not only affect overall performance and emissions but also obscure the visibility of the aged system being detected, such as reduced catalytic converter capacity, by distorting the engine control unit data stream (see Fig. 9).

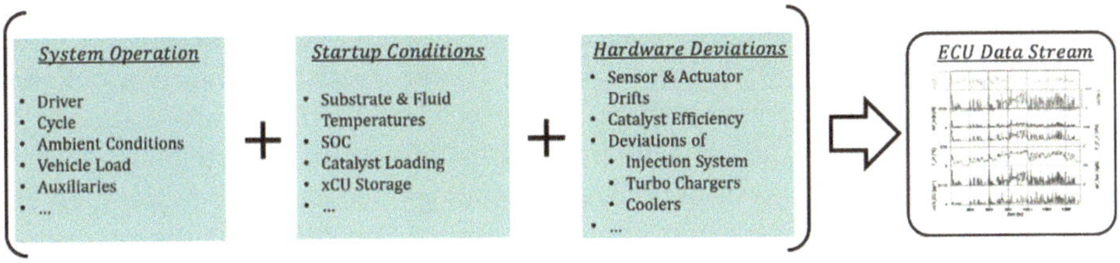

Fig. 9. Variation Parameters influencing the ECU Data Stream

Thus, to make a specific hardware deviation detectable during vehicle operation, it is necessary to identify "monitoring-channels" in the ECU data stream that respond sensitively to this deviation. Additionally, specific conditions need to be defined (i.e., release/enabling conditions) under which the identified monitoring-channels are not distorted by other cross-influences. While this approach is well-known from OBD, it still requires system knowledge and experience for its implementation, which poses a challenge, especially when new technologies evolve. As a response to that, AVL has developed the MSA approach [10]. MSA generates in a systematic and generic way release conditions and an associated monitoring function for each hardware deviation (or aging phenomenon), which can be described within a simulation environment (e.g., SiL or HiL).

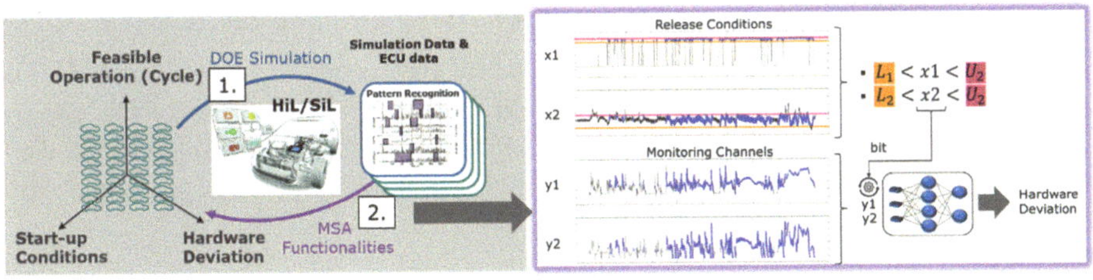

Fig. 10. MSA Methodology

As visible in Fig. 10, in the first of two steps, the feasible "real-world variability" in terms of operation, startup conditions and hardware variation is explored with an online DoE approach. Typically, 100 to 1000 experiments are simulated, where all variation parameters are varied simultaneously. In the second step, the simulated ECU data streams are processed with pattern recognition and machine learning methods. As a result, release conditions, monitoring-channels and corresponding monitoring functionalities are obtained that estimate the hardware deviations as imposed in step 1.

Thereby, it is ensured that the monitoring function remains robust against other variations and the release conditions can be frequently met during normal usage.

MSA shall be demonstrated by the following example: In a SiL environment modeling a vehicle, its engine, and its controllers, 40 hardware deviations were simulated in various combinations through approximately 1000 experiments. Among them, the hole sizes in the AGR mixture were varied between 0 and 6mm to account for different "underboost-scenarios". As shown in Fig. 11, the monitoring function obtained through MSA can robustly extract the hole size from the ECU data stream despite 39 other hardware deviations across 3 different cycles. Moreover, the extrapolation capability to other testing cycles of the MSA methodology becomes also evident. To test the obtained function against a corresponding conventional OBD monitor, a threshold of 5.5mm was located for the BPU deviation.

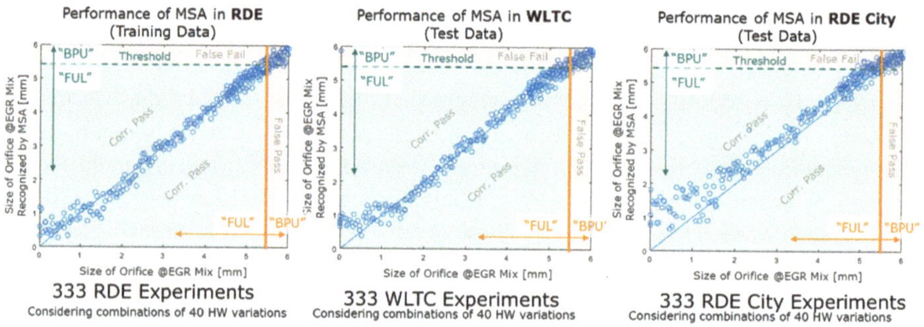

Fig. 11. Prediction and Extrapolation Capability of MSA Functionality

With the latter, the applied OBD monitor can also be evaluated based on the simulation results. The corresponding counts in Fig. 12 demonstrate that MSA enables a significantly more robust detection of the FUL and BPU states compared to OBD. For instance, pointing on the brackets, it is visible that 95 of the simulated RDE experiments, where the hole size was set below 5.5 mm and other hardware variations were considered, were classified as a "false fails" by the conventional OBD. In comparison, MSA only misclassified 4 cases.

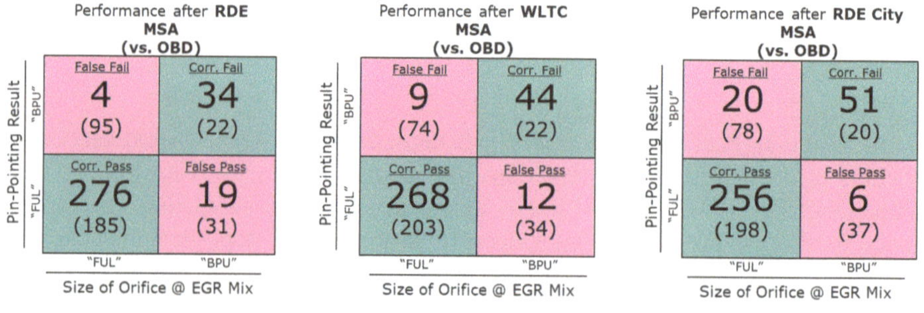

Fig. 12. Comparison of MSA Against Conventional OBD in terms Robustness

Comparable results were achieved for the detection of other emission critical hardware deviations, as well as for the identification of "intelligent tampering," on the same but also other simulated systems. Consequently, we attribute significant potential to the revolutionary MSA technology as a pin-pointing functionality for EU7 OBM. However, utilizing MSA as an OBM in-vehicle observer also necessitates a resource and runtime analysis on state-of-the-art control units. To this end, we assessed the memory requirements (Flash and RAM) and the execution time of 30 MSA functions on a 300 MHz CPU (see Fig. 13.). Depending on the CPU technology being utilized and the available resources, it is important to assess whether the integration of MSA functionalities is feasible.

CPU Core

- 32-Bit Aurix TC387, Type TC 1.6.2
- 4 Cores, 2 Checker Cores, 300MHz
- Cache per CPU: Program (32kB), Data (16kB)
- SRAM per CPU: PSPR (64kB) DSPR (240kB), DLMU (64kB)
- SRAM global: LMU (128kB), DAM (64kB)
- Flash: Program (10MB), Data (512kB)

	30 MSA Functions
Flash	~105 kB
RAM	~17 kB
Execution Time	~5.4 ms
Max CPU Load	5%

Fig. 13. ECU Load for MSA Models

Concluding, MSA functions are not only suitable for pinpointing on partial degradation in the case of an EEDWS notification, but also offer other potentials through continuous evaluation at the engine control unit. For instance, the component deviations obtained through MSA functions can be utilized to adapt the Kalman-Gain (Sect. 2.3) or virtual emission sensors (Sect. 3.2) to the aging of the individual vehicle. MSA can so be considered a key technology for implementing a digital twin, as it translates sensor signals from a vehicle into underlying hardware deviations. The latter are essential to fine-tune emission models of the target vehicle or components.

4 Summary & Outlook

As we move more towards virtual emission sensors, using digital twins for engine and aftertreatment in the ECU environment, we can get more benefits on a system level, also enabling healing functions and better predictive maintenance linked to customer benefits. However, this revolutionary approach also requires us to change the calibration process itself, since late hardware changes can create significant loops. In this case, OBM system definition and application should be seen as an integral part of the development process from concept definition to SOP—and not just as a late addition. Doing so, the virtual development environment also supports the MSA requirements, moving costly HW focused tests to SiL or HiL environment in an early development phase, increasing system robustness and lowering quality costs in the field. Looking at the big picture, the evolutionary approach of transferring more semi-physical models and statistical evaluations to the vehicle environment becomes more appealing for the mid-term future harvesting additional customer relevant benefits combined with improved system robustness. In the short term, the improved classical approach seems especially attractive for CI applications and for meeting minimum requirements combined with a shorter time to market. Stricter legal requirements could soon limit this solution.

AVL keeps improving both possible approaches to be ready for future requirements, further optimizing the cost to benefit ratio on a system level.

References

1. P. Goetschl and K. Klumaier, "On-Board Monitoring (OBM) for EU7 - Challenges and possible approaches," Baden-Baden, 2023.
2. The European Commision, "Proposal for a REGULATION OF THE EUROPEAN PARLIAMENT AND OF THE COUNCIL on type approval of motor vehicles and engines and of systems, components and separate technical units intended for such vehicles, with respect to their emissions and battery ...," 19.12.2023.
3. M. Riener and H. Herrmuth, "Advanced OBD development methodologies to meet Euro7/VII," [Online]. Available: https://mobex.io/webinars/advanced-obd-development-methodologies-to-meet-euro7-vii/. [Accessed 18 11 2021].
4. R. Gunja, H. Wancura, B. Raser and M. Weißbaeck, "Euro-7/VII-Konzepte in der Interaktion mit OBD und OBM," Motortechnische Zeitschrift MTZ, p. 2022, 11 2022.
5. The European Commision, "Commissiom Regulation (EU) 2017/1154," Official Journal of the European Union, p. 643, 2017.
6. P. Goetschl, K. Klumaier, R. Merl and T. Nishizawa, "Challenges and development approach for EU7 On-Board Monitoring," in JSAE Annual Congress (Spring), Yokohama, 2023.
7. H. Wancura, C. Kreis and S. Krapf, "Euro7/VII—Prospects of the Diesel Engine for the LCV Segment," in 12. Internationalen AVL Emissions and Energy Forums, Ludwigsburg, 2022.
8. R. Luef, K. Klumaier and R. Merl, "Model-based Gasoline Tailpipe Emission Prediction on a Real Time Virtual Test Environment as an important contribution to achieve RDE and OBD com-pliance," in JSAE Annual Congress (Spring), Yokohama, 2020.
9. M. Stockmeier, M. He, D. Stemmer, M. Wüstner, F. Eberhard and L. Schiller, "Focus on virtualization: Rapidly available model-based calibration for gasoline hybrid powertrains in a software-in-the-loop test environment," in International Symposium on Development Methodology, Wiesbaden, 2021.
10. M. Piffl, M. Schuessler, A. Danninger and H. Herrmuth, "Model-based Approach to Accurate Virtual Sensors for OBM Considering Production Tolerances and Aging," in SAE On-Board Diagnostics Symposium-Europe, Online, 2022.

Innovative and Cost-Effective Exhaust After Treatment for Lev Tier Iv Emission Legislation

Mats Laurell[1](✉), Henrik Klövmark[1], Rolf Brück[2], Katrin Konieczny[2], and Lorenzo Pace[2]

[1] Aurobay, Gothenburg, Sweden
{mats.laurell,henrik.klovmark}@aurobay.com
[2] Emitec Technologies GmbH, Lohmar, Germany
{rolf.brueck,katrin.konieczny, lorenzo.pace}@emitec.com

1 Introduction

Large scale vehicle electrification is expected to play an important role in eliminating fossil fuel dependency. However, although battery electric vehicles eliminate tail-pipe CO_2 emissions, up- and downstream emissions connected to battery manufacturing and recycling represents a major challenge. Therefore a "cradle-to-grave" perspective trough a Life Cycle Analysis (LCA) protocol becomes increasingly important [1]. In a plug-in hybrid vehicle (PHEV), low utilization batteries of BEVs are replaced by an ICE resulting in lower zero mileage CO_2 footprint and, depending on utility factor, significantly lower CO_2 emissions per km than a conventional ICE vehicle. The differences in operation scheme of the ICE in a PHEV versus in a conventional ICE vehicle does however require specific attention to hazardous emissions.

The research regarding SI ICE is mostly focusing on the cold start, considering that emissions in warmed up conditions are very close to zero. Cold start conditions are mainly defined by the catalyst heat up phase, where most of the engine out emissions pass through the catalyst untreated and are then released into the atmosphere. One typical goal of the researchers for all ICE applications (not just PHEVs) is to find ways to decrease the time for the heat up phase of the catalyst to a minimum. One possible solution to accomplish this, is to use an Electrically Heated Catalyst [2] or Disk [1, 3] to heat up the catalyst during a number of seconds after the engine cranking or even before the cranking starts itself.

Another important aspect to consider, is that after engine cranking, the idling time will be reduced or even eliminated in the Emission Legislations being enforced in the coming years. The reduction of the idling time may have a massive impact on the tailpipe emissions [5, 6] and on the system layout. Moreover, the possibility to apply high loads on the engine right after the cranking, has the consequence that the catalyst

A. Heintzel (Hrsg.): Int.Engine.Congr. 2024, Proceedings, S. 79–105, 2025.
https://doi.org/10.1007/978-3-658-47966-4_7

volume above light off temperature must be increased. This is needed to treat the higher mass flow rate, compared to common case.

For those cases in particular, the active heating has massive benefits when it is coupled with auxiliary air (AAI). This is to transfer as much of the generated heat as fast as possible from the heating disk to the catalyst behind the active heating itself.

This paper investigates pre-heating strategies in combination with the introduction of AAI, primarily in a PHEV context. The starting point for the discussions is the future legislation and a PHEV use-case approach. The PHEV use-cases depend on a combination of electrical and ICE propulsion and therefore differ a lot from ICE only use-cases. The paper continues to discuss the unique packaging challenges with PHEV and what impact this has on EATS design. This then leads to the introduction of emission system design with EHD as active heating and AAI. After this, the realization of emission tests using this design and methods for evaluation are discussed. The result chapters towards the end are divided into two parts, where the first presents the emission test results and the latter part discusses methods and results and deals with how to design and evaluate AAI flow conditions with a closed coupled catalyst.

2 Development in US Legislation regarding PHEV

A paper from CARB [4] highlighted the unique emission challenge for PHEVs

during customer operation that was then named high power cold start (HPCS). In the investigation CARB defined five different versions of highway entrance ramp-up accelerations. Three different PHEV vehicles at that time available on the market were tested for the various HPCS challenges. The different vehicles behave differently as a consequence of the available power from the two different propulsion system, electrical and ICE. All vehicles have an emission challenge when the propulsion system switches to the ICE system with a sudden engine start combined with a direct high-power demand.

In the proposed upcoming LEV IV legislation CARB has updated its legislation including PHEV to consider the HPCS challenge. For HPCS the new PHEV updates means that the so called US06 cycle should start with a cold vehicle (+23 °C) and full SOC level. Considering that the US06 includes higher power demand, it can be possible that the ICE has cold start with direct high load. Depending how the vehicle is designed and calibrated this challenge for the vehicle can be more or less severe.

Other proposed updates for PHEVs regard the strict rules that define what type of vehicle can qualify as a ZEV: BEVs of course, and certain PHEVs with minimum all electric range (certification range value of greater than or equal 70 miles). In addition, PHEVs which qualify for ZEV status must be certified to SULEV 30 or lower emission standards, have extended warranty (15 years, 150,000 miles), certain battery labelling and warranty requirements, and have a minimum US06 all-electric range of 40 miles. The PHEV portion of the total ZEV requirement is capped at 20% of each vehicle manufacturer's production. On this paper we exclude considering potential ZEV benefits.

To investigate this challenge with regards to catalyst systems, Aurobay used a Volvo PHEV test vehicle. Various calibrations of the engine and vehicle software were introduced, which altered the control and balance of energy and torque from ICE and electric motors. This enabled different variants of sudden ICE cold starts during the US06 cycle. The first vehicle test campaign resulted in 10 different variants. They have all different behaviors in ICE usage in the cycle. This of course results in different emission challenges as can be seen by 4 examples in Fig. 1.

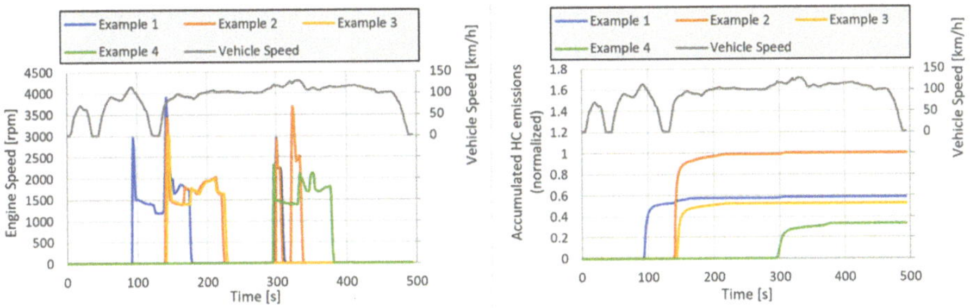

Fig. 1. Four examples of PHEV applications, US06 with different ICE HPCS behavior

3 Case Study with PHEV Vehicle

3.1 Cold Start Use-cases

Already from the early stage of PHEV applications development, the sudden engine cold start was raised as an upcoming challenge to be taken into consideration. The sudden transition from pure electrical to ICE propulsion, due to extra power request or low battery charge status, is very typical for PHEVs.

In general, the powertrain layout of the PHEVs has changed over the past years, from mainly high-end vehicles with high performance, to more common vehicles with extended electrical drive range. Nevertheless, the so-called high power cold start is always a critical use-case [1, 5, 7], even if probably less frequent on later PHEVs generation due to more mature and sophisticated technologies combined with the balance towards customer expectations.

Typically, emission test cycles like FTP75 or WLTC, in contradiction to sudden ICE start with torque demand, consists of an idle phase at the beginning of the test. During this idle period usually, the engine is calibrated with significant ignition retardation combined with high idle speed and a slightly lean lambda. This combination has been proven to allow optimal condition for low engine out emissions, steep temperature ramp for early light off and sufficient oxygen for exothermic reactions in the catalyst.

In the upcoming US legislation, the idle period for FTP75 is reduced from 20 seconds to 8 seconds. For PHEVs, where there could be no pre warning at all for an ICE start, US authorities now have updated the test procedure and included the US06 with cold engine start.

3.2 The Electrically Heated Catalyst

A typical electrical heated catalyst consists of a relatively short heating disk connected nearby to a more standard type of catalyst substrate (Fig. 2). The electrical power that is typical for those disks is in the range of 4–6 kW on 48V applications.

This available power, that is converted in available heat, must be transferred as quickly as possible to a large volume of the catalyst, to cope with the PHEV high-power cold start challenge. Even better would be to have a large volume of preheated catalyst before this sudden ICE start.

An extensively discussed challenge is of course to know when pre-heating should be started since the vehicle can´t predict exactly the future torque request. It has been discussed the use of predictive software using on board cameras and maps connected with GPS. This very interesting topic will not be discussed further in this paper, and the investigation will be carried out under the assumption that the ICE start can´t be predicted.

The requested high electrical power to bring a big catalyst volume above the light off temperature is a challenge for the heating disk, both from a durability point of view, considering the high temperature changes, but also from a heating transfer point of view. This second aspect is strictly related to the uniformity of the flow in front of the disk itself. This is even more challenging when considering the tight space available for the installation of the EHC in front of the catalyst and the potential need of an auxiliary air introduction (AAI) to pre-heat the catalyst itself. The auxiliary air introduction has a positive effect on tailpipe emissions, as discussed in [11]. Possible positions of AAI are the exhaust manifold tubes in front of the turbo charger and the inlet cone in front of the catalyst. The optimization of the flow, coming from the turbocharger, with open and close wastegate, and coming from the nozzle, is a challenge and sometimes it is necessary to accept compromise flow distribution from the engine or from the nozzle due to space constraint.

Fig. 2. Electrically Heated Catalyst

3.3 Catalyst Installation and EHC Installation

The company Aurobay, formed in the year 2021, is having design responsibility for catalytic converters in Volvo vehicles.

The most common of Volvo gasoline PHEV applications, introduced 2016 [9], had a catalyst system developed in collaboration with Emitec. The development of the so-called Compact Catalyst® is explained in [8] and also how this catalyst concept within the collaboration (Fig. 3) ended into a patented solution as a two-brick system. It also explains the unique feature with the possibility to package a large catalyst volume in a small space in close coupled position. The innovation hereby could contribute to challenging battery installations. The installation is in this paper further on named as Generation 2 (G2).

Fig. 3. A common Volvo PHEV Exhaust System, in this paper named Generation 2 (G2)

Today, the most common versions of engines from Aurobay, have a high turbo installation. They belong to Generation 3 (G3) engine family [10]. Until today those engines are mounted in mild hybrid vehicles. The design is enabling space for catalyst installation in a different way. The design is scalable and also enables large substrate diameters supporting decreased back pressure. It is optimized for the best overall compromise for current applications and the flow distribution is optimized for catalyst function over normal operation. The turbo design, the inlet cone and the first substrate design are therefore tuned together. The catalyst design is shown in Fig. 4.

In the flow distribution investigation described in this paper the electrical heated catalyst with auxiliary air is positioned at the same position as the front face of the standard substrate on G3. In Sect. 5 the challenge of uniform flow distribution with AAI installed in a close coupled position on the G3 design, will be discussed. So far, this G3 layout with AAI is not validated for emission tests. Instead, for the emission investigation, to show the potential of the EHC and AAI, the tests were done with another layout which (Fig. 5) was not aimed for PHEV though. The relatively long inlet cone and the straight part of it allow for an optimal auxiliary air introduction with a simple nozzle geometry.

Fig. 4. Generation 3 (G3) Volvo Exhaust System

3.4 The Electrically Heated Disk

In previous papers [1, 7] it has been discussed that the most beneficial position of the heater element is in the center of the catalyst, even if there are applications with the heater element in front of the first brick. One reason for that, was that the heating disk was coated and by having it beyond a first brick, the coating is protected from poisoning and ageing. The alternative, to have an uncoated heater disk is investigated in this paper and in this particular case, the aging of the heating disk is not an issue at all.

The Electrical Heated Catalyst (EHC) is a well-known and field-proven component and has already been in serial production in vehicle applications. The Electrically Heated Disk (EHD) is a new product, based on the same field proven components as the EHC and is by definition not coated, see Fig. 6. According to extensive voice-of-customer interviews, a new stand-alone heater was developed to allow flexible installation at different locations in the exhaust aftertreatment system. To achieve this, the support catalyst matrix was removed, and replaced with a stainless-steel frame at gas inlet side, also called the support structure. All other key components were carried over from the EHC to design a robust heater while keeping the number of new components to an absolute minimum [11].

Fig. 5. System (including AAI inlet) used for Emission test

In the flow distribution investigation described in this paper the electrical heated catalyst with auxiliary air is positioned at the same position as the front face of the standard substrate on G3. In Sect. 5 the challenge of uniform flow distribution with AAI installed in a close coupled position on the G3 design, will be discussed.

An alternative version of this technology, that has also been marketed from another supplier [13], consists of a single electrical heating disk with a different support design.

From an application point of view, both the catalyst ageing and the OBD2 of an electrical heated catalyst are often raised as particular challenges. The first OBD2 question normally discussed is how to judge that all the heat energy that is supplied to the EHD comes to an effective emission advantage. This question is the same for any EHC or EHD solution. However, the issue about efficiency degradation, both for the OBD limit thresholds as well as for emission full useful life performance, is somewhat less severe in the EHD case compared to a coated EHC, this considering that the EHD does not have any coating on top, that can degrade over time. It may simplify a potential industrialization project, since many activities during the verification process relies on rapid bench aged parts, which would need to include an EHC but not an EHD. Less complexity simply means more straightforward development process, which likely helps achieving high robustness in the final product. Another aspect that can be seen as an advantage of an uncoated EHD compared to a standard layout without any EHD, is that during transients or extreme high loads, which are most critical for catalyst ageing, the EHD (with no electrical heating applied) will to some extend act as a temperature shield and protect the catalyst brick behind (Fig. 7).

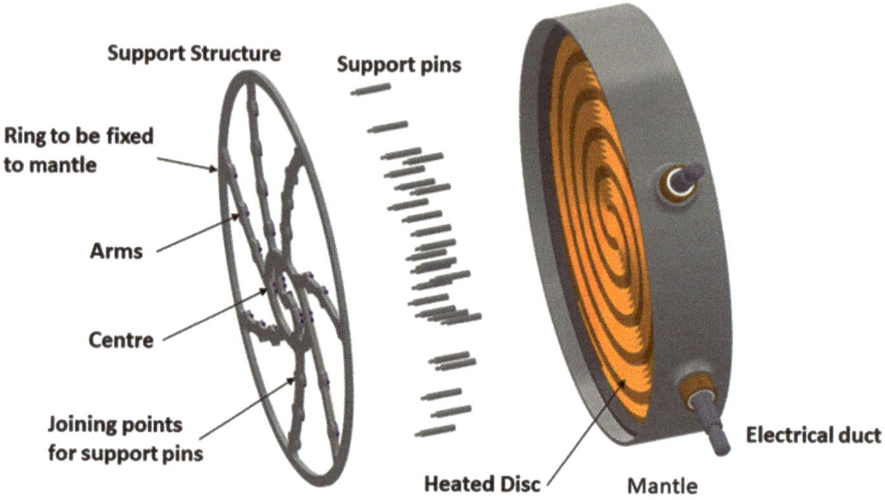

Fig. 6. Electrically Heated Disk

The effect can be understood since the (switched off) EHD reacts as a heatsink before the first coated substrate. Considering that the main goal of an EHD is to improve cold start, it is also a benefit for aging behavior of the first coated brick.

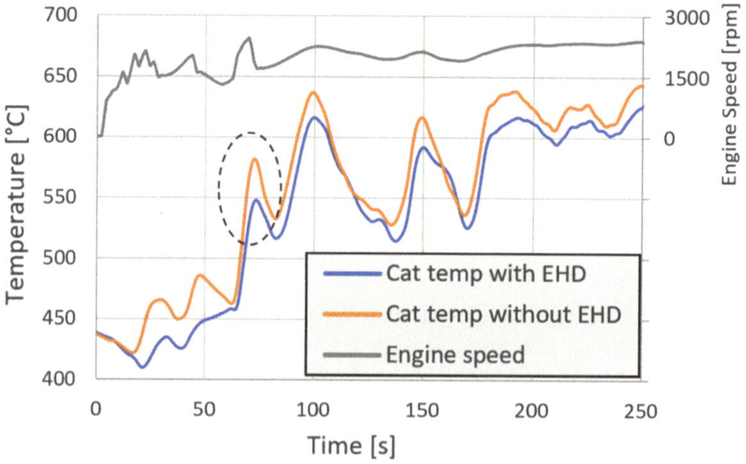

Fig. 7. As example dashed encircled shows 40 °C lower temperature 8 mm from front face in first coated brick during one transient (Here a portion of WLTC cycle measurements)

3.5 Cost Efficiency and development of EHD System

The addition of an EHD to the catalyst system will increase costs. The potential here is to compensate these added costs, of the EHD itself, the cables and the controller, with a reduction in the catalyst PGM loading. It can be assumed that there is sufficient electric energy storage in the vehicle for any PHEV configuration.

Furthermore, from a system perspective, the desired emission performance together with overall complexity needs to be balanced. Previous system designs are gathered in Fig. 8. In one study [1] system C was evaluated for cost and sustainability in a zero-emission impact context. Here a cost reduction is valid for the reduced PGM load. It can also be an attractive argument for contributing to the sustainability of ICE vehicles through improved air quality and lower mineral depletion of critical materials. Utilizing significantly less PGMs in the design, is an enabler for a higher portion of the PGMs from recycled catalysts. This low-cost catalytic substrate system had a 60% PGM reduction compared to a standard Eu6d system. It performed well, with over 90% cold start emission reduction compared with the Eu6d system. Compared with system D that was decided to test in this paper, system C had a second EHD. Furthermore, this system had its second substrate positioned under-floor. Both systems C and D utilize the same low cost TWC substrates.

The extra feature for system C, besides the extra heat from the second EHD, is the ability to convert condensed emissions. Hydrocarbons are solvable in water and when hot exhaust gas contacts the cold exhaust system walls, the water in the exhaust gas condenses, leading to the condensing of pollutants to the pipe walls for a period of time until the water later evaporates when those surfaces are heated. With system C, cold start emissions that slips through the front part of the system have an extra chance to be converted in the second part of the system, following evaporation from the cold walls inside the exhaust line. The experience and knowledge on how to utilize this effect was inherited from system B. System B was designed to reach a zero-emission impact target and evaluated [12] with less priority on cost and sustainability than system C. Other specification differences included a HC adsorber that was mounted in-front of the second electrical heater.

Experience has shown that it is possible to take advantage of emission condensation without a HC adsorber material when a heater and a catalyst are placed behind cold surfaces. An extreme version that utilizes this feature is exemplified in system E [14]. It has until now only been tested with the condense-adsorber-device part, not having the heater as well, relying on that the hot exhaust gases manage to heat the catalyst enough before the condensed cold start emissions evaporate from the condense-adsorber-device. Combining geometries with heaters and cold surfaces can be an efficient approach to control and achieve extremely low emissions levels. Here preheating plays an essential part.

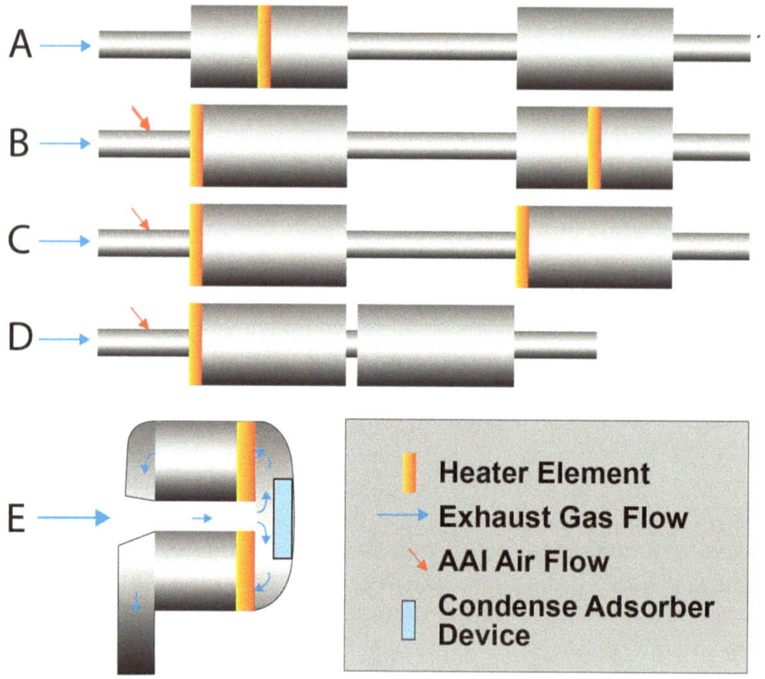

Fig. 8. Schematic representation of considered system design alternatives

Therefore, the system A [7] represents a good compromise also without AAI. When the heater in this case was embedded between two bricks, due to the insulation of the adjacent substrates, a large portion of the front catalyst avoids the delay of catalyst heating due to the already overcome dew point in the substrate. For a PHEV with it's typical packaging restrictions, as mentioned before, it was decided to aim for system D, judged as to be the best balance. Thereby also taking advantage of the ease to industrialization discussed in Sect. 3.4. Combined with pre-heating and AAI, this system maximizes early conversion of gaseous and condensed cold start emissions.

3.6 Test Set Up Description

Typically EATS hardware evaluations are done on engine test bench or on complete vehicle in an emission test cell. A PHEV vehicle is more complicated than a standard ICE or MHEV vehicle with its dual propelling force sources (ICE and electric motor). It is a continuous interaction between the ICE power source and the electrical power source. The driver behavior and the state of charge of the battery are factors that effects the power demanded from the ICE. This becomes a source of spread in the evaluation of results from PHEV vehicles. It is then clear that especially a PHEV hardware investigation will benefit from being performed in a successful engine test cell setup.

Several variants how to reproduce tests from vehicle to rig have been judged. Aurobay have concluded to apply a method based on an add on for open loop torque control via CAN. A special engine management software that controls engine output torque according to a recorded trace is needed, with as few as possible interim steps that could alter the controlled torque output of the engine. With this approach, Aurobay have been able to replicate an almost 100% exact behavior of load, speed, ignition, fuel cut events and lambda as in the test from the car where it was recorded. One of the most important aspects of this method is that the variation test-2-test (often called repeatability performance) is excellent. Another thing that is important, is that this methodology allows to keep the original software calibration in the ECU. Otherwise, there is often quite some work to recalibrate, and by that it is not known if the test results are representative or not in the end. In addition to the above-mentioned advantages, the engine initial rev-up from stand still from the vehicle application, can easily be replicated thanks to original calibration designed with ISG starts.

Alternative test cell control options of the test cell as open loop or closed loop control of the throttle (or pedal) does e.g., not give representative behavior of fuel cut events. Attempts with a driver model and simulated gearbox and vehicle have not been even close to as representative with a reasonable work effort.

In this study one further step was investigated: the possibility to use an updated and newly developed engine than the one used in the test-vehicle. In the car, a G2 engine is used (G2 and G3 explained in Sect. 3.3), while in this study, a G3 engine has been tested. But without having a PHEV vehicle already equipped with a G3 engine, no vehicle test was possible to perform. Therefore, the open loop torque control via CAN was the perfect way for this investigation, as it solves the interest in investigating how a desired EATS would behave in circumstances with a later developed engine.

To allow for several tests per day, a fast-cooling system for both the engine as well as for the exhaust aftertreatment was installed. It was also decided to reduce the test duration to only cover the most interesting parts. For the standard WLTC, the beginning and a total of 390 seconds portion was suitable, as there is a natural start/stop event at 390s in the WLTC cycle (Fig. 9). The shortened cycle and quick cooling time allowed for a reduction in test time, resulting in the possibility to perform up to 8 tests per day in the engine test bed.

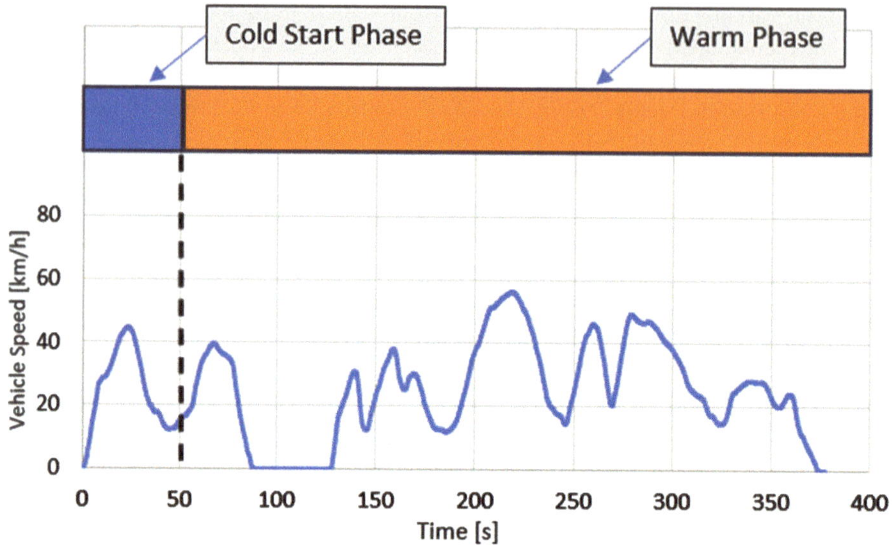

Fig. 9. WLTC Emission Cycle: Cold Start and Warm Emission Portion

For HPCS, the two most representative cycles regarding challenging sudden ICE starts were selected from the 10 cycles measured on the PHEV vehicle mentioned earlier. These were then used for the engine bench emission tests, reducing the test campaign complexity. These two cycles are named HPCS-A and HPCS-B and were also chosen with the respect of fitting in within the same duration and boundaries to keep the same test efficiency heritage as with the WLTC testing. Emissions were evaluated after 50 seconds to represent the cold start and take-off emissions. The results are explained in Sect. 4.

Emission calculations are the same as those from vehicle emission testlab (VETSONE), however a CVS-tunnel is not used in the engine test cell, instead calculated total exhaust flow from the ECU is used.

3.7 Lambda Offset during Cold Start

Tail Pipe emission measurement repeatability is of course very important for evaluation of catalyst systems. It is well known that the 3-way catalyst, by switching between rich and lean value, can reach a conversion efficiency in warm conditions above 99.5%. A typical production calibration is to start the lambda control as early as possible. However, even though this is the best choice for the individual application in production, it may not be the best option during catalyst hardware evaluation.

During the screening of different technologies and installations it is not possible to manage a full calibration development for each technology or installation. In this case, if the calibration is fully developed to the optimal for each solution, to do a fair comparison, more tests and further calibration work would be needed.

In our test campaign we decided to solve this issue by applying a special calibration strategy. This allows for a fair comparison of the different technologies, but with the drawback that NO_X emissions cannot be evaluated. We believe on the other hand, that it is sufficient to evaluate HC and CO in a first step and then we know by experience that we later can fully develop the control strategy to also optimize the conversion of NO_X.

The traditional calibration strategy during cold start is to start at lean lambda, then after some seconds perform a rich catalyst neutralization step. From this point on, in case of the production calibration, the lambda control would switch between the two target lambda values. But for this test campaign, a steady target lambda value was set to 1.02 (Fig. 10), by using a special ECU software. Hereby, of course, there will soon be a poor NO_X conversion of the system.

One other positive aspect of this method is that the catalyst neutralization step after the lean cold start, is carried out as is typical done in final production applications. This helps to keep the catalyst coating material in a state that is as normal as it is possible, with the special (lean) lambda control approach. As comparison, a completely lean start combined with a lean take off, without the rich neutralization step, is judged to be less representative for how precious metals and oxygen storage are treated in a standard calibration.

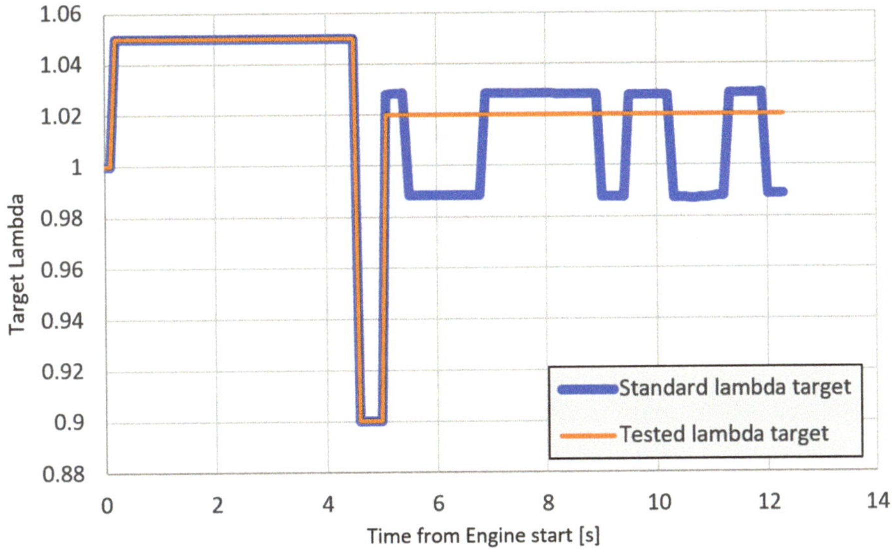

Fig. 10. Target Lambda Value: Standard and Test Approach

4 Emission Test with the Aurobay Zero Emission Engine

4.1 Comparing Catalyst Challenge WLTC vs HPCS

The High-Power Cold Start cycle is challenging and somewhat unique. There is an instant high load and high engine speed increase during a short phase just after engine start, which is highly challenging for emissions control. This is directly followed by a moderate load phase since the E-drive takes over the propulsion of the vehicle as soon as it is able or allowed to do so. In this phase the ICE is mainly charging the battery. This type of PHEV start means, in our assumption, a very instant need of active catalyst volume for conversion.

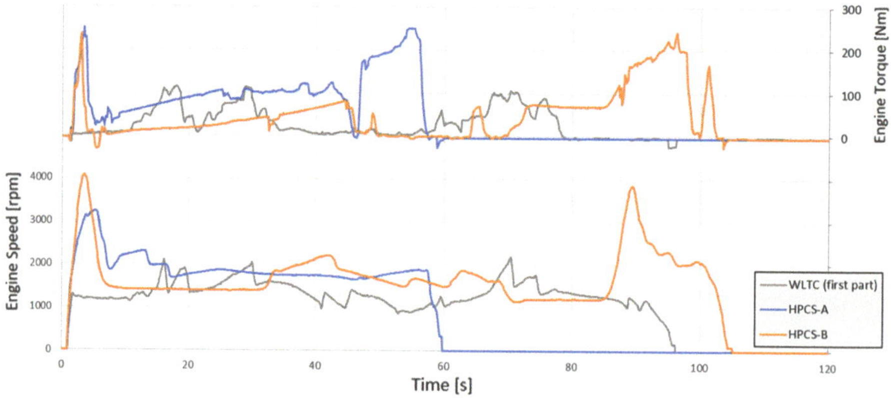

Fig. 11. Torque and Engine Speed Traces during the Test Cycles (first 120 seconds)

The test cycles were performed by applying the described approach, using the trace replication method, in conjunction with the special lambda control strategy, described in earlier chapters. A comparison of the cycles, the torque and the engine speed, are shown in Fig. 11. The picture shows the two PHEV cycles: HPCS-A and HPCS-B, in relation to the beginning of a mild hybrid (MHEV) WLTC cycle.

For a more detailed comparison, a zoom in on a few other signals during the first 30 seconds, explains the preconditions for the catalyst light-off challenge, see Fig. 12.

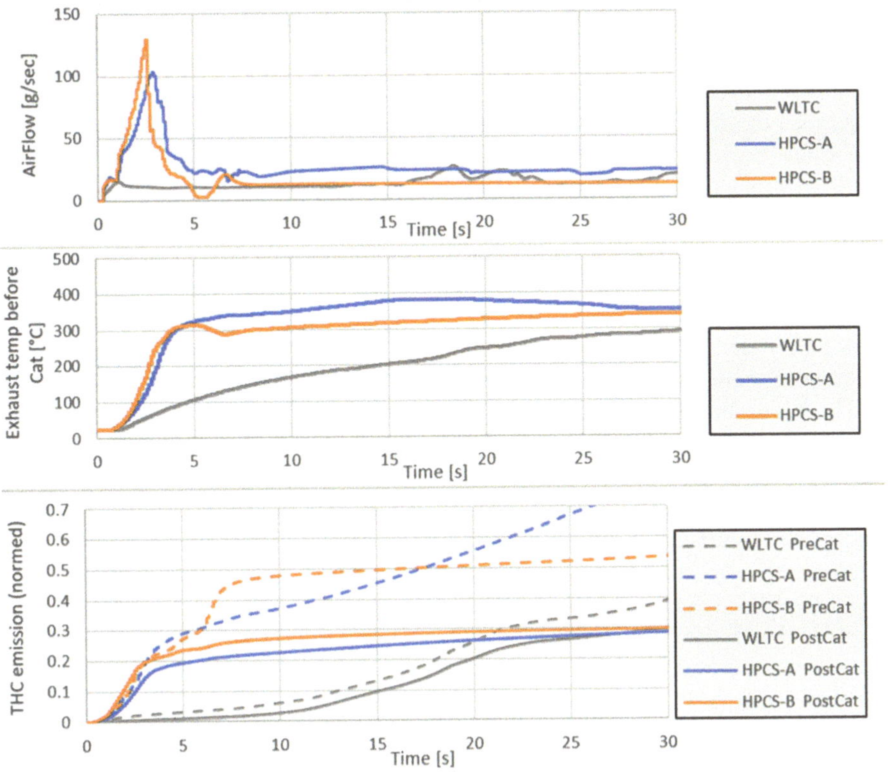

Fig. 12. THC emissions, catalyst inlet temperature and engine inlet air mass flow during the first 30 seconds

It can be seen in Fig. 12, how both HPCS versions bring a significantly increased catalyst challenge regarding the cold start emissions compared with the WLTC on a mild hybrid. The THC engine out emission and air mass flow peaks just after engine start, confirms this. An observation here is also that the accumulated THC tailpipe emissions from both HPCS cycles, reaches a similar level to the WLTC cycle after around 30s seconds for this catalyst system setup. By that even with higher engine out emissions, some compensation is achieved when the catalyst light-off occurs considerably earlier due to the increased energy flow from the instant high-power output of the ICE. In addition, it should be noted that the total emissions from a PHEV during harsh driving compared to a standard mild hybrid with a WLTC cold start, are comparable.

The HPCS-A and the HPCS-B cycles are quite different to each other. The HPCS-A cycle has in comparison, a relatively high load after the start period. On the other hand, the HPCS-B has a much more aggressive start. Compared with a mild hybrid during harsh driving (example US06), the mass flow typically stays lower in the PHEV use-cases. Those are interesting observations to keep in mind for catalyst design discussions.

4.2 Emission Test results with and without Auxiliary Air

Even though emissions are not severe from HPCS, in comparison to WLTC as discussed in previous chapter, it is of course interesting to compare what the system design uniqueness means for PHEV's. This since possibilities for newer stricter legislations can be needed to consider. For the test campaign discussed here, only WLTC and HPCS at ambient temperature, +23 °C, has been investigated. The aftertreatment system has a high degree of added complexity when it comes to mechatronics. One EHD together with an auxiliary air system brings a high degree of freedom for parameter settings and in turn, automation application work. In this case, the electrical power of the EHD and the air mass controller for the auxiliary air, were both controlled directly from the test cell. The EHD power unit offered full flexibility of voltage and current for a power range up to 10kW.

The base catalyst system chosen is named LOW SPEC in Table 1 and is compared with one of the currently most common production variants designed for EU6d and ULEV70 regulations. The LOW SPEC system was designed with the target to reach a 70% decrease in PGM mass compared to today's standard EU6d vehicle. This PGM target relates to the fact that 30% of the used PGM today in production for new vehicles are being recycled. Furthermore, it was also desired to compensate the increased cost of the EHD system at least partly. To achieve a low-PGM system it was critical to minimize the catalyst specification as much as it was practically possible. A total catalyst volume of 2 liters was chosen (Table 1). From available parts, we decided to stay at 60% decrease of PGM.

Table. 1. Properties of the tested catalyst system (production system for comparison)

	Production system Eu6d/ULEV70	Low Spec
Number of TWC bricks	2	2
System PGM loading	100%	40%
Celldensity (cpsi)	600 + 400	400 + 400
Volume (liter)	0,8 + 1,2	1 + 1

Three different heating strategies with EHD have been investigated (Fig. 13). One of the heating strategies combines pre-heating and AAI.

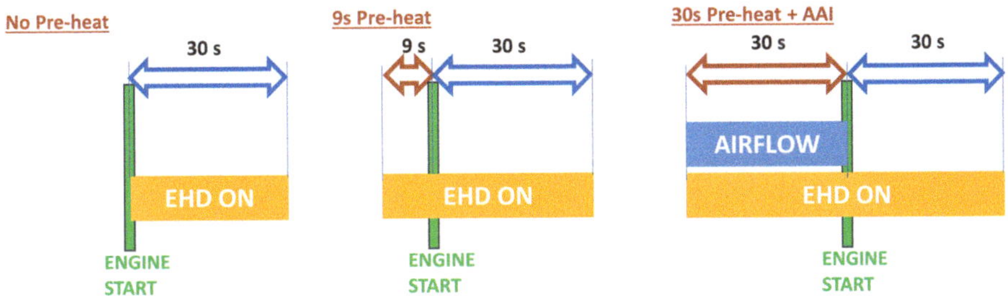

Fig. 13. Tested Heating Strategies

For the reference without heating, the EHD was taken out from the catalyst system, otherwise it would have acted as an unrealistic heat sink, that would have deteriorated the reference emission level in a non-representative way. All tests with EHD heating, used a post-engine-start heating period of 30s with 4kW EHD power. Pre-heating without AAI had a duration of 9 seconds and pre-heating with AAI had 30 seconds duration, in this case the heating power was 5 kW with an air mass flow rate of 30 kg/h.

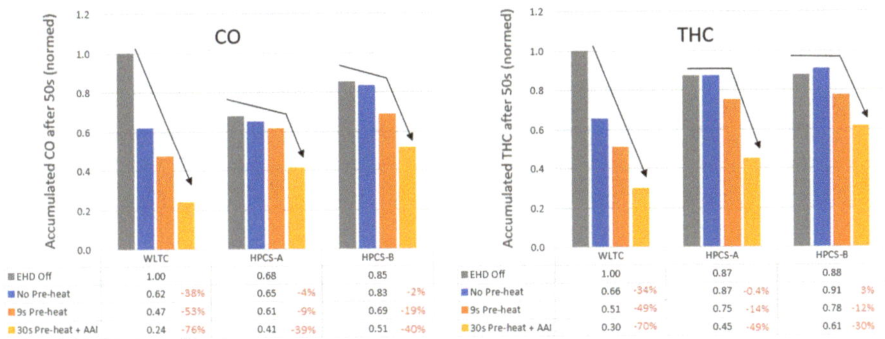

Fig. 14. Emission Results

All results in Fig. 14 are the averages of 3 tests with near negligible individual variation. The difference between the strategies is to some extent dramatic. The WLTC results can benefit significantly already from the mildest version of EHD strategy without any pre-heating nor AAI. The HPCS versions needs at least some pre-heating, but it is only with the AAI strategy, that the emission reduction is significant.

The 9s preheating case shows approximately 50% benefit for WLTC, whereas 20% for the HPCS cycles, which can be judged as only a moderate improvement. Lastly the pre-heating with AAI has almost 80% benefit for the WLTC and in this case we also see a strong benefit on the HPCS cycles with a 40–60% improvement.

It can be stated, considering the overall performances of the three heating strategies, that only EHD Pre- and Post-heating is not as effective in HPCS cycles as it is in the WLTC cycle. AAI has a great beneficial effect in the HPCS cycles. So, one could argue that PHEV need to consider AAI if considering an electrical heater solution. Then this of course depends on the overall system specification, together with the level of drivability quality that the final product aims for. It is mentioned earlier that the WLTC looks really promising with the simplest heating strategy. A PHEV have a big potential, especially in the case when the customer attributes like drivability and customer convenience is sacrificed. Then a cold start is more balanced towards emissions and will thereby avoid the unique challenges from a HPCS, that is described in this chapter.

5 Optimization of Auxiliary Air Nozzle Position for G3 catalyst geometry

The auxiliary air injection in the inlet cone has been investigated in a previous work [6] where it has been demonstrated how a uniform velocity distribution is a key factor to have an efficient heat transfer for the heating device to the catalyst. A uniform velocity distribution maximizes the heat transfer from the disk to the air or the exhaust gas, avoiding the presence of hot spot, that could limit the maximum power applied to the EHD/EHC itself.

The focus of previous investigation was on the influence of the velocity distribution on the pre-heating performances, without paying too much attention to the design of the nozzle or the possible use of the nozzle design for serial production.

These tests have been carried out using an available EHC. No differences in flow distribution and velocity distribution are expected according to the same layout of the disk on the EHC and EHD.

5.1 Comparing Catalyst Challenge WLTC vs HPCS

In an early test campaign, the auxiliary air introduction point was positioned in front of the turbocharger. In this case, the waste-gate was kept close. The flow distribution measurements have been carried out at room temperature with a mass flow rate of 10 and 30 kg/h. Both are representative of mass flow rate with market available air pumps.

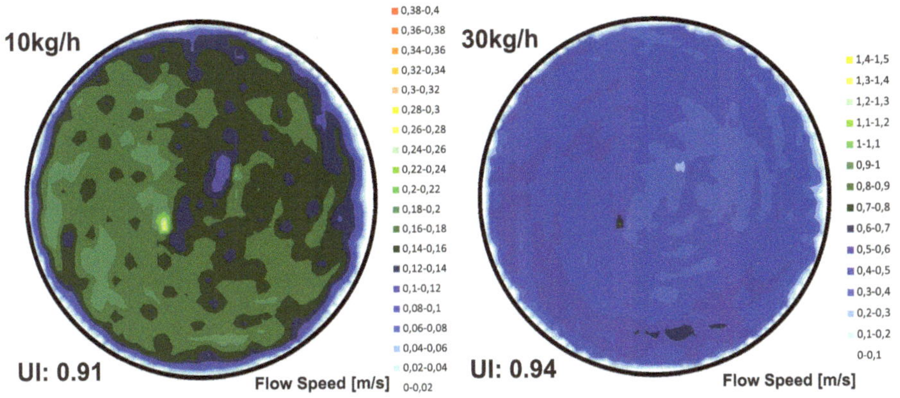

Fig. 15. Flow Distribution with AAI in front of the turbocharger (please take into consideration the different velocity scale)

While the uniformity index is a commonly used parameter to judge how good is the flow distribution on a surface, the velocity distribution (Fig. 16) is used here to better explain how fast the heat generated on the EHC/EHD heated surface can be transferred to the air and then to the catalyst.

If the distribution of velocity is concentrated around the mean value, it means that in a very large part of the surface the flow velocity will be the same or similar to the mean value, increasing in this way the heat exchange area between the hot EHC/EHD surface and the cold auxiliary air flow.

On the other hand, if the shape of the flow velocity distribution is "flat", it means that on the surface there will be areas with higher velocity, where the heat exchange is good, but also areas where the flow is slower, and the heat exchange is worse. This is a situation that must be avoided, to prevent overheating areas.

Both uniformity indexes (Fig. 15) and velocity distribution (Fig. 16) are quite good, in this case.

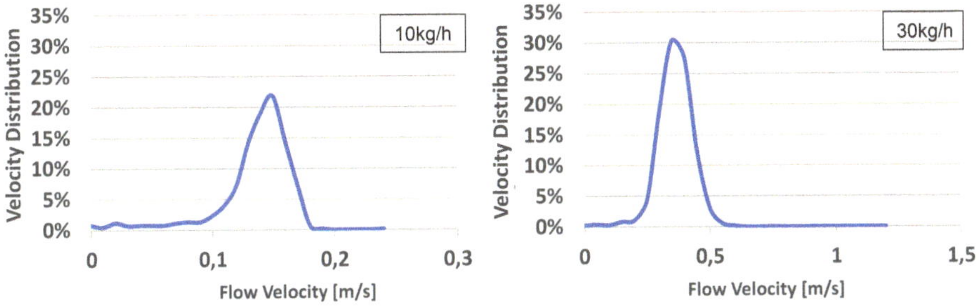

Fig. 16. Velocity Distribution with AAI in front of the turbocharger

In this test campaign, the goal was to check how long it was possible to heat up with 4 kW before reaching the max temperature (900 °C) in the hot spot. With 10 kg/h it was possible to heat up 30 seconds, while with 30 kg/h it was possible to heat up 90 seconds.

In the next step the position of the AAI was moved to the inlet cone, considered that this could be a more viable solution for a serial production. It was also decided to keep the heating time within 30 seconds.

5.2 Test with AAI in the Inlet Cone

For the present investigation, a potential intended serial production nozzle design has been used. In order to carry out flow distribution measurements with the nozzle in different positions, the inlet cone and the nozzle itself have been 3D-printed.

This allows for a quick modification of the nozzle position and inclination, as shown in Fig. 17.

Fig. 17. 3D-printed Nozzle and G3 Inlet Cone, connected to the turbo charger, baseline 0° (left) and 120° (right) Nozzle Position

Measurements were also done with a mass flow of 10 and 30kg/h at room temperature. The inlet cone has been connected with the turbo charger to replicate the flow conditions of the real application.

The flow distribution is measured using a hot wire with a mesh resolution of 5 mm, shown in Fig. 18 as a schematic illustration.

Fig. 18. Test Set-up for the uniformity index flow distribution

To assess the effect of the flow distribution on the heating performance of the EHC, a set of infra-red measurements have been carried out.

5.3 Flow and Temperature Distribution with 10 kg/h Mass Flow Rate

The first tests have been carried out with a mass flow rate of 10 kg/h at room temperature (20 °C).

Figure 19 show the influence of the nozzle position on the Uniformity Index (UI) and generally on the position of the high flow velocity area.

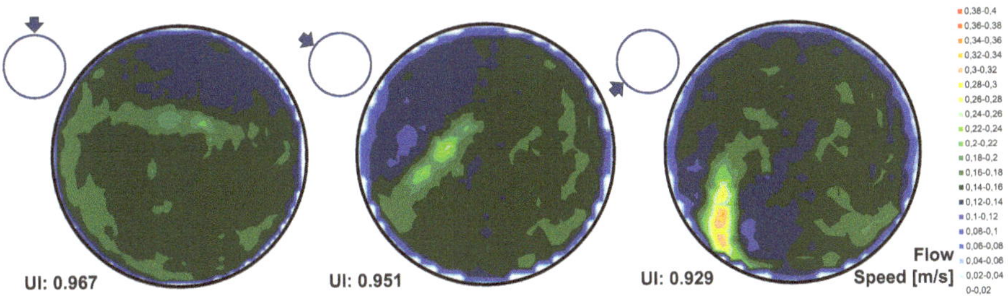

Fig. 19. Flow Distribution and Uniformity Index (UI) for the three tested nozzle positions

The flow velocity distribution, as reported previously, is a key parameter to better understand if the EHC/EHD can reach its maximum power with that particular flow field. The flow velocity in an empty tube under the given boundary conditions would be 0,146 m/s. The flow velocity distribution (Fig. 20) of the layout 1 (with 0° nozzle position) is very promising, the flow velocity distribution curve is very narrow. Layout

2 (with 60° nozzle position) and layout 3 (with 120° nozzle position) have a lower percentage of the velocity flow at the theoretical value, this means that they have also more areas with very low flow, resulting in the heat exchange coefficient of the layout 2 and layout 3 to be lower than the one of layout 1.

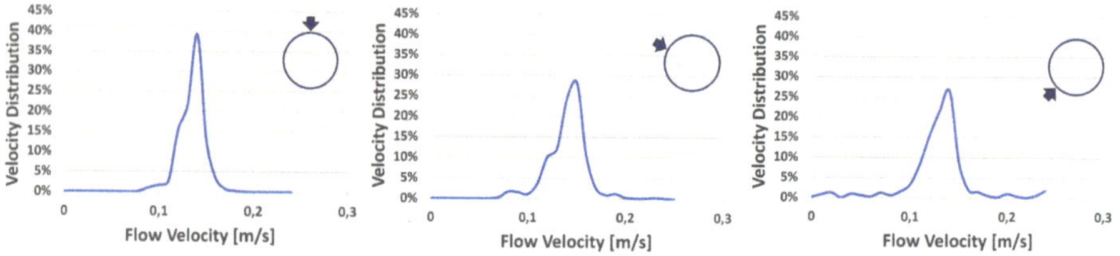

Fig. 20. Flow velocity distribution for the three tested nozzle positions, 10kg/h mass flow rate

Layouts 1 and 3 have been chosen for the Infra-Red measurement. Layout 1 as the more promising and layout 3 as the less promising.

5.4 Infra-Red Measurement with 10 kg/h Mass Flow Rate

The test has been carried out applying the maximum possible power for 30 seconds with the limit of the 900 °C at the hot spot and 10 kg/h mass flow rate at room temperature.

Figure 21 shows the results of the Infra-red measurements: both configurations perform quite well. Layout 1 has a higher mean temperature, and 3.3 kW can be applied, while layout 3 has a lower mean temperature and 3.2 kW can be applied. The small advantage in maximum applied power and mean temperature of the layout 1 confirms the importance of a good AAI flow distribution even if it should be noted that the Hot Spot is near the mantle, where all three configurations have similar flow conditions.

A possible, further, optimization could be, to increase the flow conditions near the hot spot.

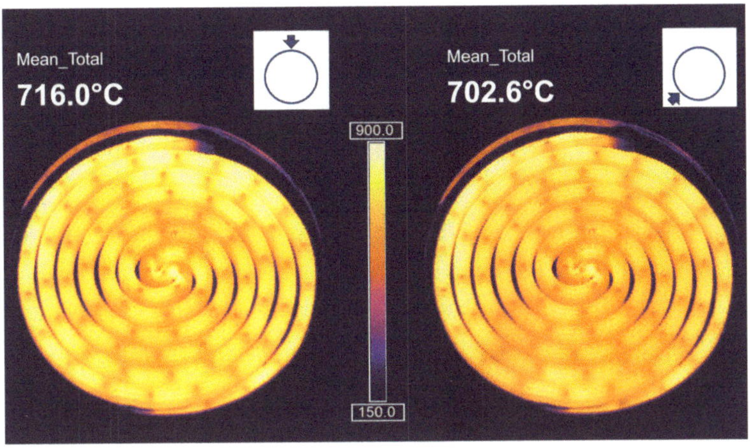

Fig. 21. Infra-Red Measurement of Layout 1 and 3, Max (Picture) and Mean Temperature after 30 seconds

5.5 Flow and Temperature Distribution with 30 kg/h Mass Flow Rate

The second set of tests has been carried out with a mass flow rate of 30 kg/h at room temperature (20 °C).

Figure 22 shows the influence of the nozzle position on the Uniformity Index (UI) and generally on the position of the high flow velocity area.

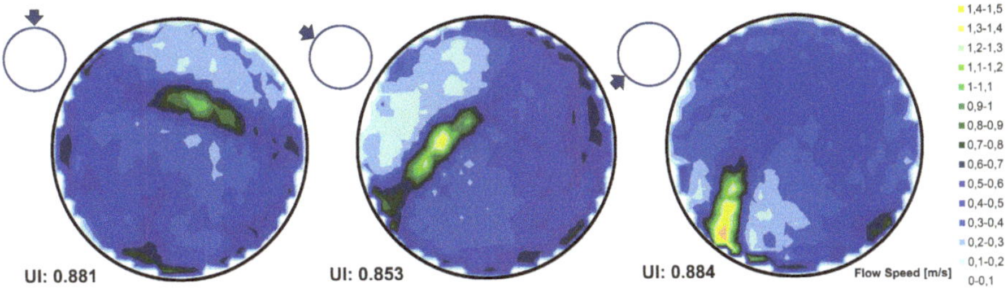

Fig. 22. Flow Distribution and Uniformity Index (UI) for the three tested nozzle positions

Figure 23 shows the flow distribution of the three tested nozzle positions.

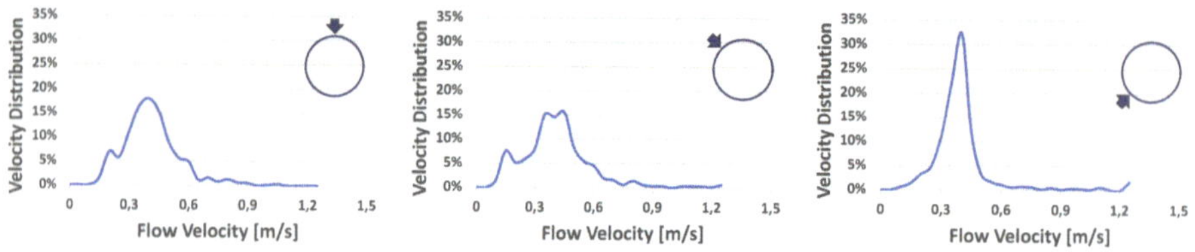

Fig. 23. Flow Velocity distribution for the three tested nozzle positions, 30kg/h mass flow rate

Comparing the velocity distribution, it is quite evident that the layout 3 has the highest potential, even if the flow distribution of layout number 1 could be also promising.

For the infra-red measurement, the layout 1 and the layout 3 have been chosen.

5.6 Infra-Red Measurement with 30 kg/h Mass Flow Rate

The test has been carried out applying the maximum possible power for 30 seconds with the limit of the 900 °C at the hot spot and 30 kg/h mass flow rate at room temperature.

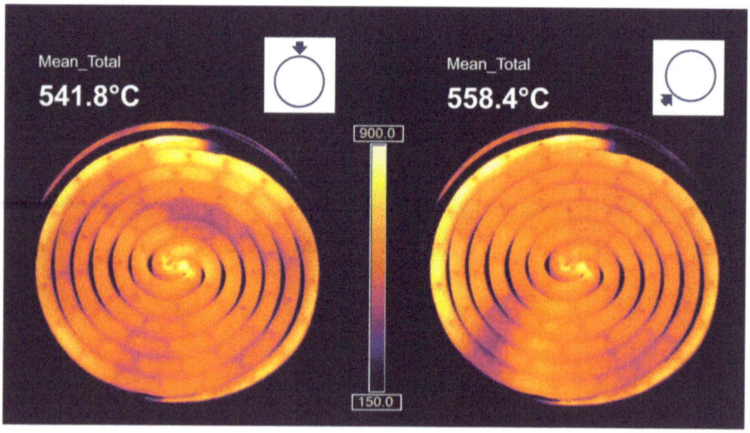

Fig. 24. Infra-Red Measurement of Layout 1 and 3, Max (Picture) and Mean Temperature after 30 seconds

Figure 24 shows the results of the Infra-red measurements, the cold area where the flow speed is higher are very well visible. The layout 1 reaches its maximum temperature after 30 seconds when 3.8 kW are applied, while it is possible to use 3.9 kW on the layout 3. The layout 3 has a slightly higher mean temperature. The same considerations as for the case with 10 kg/h on the position of the hot spot can be made also for this use-case: an improvement of the flow conditions near the hot spot could led to increased applied power, even if 3.9 kW can be seen as a very promising result.

5.7 Test Results of AAI Influence

With low mass flow rate, the flow distribution of the two nozzle positions (in front of the turbocharger and in the inlet come) are very good, but the applied power is higher with the nozzle in front of the turbocharger. This is due to the fact that that flow is well distributed around the hot spots, not only in the center of the EHD. With higher mass flow rate the position in front of the turbocharger shows a clear advantage, but a further optimization of the nozzle position in the inlet cone could improve the situation and allow even higher electrical power.

6 Discussions Test Results

In previous investigations [1, 12], the combination of AAI and EHC/EHD has proven to be capable to reach a close to zero emission impact level during WLTC. That was with EHC/EHD both in closed coupled and under floor position. In this investigation the system was more moderate equipped with a single EHD. This together with low specified substrates provides an overall compensated approach regarding cost.

It is expected, with continuously increasing market share of PHEV at least within ICE equipped portion of vehicles, that the test on WLTC will be not sufficient to judge the efficiency of an EATS during RDE. Moreover, PHEV have further unique use-cases and in the U.S. the HPCS will be soon introduced as part of the emission certification. Then there is a need to investigate those use-cases in early development. As predevelopment investigation we could apply the use-case from an existing PHEV into a newer engine by simulating the electric drive in the test bench with reliable vehicle correlation. In comparison to WLTC: a large catalyst volume during a very short time period, just after the internal combustion engine start, is instantly needed to prevent high tailpipe emission. The exact use-cases will of course depend on the final PHEV product. Still, early knowledge from the first assumption of upcoming application is of strong value.

The work done shows how the AAI in combination with EHD and pre-heating might be a key technology to achieve a near-zero tailpipe emission even with PHEV during HPCS.

AAI must be optimized to transfer the heat generated over the EHD to the catalyst, taking also into account that during the HPCS a large catalyst volume above the light off temperature is needed, and allow a high power to be applied on the EHD itself.

The authors have used an experimental methodology to measure the flow uniformity index and the flow velocity distribution in order to assess if a given nozzle position can be used in an EATS in combination with an EHD. The goal is to use, during pre-heating, as much as possible the available heating power, without having overheating areas, which can be dangerous for system durability.

7 Conclusion and Outlook

It has been demonstrated how the combination of AAI and EHC/EHD, in particular for the PHEV use-cases, have the potential to balance customer attributes with system cost and legislation.

PHEVs have unique possibility to integrate preheating thanks to extended electric drive capability and thereby take advantage of AAI.

Nozzle optimization is crucial to take full advantage of EHC and EHD in this context. Development of an optimized nozzle would have a positive effect on cost, environment, and sustainability. In a next investigation, CFD model will be validated using the results of this test campaign and new positions of the nozzle will be investigated. A possible further development of the measurement could be the use of a glass window on the inlet cone to measure the temperature distribution of the EHC/EHD during preheating and auxiliary air introduction.

Selecting in particular an EHD solution (instead of EHC) may have many benefits regarding aging of the first coated brick and OBD2.

Alternative fuels that are non-fossil are also crucial for the future of ICE and therefore PHEVs. The eFuel mainstream is not yet defined, but at least some of todays (methanol, ethanol) fuel compositions can bring additional cold start challenges and the investigation of EATS with AAI and EHD can bring also a possible solution for these challenges.

References

1. How to Accomplish a More Sustainable Internal Combustion Engine Using LifeCycle Analysis: A Demo, Joshua Paul Dudley, Mats Laurell, Christoffer Thuve,Henrik Klövmark, Aurobay; SAE Energy & Propulsion Conference & Exhibition, Greenville, USA, 2023
2. Unlimited Clean Mobility Emission Management for EU 7 and China 7 Powertrain Concepts Rolf Brück; Ines Lienou L.; Dr. Axel Schatz, Vitesco Technologies Emitec GmbH; Lohmar, Germany; Dietmar Ellmer, Vitesco Technologies; Regensburg,Germany; Aachen Kolloquium, Germany, 2020
3. The Exhaust Aftertreatment for Future World-Wide Internal Combustion Engines;From Passenger Car to Trucks, Rolf Brück, Dr. Manuel Presti, Lorenzo Pace, Frank Bohne, Jan Hodgson, Jan Kramer, Emitec Technologies GmbH, Lohmar, Germany;Aachen Kolloquium, Germany 2023
4. Characterization of Gaseous Emissions from Blended Plug-In Hybrid Electric Vehicles during High-Power Cold-Starts, Allen Pham, Marko Jeftic, California Air Resource Board, SAE Technical Paper 2018-01-0418, 2018
5. Exhaust Aftertreatment Technologies for Hybrid Electrical Vehicles, Lorenzo Pace and Rolf Brueck, Vitesco Technologies Emitec; Paul Rodatz and Peter Senft, Vitesco Tech-nologies; SAE ICE, Capri, Italy, 2021
6. Innovative Catalyst System to Achieve EU7 Legislation for Electrified Powertrains, Rolf Brück, Katrin Konieczny, Vitesco Technologies Emitec GmbH, Lohmar; Interna-tional Vienna Motor Symposium, Wien, Austria, 2022

7. Strive for zero emissions impact from hybrids, Mats Laurell and Fredrik Ekström, Volvo Car Corporation, Katrin Konieczny and Lorenzo Pace, Continental Emitec; SAE ICE, Capri, Italy, 2019
8. The innovative exhaust gas aftertreatment system for the new Volvo 4 Cylinder En-gines; a unit catalyst system for gasoline and diesel cars, Mats Laurell, Johan Sjörs, Staffan Ovesson, Mats Lundgren, Volvo Car Corporation, Rolf Brück, Manuel Presti, Emitec GmbH; Aachen Kolloquium, Germany, 2013
9. Fleiss, M., Thomas M. Müller, Martin Nilsson, Jonas Carlsson, "Volvo Powertrain Integration into Complete Vehicle", ATZ 03/201
10. The New Aurobay 250 PS, 2.0 L Miller Engine, Michael Denny, Andreas Berntsson, Ayolt Helmantel, Aurobay, Gothenburg, Sweden; Aachen Kolloquium, Germany, 2023
11. The Exhaust Aftertreatment for Future World-Wide Internal Combustion Engines; From Passenger Car to Trucks, Rolf Brück, Manuel Presti, Lorenzo Pace, Frank Bohne, Jan Hodgson, Jan Kramer, Emitec Technologies GmbH; Aachen Kolloquium, Germany, 2023
12. The sustainable engine; towards a provider of power and torque with zero resulting en-vironmental impact, throughout the complete life cycle, Tomas Johannesson, Joshia Dudley, Mats Laurell, Aurobay; Aachen Kolloquium, Germany, 2022
13. Technology options to reduce cold start – electric catalyst heaters, Florian Kunath, Dominik Rose, Bertrand Coulet, Per Nicolin, Thorsten Boger, Thomas A. Collins, James M. Grochocinski, Mrinmoy Dam, Corning; Aachen Kolloquium, Germany, 2022
14. US20170184001A1 Volvo Cars Patent: Catalytic Converter and Exhaust Gas Aftertreatment Arrangement Catalyst Substrate Innovation for Achieving RDE and SULEV 30 Emission Legislation, Rolf Brück, Peter Hirth, Bin Hu, Christian Schorn, Continental Emitec GmbH; International Vienna Motor Symposium, Wien, Austria, 2016

Hydrogen Internal Combustion Engine with Highest Efficiency

André Ferrarese[1(✉)], Elio Augusto Kumoto[1], Ralf Marquard[1], Martin Wieser[2], Anton Arnberger[2], Ulf Lundqvist[3], David Mumford[3], and Jake Walker[3]

[1] Tupy SA, R. Albano Schmidt, Joinville-SC, Brazil
andre.ferrarese@tupy.com,
{elio.kumoto,marquard.ralf}@mwm.com.br
[2] AVL GmbH, Graz, Austria
{martin.wieser,anton.arnberger}@avl.com
[3] Westport Fuel Systems, Vancouver, Canada
{ulf.lundqvist,david.mumford,jake.walker}@wfsinc.com

1 Commercial Hydrogen Engine Combustion Concepts

Long term Green House Gases (GHG) and CO_2 targets are linked to carbon neutral transport solutions. On commercial vehicles three main propulsion systems are in a strong position as zero emission: battery-electric, fuel cells and hydrogen internal combustion engines (H_2-ICE). In a holistic approach, the energy and emission demand on new batteries cells are too high to produce new vehicles or maintenance for durability. It is given by the high dependence on critical minerals. Due to that, H_2-ICE is becoming a strong complementary solution, especially for high load applications [1–3]. Once H_2-ICE is the one with the lowest TCO (lowest acquisition and maintenance costs) and highest durability (minimum 3 times higher) than other zero emission solutions, it has becoming focus on commercial vehicles.

The first generation of hydrogen engines for commercial vehicles follow spark ignited engine combustion processes, which are operated lean with intake manifold or low-pressure direct injection [2–5]. Although these engine concepts are characterized by low injection pressures which means advantages regarding the necessary hydrogen storage concepts, the achievable efficiencies and power densities are limited due to the combustion process. That demands an engine design with higher volumetric displacement to achieve the required power level and transient performance [6].

A. Heintzel (Hrsg.): Int.Engine.Congr. 2024, Proceedings, S. 106–116, 2025.
https://doi.org/10.1007/978-3-658-47966-4_8

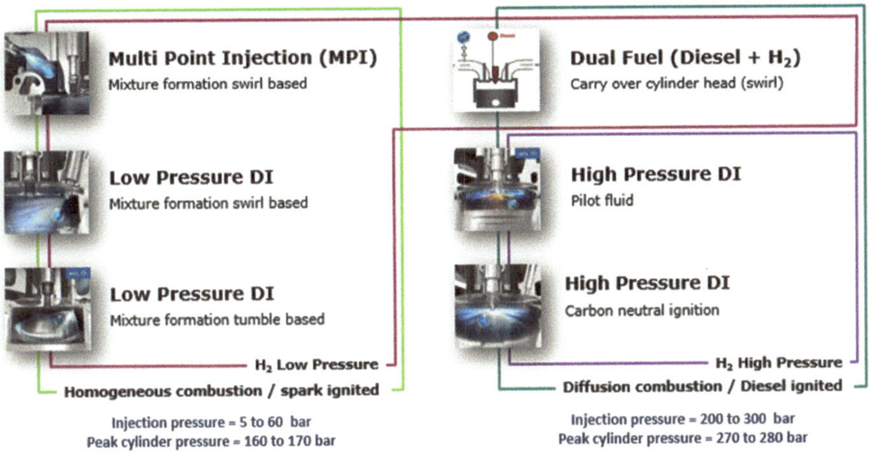

Fig. 1. Different injection strategies for H_2 internal combustion engines. Adapted from [2]

To achieve a significant increase in efficiency, an alternative combustion process is required that comes close to Diesel engine combustion. To accomplish this, the hydrogen must be injected under high pressure near to top dead center, which is enabled by high pressure direct injection systems [7, 8], such as Westport's HPDI™ product. As shown in Fig. 1, the combination of hydrogen with a hydrocarbon as ignition fuel is a possibility to have a robust system. In the same way for a zero emission option at a complete cycle, Diesel can be substituted by HVO. For low carbon strategies, Diesel cycle combustion of hydrogen is possible in two main ways: Hydrocarbon pilot (minimum pilot fluid content to support mixture ignition) or an alternate ignition source like a glow plug. The use of Diesel cycle to burn hydrogen can enable efficiency in a combustion engine exceeding 50% BTE [6 and 9].

In terms of material development, spark ignition H_2-ICEs will need solutions to reduce weight and size. As higher volumetric displacement is necessary to deliver the required power and torque, packaging can become an issue. Mechanical and thermal loads are below Diesel levels, which indicate an easy carry over of materials. On the other hand, the HPDI design is more demanding from mechanical and thermal perspectives, e.g. peak firing pressure can be up to 280 bar and injection pressures 300 bar. This brings the material selection into high focus for HPDI design.

The presence of hydrogen in the combustion chamber creates a risk of hydrogen embrittlement on the exposed parts of the chamber. Improved cast iron and stainless steel alloys are being developed to enable same high durability to Hydrogen than Diesel [10]. However, the diffusion flame combustion of the HPDI™ system should allow the H_2 to be fully combusted before any significant engine component exposure considering the pilot fluid concept as illustrated in Fig. 2. Only the fuel system itself would be constantly exposed to H_2 [6].

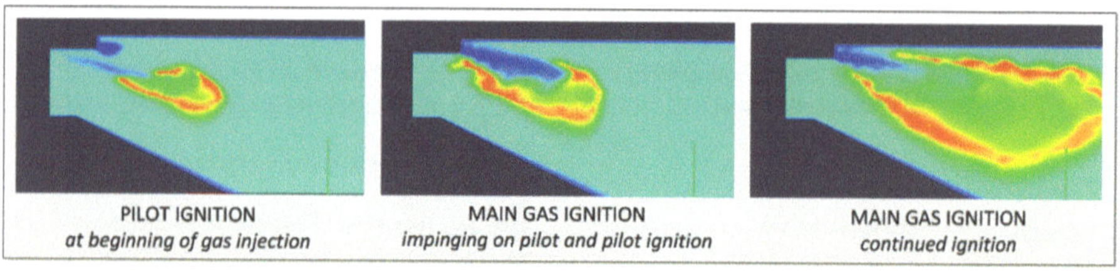

Fig. 2. Steps of ignition with the pilot fluid injection concept

2 Demonstrator Platform

It was decided to use existent Diesel platform to transform it on a HPDI H_2-ICE version, which main data is presented in Fig. 3 below:

ignition	charging	cyl.	VH	EGR	max. torque	rated power
[-]	[-]	[-]	[l]	[-]	[Nm]	[kW]
reference diesel engine		6	~13	cooled	2500	340

Fig. 3. Main data of series Diesel engine used as platform

A summary of key steps is presented in Fig. 4.

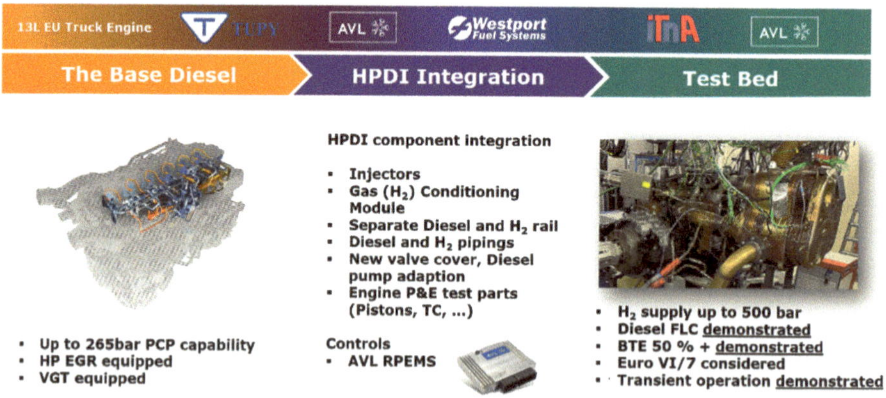

Fig. 4. Overview of the H_2-HPDI Demonstrator Program and its Contributors [9]

The selected platform was already high loaded, 265bar on peak cylinder pressure (PCP) using already Compact Graphite Iron (CGI) alloy at cylinder head. Pistons were maintained on the performance tests are original steel grade material. Only variations on compression ratio that demanded new designs, but no material change.

The demonstration of the necessary vehicle components has not been part of the program. The current approach foresees a 700bar tank system, and an integrated gas module, as a buffer tank, and a mobile compressor which extends the drive range if the tank pressure falls below the demanded injection pressure.

3 Test Results

The baseline series production 13L in-line six HD Diesel engine for the H_2-HPDI Demonstrator shows highest brake thermal efficiency (BTE) of 47.6 % before any adaptions on the engine. HPDI concept used Diesel as pilot fluid. After equipping the base line engine with Westport HPDI 2.0 fuel injection system measurements with hydrogen showed a brake thermal efficiency of 48.7 % which is an increase in brake thermal efficiency of +1.1 % (Fig. 5). The increased BTE was achieved due to additional expansion work and reduced friction because of lower injected Diesel quantity and lower injection pressure (between 200 and 350 bar). The test engine was equipped with an external Diesel supply unit with Diesel at requested rail pressure. The HP- and LP-Diesel-pump of the engine was disconnected which led to increased brake thermal efficiency since Diesel pump friction loss of around 0.3 % (evaluated in strip-down analysis) was not considered.

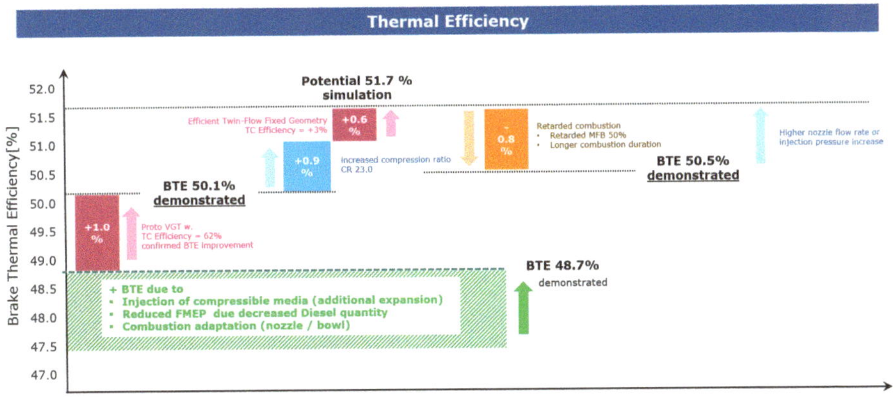

Fig. 5. Brake thermal efficiency improvement—demonstrated and potential results

To increase the brake thermal efficiency, the serial VGT turbocharger was exchanged by an improved VGT turbocharger with efficiency of 62%. With the exchanged VGT turbocharger a brake thermal efficiency of 50.1 % was demonstrated [9].

For further brake thermal efficiency increase an estimation has been conducted on base of 1-D simulations and comparable engine measurements (Fig. 9). According to the estimation a brake thermal efficiency of 51.7 % could be identified with increasing the compression ratio with 23.0 and installing a high efficiency turbocharger. The increased compression ratio was realized with a new piston with adapted piston bowl shape for best hydrogen jet and piston interaction. To improve the gas exchange, a high efficiency fixed turbine geometry turbocharger with efficiency of 68 % (twin scroll layout with wastegate) was installed and combined with a double flooded EGR system to ensure proper EGR-rates for lowest emissions. A Miller valve timing promises additional 0.1 % increase of brake thermal efficiency. However, test results showed 50.5% BTE due to retarded combustion. The use of higher nozzle flow rate or injection pressure increase could cope with this releasing the potential increase. Any how the demonstration of +2.9% on BTE compared to Diesel is an important milestone to high efficiency hydrogen combustion engines.

The hydrogen injection pressure was adjusted to about 200–300 bar, depending on the peak cylinder pressure. The injection pressure was adjusted 50–80 bar higher than the peak cylinder pressure to ensure an overcritical pressure ratio between injection pressure and cylinder pressure for the Start of Injection. Furthermore, NO_X emissions rise with a higher injection pressure, whereas efficiency has a sweet spot with falling efficiency when raising or lowering injection pressure (Fig. 6).

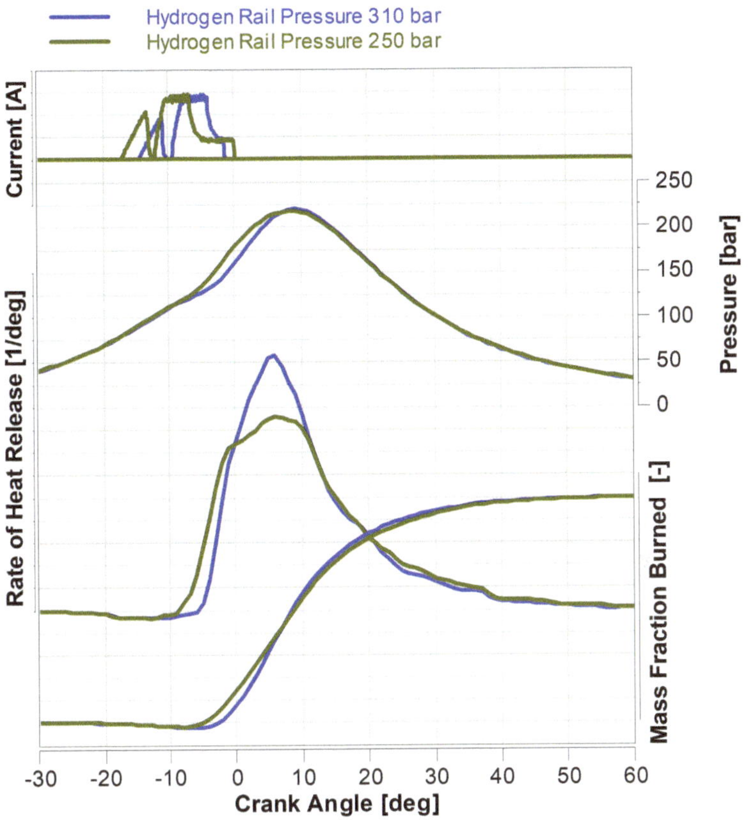

Fig. 6. Injector current, cylinder pressure and rate of heat release for 1200rpm, 21 bar BMEP from [9]

The measurement data indicate Diesel-like behavior of the H_2-HPDI combustion concept in terms of NOx emissions and exhaust gas temperature level. Consequently, the aftertreatment concept meeting Euro 7 legislation according to the Draft Agreement of 12/2023 (limit values displayed in Table 1 will look like the currently proposed Diesel system architectures (Fig. 7). As the H_2-HPDI combustion concept tends towards higher NOx emissions EAS concept 2 in Fig. 7 is the most suitable configuration to meet the emission targets and limits.

Tab. 1. Limit values in the WHTC cold/hot and RDE limits for EURO VI E and Euro 7 (based on draft agreement)

Pollutant emissions	WHTC		RDE	
	Euro 7 Draft Agreement WHTC c/h combined (1/7 vs. 6/7) 20.12.2023	EU VI Limits WHTC c/h combined (1/7 vs. 6/7)	Euro 7 Draft Agreement 20.12.2023 ISC c/h combined (1/7 vs. 6/7)	EU VI Limits ISC combined (1/7 vs. 6/7)
-	**per kWh**	**per kWh**	**per kWh**	**per kWh**
NOx in mg	200	460	260	690
PN_{10} in #	6x10^11	6x10^11 (PN 23)	$9*10^{11}$	$9.78*10^{11}$ (PN 23)
NH_3 in mg	60	10 (ppm)	85	N/A
N_2O in mg	200	N/A	260	N/A
CH_4 in mg	500	500	650	750
CO in mg	1500	4000	1950	600
NMOG in mg	80	160 (NMHC)	105	240 (NMHC)
HCHO in mg	N/A	N/A	N/A	N/A
PM in mg	8	10	-	N/A

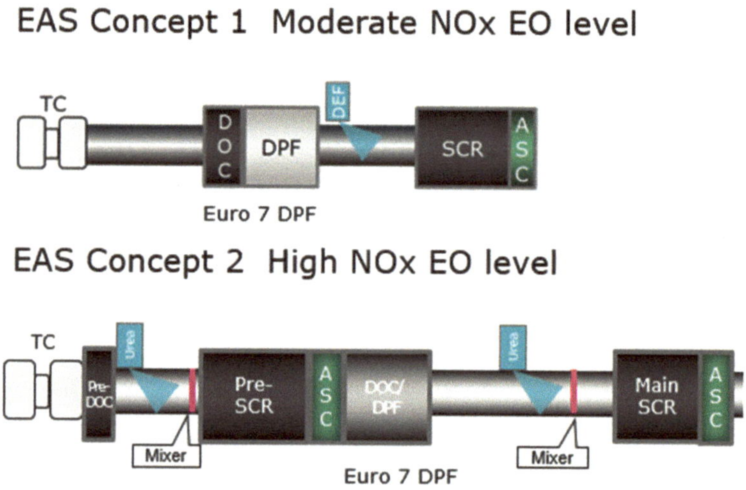

Fig. 7. Possible system architecture for a Euro 7 compliant aftertreatment system for the H_2-ICE HPDI combustion concept

The Euro 7 PN10 limit (Table 1) requires the integration of a particle filter in the exhaust aftertreatment system even for the HPDI concept. However, the expected low level of soot emissions eliminates the need for active soot regeneration. SCR desulfurization and Urea deposit removal procedures can be performed by exotherm HC oxidation on the Diesel Oxidation Catalyst (pre-DOC) respecting the temperature limits of the downstream components. Typically, in a Diesel exhaust aftertreatment system with two SCR systems, the dosing of the AdBlue dosing systems is varied to enable passive soot oxidation on the DPF when the operating conditions are favorable and usually the pre-SCR catalyst volume is lower compared to the main-SCR volume. As the soot emissions for the H_2-HPDI combustion concept are considerably lower compared to the Diesel combustion the above-mentioned design and operation guidelines do not necessarily apply for an exhaust aftertreatment system specifically designed for an H_2-HPDI engine. In the recent measurement campaign at AVL an aftertreatment system corresponding to the concept 1 configuration in Fig. 7 was applied and its performance was measured in the WHTC c/h. In Fig. 8 the results are graphically compared to Euro VI and Euro 7 emission limits. While the Euro VI NOx emissions are already reached the Euro 7 limits could be reached by adapting the EAS configuration and implementing heat-up strategies on the engine side.

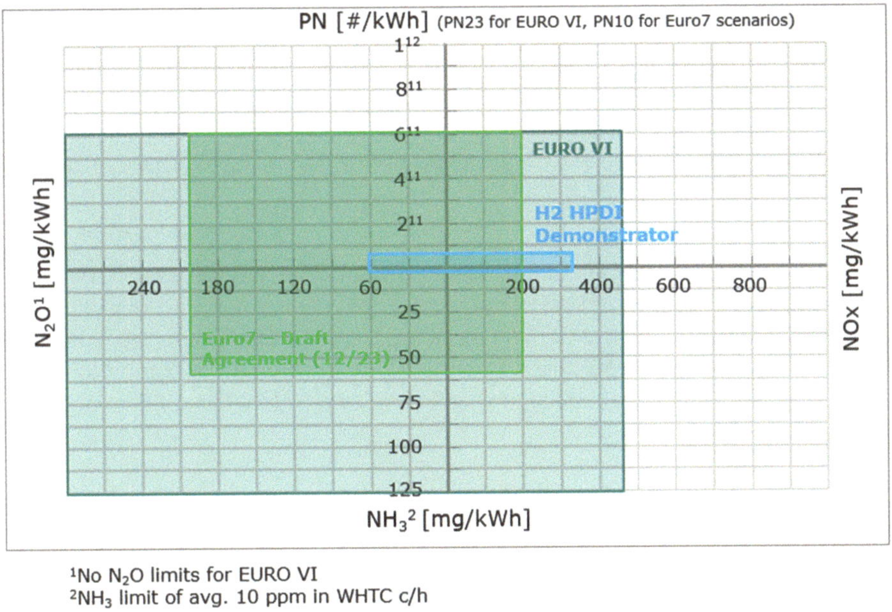

Fig. 8. NO_x, N_2O and NH_3 emissions in WHTC c/h

In terms of max. power and max. torque, results are illustrated in Fig. 9.

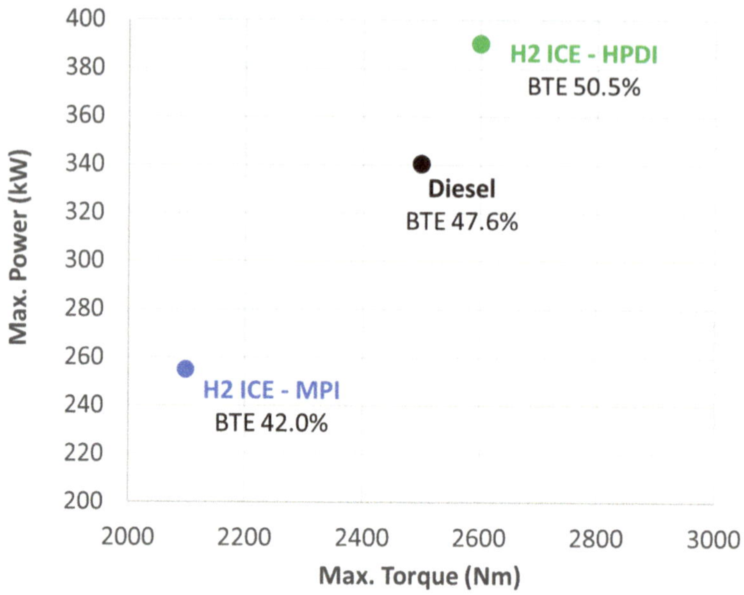

Fig. 9. Power and torque comparison with different powertrain concepts (same 6 cylinder 13L design)

The tank to wheel (TTW) CO_2 reduction from Diesel Engine was 97.5% in full load conditions. Use of sustainable paraffine fuel (e.g.: HVO, e-diesel) is also possible as pilot fluid for well to wheel (WTW) further reduction. These results are in line with the proposed limits of Zero Emission Vehicle (ZEV) in Europe of 5 g/ton-km CO_2e. H_2-ICE with HPDI shows potential to attend even lower values like 1 g/ton-km CO_2e.

4 Conclusions and Next Steps

H_2-ICE is a promising carbon neutral solution for commercial vehicle transportation. Its high performance at full load conditions combined with lower cost and much higher durability are factors for a significant advantage over other carbon neutral solutions.

In a common project of AVL, TUPY, Westport and ITnA (TU-Graz) a H_2-ICE with HPDI demonstrator was designed, built up and tested under transient conditions. A brake thermal efficiency above 50% was demonstrated, with further hardware upgrades thermodynamic calculations promise almost 52% BTE [11]. The demonstrated full load curve is identical to a typical Diesel, enabling even higher power levels with transient torque built up calibrated to be identical to series. H_2 combustion can release 24% more energy compared to diesel for a given airflow. It combined with the fact that H_2 combustion do not produce smoke, the smoke limiter is no longer needed. Then transient torque builds up time can be reduced with approx. 50% compared to the baseline diesel.

Fuel efficiency is a major contributor to optimize total cost of ownership for heavy duty vehicles, therefore high-pressure direct injection is an enabler to achieve carbon neutrality with highest efficiency for high load operation profiles combined with the robustness needed for commercial vehicle applications.

The H_2-ICE HPDI engine did run on a hydrogen energy ratio of 97.5% in full load conditions. A liquid pilot has turned out to be the only robust measure to ensure stable and robust conditions for an auto ignition of the hydrogen. It allows GHG reduction in line with proposals of ZEV in Europe.

A further degree of freedom and therefore potential for future combustion system development regarding NO_X-emission and max. cylinder pressure is seen in application of a combustion control with moderated injection rate.

For the P&E demonstration on the engine test cell all engine related HPDI components did show adequate reliability (engine run time limited to several 100hrs). The vehicle must foresee a 700bar Tank system, an integrated gas module, as a buffer tank, and a mobile compressor which extends the drive range if the tank pressure falls below the demanded injection pressure.

Further evaluations are focused on durability tests. It can be on vehicles or dyno and further project organization is being planned.

References

1. Mueller, M.: Sustainable Drivetrains for Off-Highway Applications. 43rd International Vienna Motor Symposium, 2022–26, Vienna, 2022.
2. Mayr, K., Hofer, F., Ragowsky, G., Gruber, W. and Arnberger, A.: Hydrogen: systemic comparison between fuel cell and engine. 9th International engine congress, Baden Baden, 2022.
3. Arnberger, A., Eder, M., Grabner, P., Raser, B.: Hydrogen Combustion Engine for CO_2-neutral Commercial Vehicles, ATZ Heavyduty Worldwide, number 03, 2021, pages 36–40.
4. Rieger, D., Mayer, F., Weller, F., Schneider, S.; Stiehl, R.: Experimental investigation of a hydrogen powered heavy-duty truck engine. 9th International engine congress, Baden Baden, 2022.
5. Korn, T., Ebert, T., Vonnoe, M., Tala, H., Lang, M.: Hydrogen Engines Strong Case – New Performance Benchmarks with Hydrogen Direct Injection. 43rd International Vienna Motor Symposium, 2022–11, Vienna, 2022.
6. Kufferath, A., Naber, D., Cornetti, G., Grzeszik, R., Kueger, M., Gaballo, M.: Development of combustion process and operating strategy for a low-emission hydrogen engine. 31st Aachen Colloquium Sustainable Mobility 2022, Paper Nr. 12, Aachen, 2022.
7. Mumford, D., Baker, S., Munshi, S.: High Performance Hydrogen Engine Applications Using Westport Fuel Systems' Commercially Available HPDI Technology. 43rd International Vienna Motor Symposium, 2022–29, Vienna, 2022.
8. Dober, G., Hoffmann, G., Piock, W., Doradoux, L., Meissonnier, G., Ouali, E., Kewley, J., Cardon, C., Coster, S., Wray, J., Muenz, S., Weiske, S.: Application of H_2-ICE Technology on Commercial Vehicles. 31st Aachen Colloquium Sustainable Mobility 2022, Paper Nr. 13, Aachen, 2022.

9. Arnberger, A., Wieser, M., Ferrarese, A., Christoforetti, P., Grabner, P. and Eichlseder, H.: Hydrogen Engine with 50% BTE – Benefits of High-Pressure Direct Injection. 19th Symposium "Sustainable Mobility, Transport and Power Generation", Graz, September 2023.
10. Ferrarese, A., Kumoto, E., Marquard, R., Wieser, M., Traxler, C., Lundqvist, U., Mumford, D. and Walker, J.: Resistant Cast Iron for Hydrogen Engine exceeding 50% BTE, 10th International engine congress, Baden Baden, 2023.
11. Eichlseder, H.; Grabner, P., Schutting, E.; Schneider, M.: Hydrogen ICE: State of the Art and Future Potential for real-world Application. 32nd Aachen Colloquium Sustainable Mobility 2023. Aachen, October 2023, paper 068.

Optimized Hydrogen Injection for Lowest NO_X Emissions

Olaf Weber[1(✉)], Jan Leberwurst[1], Jochen Broz[1], and Jens Steinmill[2]

[1] Schaeffler Technologies AG & Co. KG, Herzogenaurach, Germany
{weberoaf,leberjn,brozjch}@schaeffler.com
[2] Schaeffler Engineering GmbH, Werdohl, Germany

1 Introduction

1.1 Regulation

The European Union has recently published an updated road map for CO_2 reduction for commercial vehicles targeting CO_2 neutrality in 2050. Originally, only battery electric vehicles or fuel cell powered vehicles were considered to be zero emission vehicles. The new proposal also considers commercial vehicles, driven by an Internal Combustion Engine, emitting less than 1g of CO_2/t-kilometer, as zero emission vehicles.

The EPA also has proposed an update for the nationwide greenhouse gas regulation on commercial vehicles in 2023. Tailpipe CO_2 emissions from battery electric vehicles, fuel cell electric vehicles and vehicles with engines, fueled with neat hydrogen, are defined to be zero emission vehicles. No CO_2 related testing is required for these vehicles /8/.

China already has introduced a numerous fleet of hydrogen powered vehicles using fuel cells. It is expected, due to the technology open environment, that internal combustion engines will also be used to decrease the amount of CO_2 emitted by burning green hydrogen.

1.2 Markets

As the regulators step by step allow a high variety of machines converting hydrogen into motion, battery electric vehicles and fuel cells remain in the discussion. In addition, internal combustion engines using hydrogen are gaining share. Remote regions and locations, difficult applications requiring robust, low cost and low risk solutions provoke new approaches with internal combustion engines.

The market seems to be starting on the mid duty and heavy-duty engine side with displacements up to 16 liters. But also light commercial vehicle applications are subject to hydrogen combustion engines as refueling and service are easy compared to the huge advantage being counted as a zero emission vehicle. Even passenger vehicles

A. Heintzel (Hrsg.): Int.Engine.Congr. 2024, Proceedings, S. 117–129, 2025.
https://doi.org/10.1007/978-3-658-47966-4_9

are not the exception. Numerous manufacturers are testing 4-cylinder passenger car engines and their ability to get converted into hydrogen engines competing with fuel cell powered vehicles.

1.3 Technologies

Hydrogen and its properties regarding combustion initiate some changes in the internal combustion engine /1/. Ignition systems and piston ring packages have to be adjusted to the ignitability and fast combustion characteristics. The blow-by, covering unburnt hydrogen as well as burnt hydrogen in form of water, increases the partial pressure of water vapor in the crankcase significantly. Investigations are ongoing in how far the higher amount of water in the crankcase ventilation and/or lubricating oil causes different design efforts to keep the combustion engine as reliable as before.

The introduction of hydrogen into the engine and its mixture formation is a subject of a huge amount of investigations. As the markets are demanding clean solutions in a short amount of time, engine manufacturers are looking for practical solutions. The port fuel injection of hydrogen might be the first mixture formation system introduced to the marketplace. Replacing the Diesel injector by a spark plug, adjusting the compression ratio by a CNG based piston and applying CNG like injectors into the induction port seems to be realistic to face a soon SOP. Disadvantages like limited torque output and the potential of back firing during the gas exchange process are well known and taken into consideration. The homogeneous mixture between hydrogen and air guarantees NO_X free combustion beyond the air fuel ratio of 2.5.

2 Hydrogen Direct Injection

Hydrogen, directly applied into the intake port causes naturally a higher compressor pumping work. The air and the supplied hydrogen have both to be pushed into the combustion chamber. This leads to higher boost pressures or and smaller amounts of air facing the same air fuel ratio ending up in a lower engine torque. In addition, the high energy level of the hydrogen applied due to a pressure between 20 and 60 bar is not being used at all.

As the hydrogen combustion engine compared i.e. to its brother the Diesel engine is facing an air fuel ratio limit due to the homogeneous combustion avoiding NO_X emission, lean operation will be preferable in the entire engine map. This limits the boost pressure in addition under steady state conditions. Replacing a reliable port fuel injection system by a new direct injection system is therefore necessary to keep the hydrogen engine in a reasonable torque range. On the other side, also seen in Fig. 1, the mixture formation in the combustion chamber might not lead to a perfect homogenization.

We do then need a higher average air fuel ratio to avoid the creation of NO_X emission in the engine map leading again to a lower available torque. The capability of a direct injection system in terms of homogenization is therefore crucial for the performance of a hydrogen combustion engine.

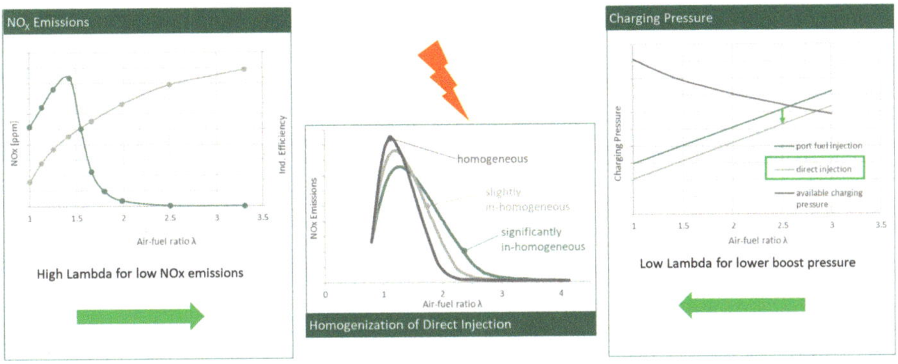

Fig. 1. Lean Running Combustion Engines targeting homogenous Mixture Formation /9,10/

2.1 Published Solutions

Based on these challenges a lot of investigations have been already done. The usage of the Schlieren method as a visualization tool for hydrogen flow is also understood as a tool to calibrate and understand CFD based methods. Figure 2 shows some investigation examples/3,4/. The injector nozzle is applied to different vertical positions on the cylinder axis. Different positions cause different kind of flow patterns. In some cases like on the left hand side, with a nozzle already penetrating the combustion chamber, the flow pattern is flat and horizontal, towards the right hand side, where the position of the nozzle is more in a recess, the flow pattern goes straight in parallel to the cylinder axis towards the surface of the piston. The back pressure in the combustion chamber is also influencing the flow pattern but seems to be a minor factor compared to the injector position.

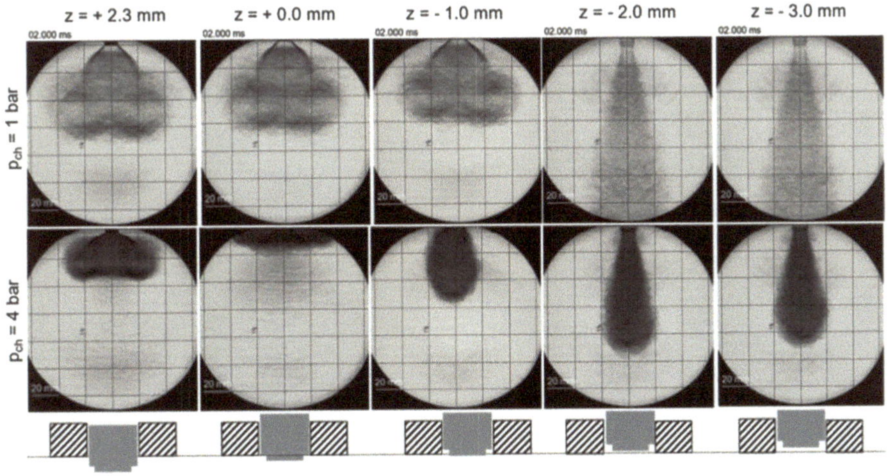

Fig. 2. Schlieren Images for Injector Mounting Positions at different Backpressures /4/

To create a more directed flow /5/ into the combustion chamber, flow caps, spray caps or deflector caps are being used. Those caps take the out of nozzle flow and direct it into a certain position in the combustion chamber, Fig. 3. A lot of investigations have been made to determine the best orientation and position of those caps. Tremendous improvements can be made by varying these positions for certain operating points. Figure 4 shows the reliable CFD simulation when compared with the Schlieren method using different kinds of caps for different purpose. However, the mixture formation process is still uncomplete and the design parameters of the injection equipment is not totally understood.

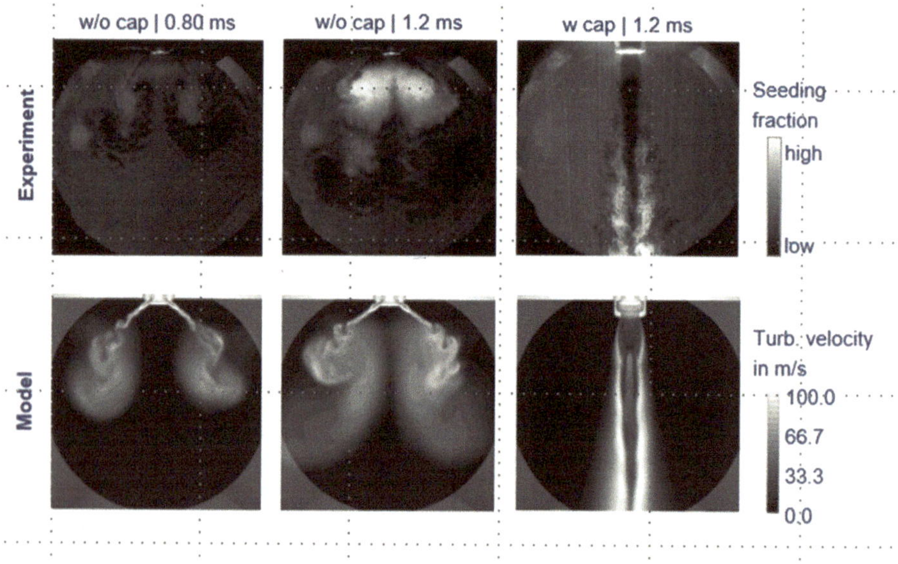

Fig. 3. Comparison of Turbulence Distribution with and without Spray Caps /5/

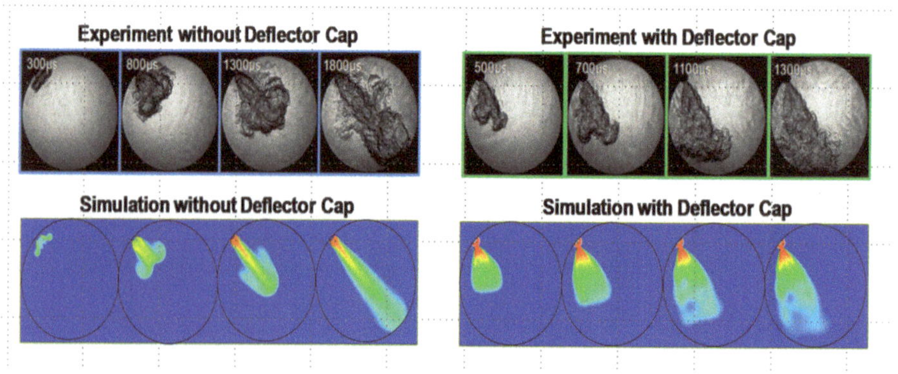

Fig. 4. Correlation between Schlieren Spray Images and simulation Results /5/

Flush mounted injectors can also generate the so-called Coandă effect /5/. Other effects lead to a close to the surface flow which inhibits the penetration of the combustion chamber by the H_2 spray. The majority of the investigations try to avoid this Coandă effect as it seems to be a more unpredictable phenomena disturbing any directed development of a new product. The occurring Coandă flow patterns are understood as they are accepted (Fig. 5).

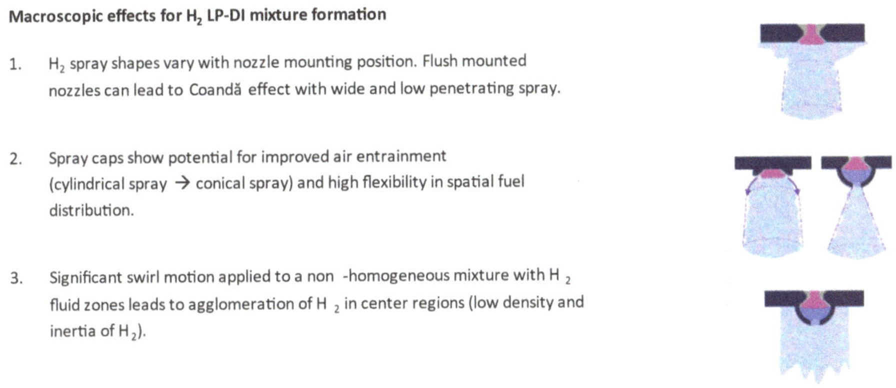

Fig. 5. Coandă Effect and how to avoid it /5/

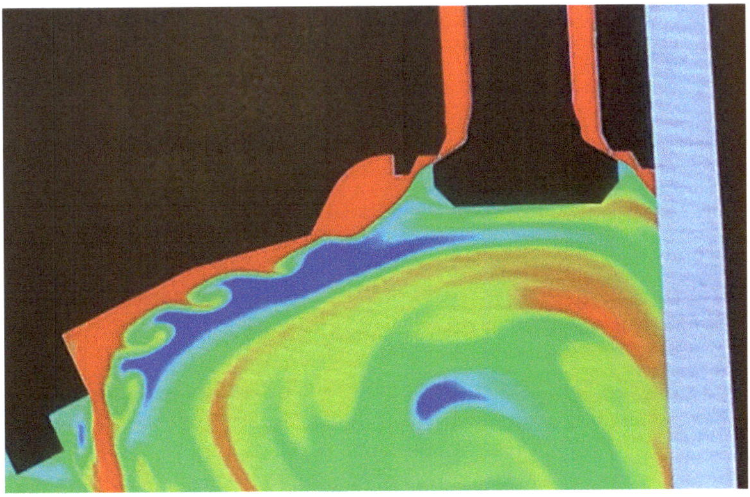

Fig. 6. Coandă Effect according to /7/ on overexpanding nozzles: Hydrogen (red) flows on the cylinder head surface reaching the spark plug on the left hand side

But the Coandă effect is also used to create stratified mixture formation processes like in /7/, Fig. 6, due to the pressure conditions around an overexpanded flow caused by a supersonic nozzle. The Coandă effect ensures a strict stratified mixture formation process where the hydrogen is flowing on top of the combustion chamber surface straight to the spark plug and ensures a proper ignition and later on stratified combustion leading to overall low NOx and reliable combustion /7/.

As a summary, the position of an injector and the upcoming Coandă effect are influencing the flow patterns significantly. A viable solution for an injector should comprise these two parameters (Fig. 7).

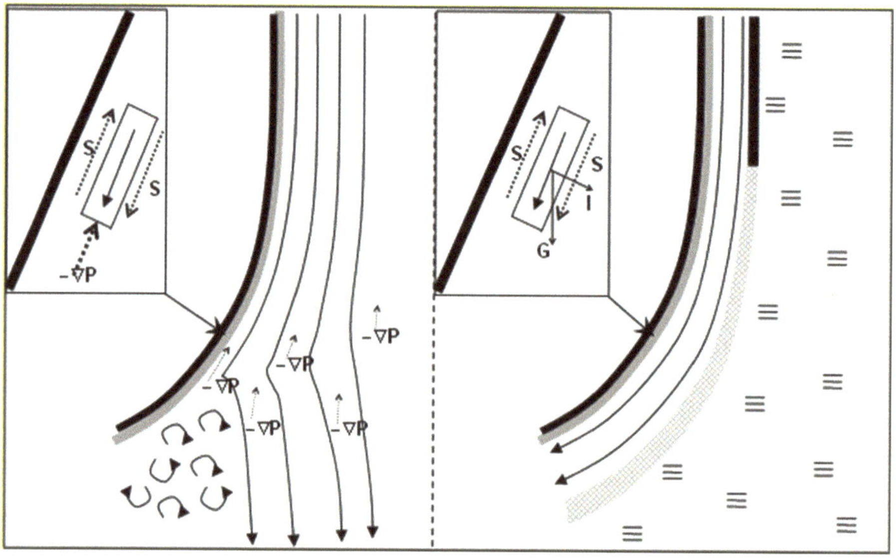

Fig. 7. Coandă Effect according to /6/

2.2 A New Approach to improve the Mixture Formation Process

Intensive development work led to a variable supersonic nozzle which controls the amount of hydrogen and the flow pattern at the same time. In addition, the pressure ratio across the injector is also decisive to direct the H_2 flow into the combustion chamber, Fig. 8.

On top of Fig. 8 the nozzle is shown with two different operation conditions. The left-hand side shows the jet flow where all the hydrogen flows parallel to the cylinder axis into the depth of the cylinder. On the right hand-side we can see the hydrogen flowing as a Coandă. The hydrogen is flowing parallel to the surface of the flat cylinder head. Figure 8 is showing two different flow patterns at the same time although the nozzle will of course create a rotational symmetrical flow pattern. The lower part of Fig. 8 explains the conditions leading to the different flow patterns: the H_2 pressure in the injector is plotted against the cylinder pressure. The drawn angel bisector divides the figure into two main parts: on the top left side we have an ordinary jet flow into the cylinder. On the righthand side of the angel bisector the flow is just reversed, hydrogen flow from the cylinder into the injector occurs.

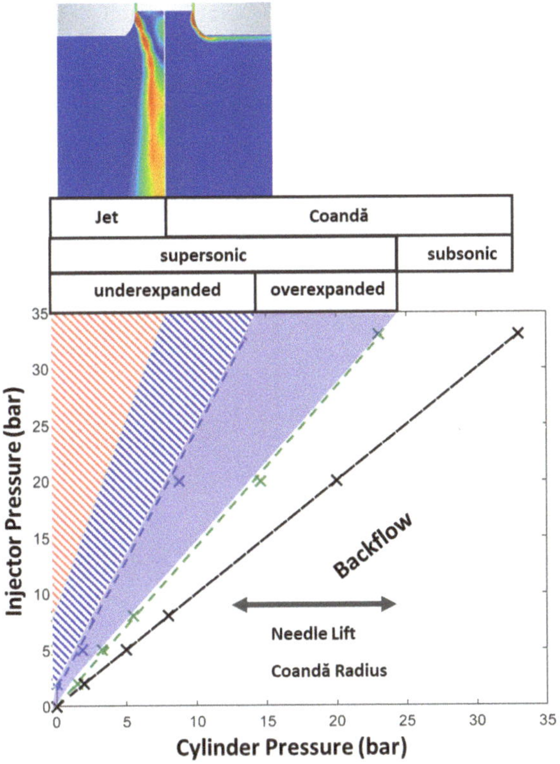

Fig. 8. Schaeffler Coandă Control of a supersonic injector

The dividing line between Coandă and jet flow is located on the very lefthand side. The Coandă and the jet are both flowing under supersonic, underexpanded conditions. The hydrogen, arriving at the nozzle exit, can still expand into the gaseous environment in the combustion chamber beyond the pressure at the nozzle exit.

The area of the over expanded Coandă flow is reached when the pressure in the combustion chamber reaches the pressure at the nozzle exit. The Coandă flow remains, even when the hydrogen flow changes to subsonic conditions.

The switching between the Coandă flow and the jet flow can now be controlled by all these parameters plotted in Fig. 8. Increasing combustion chamber pressures will create a Coandă flow and decreasing combustion chamber pressures at one point of time a jet flow. The reverse effect can be seen for example at constant combustion chamber pressure by raising the injector pressure. This will cause a change from

Coandă to jet flow. The method to combine all these engine parameters like cylinder pressure, injector pressure and needle lift leads to a wide field of opportunities to control the flow into the combustion chamber.

The Coandă radius of the injector is constant and can be seen with the same rigor as the design, orientation and position of a spray cap. The injector needle lift can be seen as the above-mentioned vertical position of the injector, flush or in recession mounted. The before mentioned in engine operation fixed parameters of spray caps and injector position and orientation can now be controlled by the new injector.

An example, how jet flow and Coandă flow can fill up a combustion chamber with hydrogen is plotted in Fig. 9. The Condă flow towards the end of compression fills up the "lean" combustion chamber area.

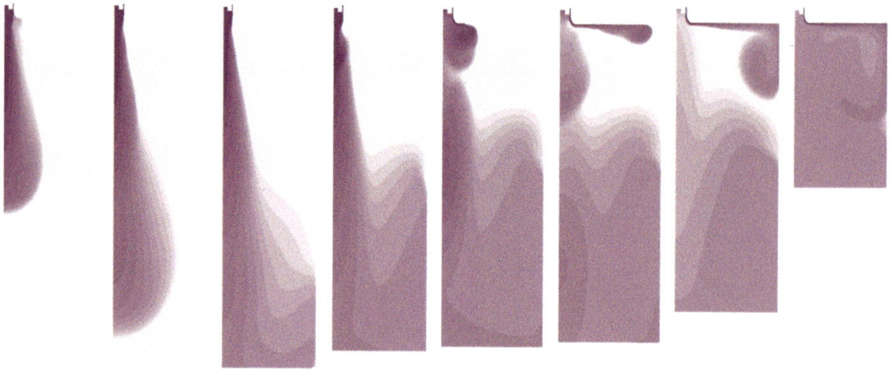

Fig. 9. Jet and Coandă flow share versus crank angle during the compression stroke. The compressing piston cuts the lengths of the snapshots from the bottom

This concept can now be applied to a combustion engine, at first with a simulation tool. We applied tools like Fluent and Converge depending on the task and its complexity. Figure 10 shows the Lambda distribution in a combustion chamber at FTDC. The plots show a cross section of half of a combustion chamber, the piston bowl can be recognized, the central injector is placed on the righthand side of the plots, the cylinder wall is, consequently, located on the lefthand side. Data for the start of injection, SOI, the homogeneity index HI, and the injected mass distribution between jet and Coandă is being varied and caused by the start of injection.

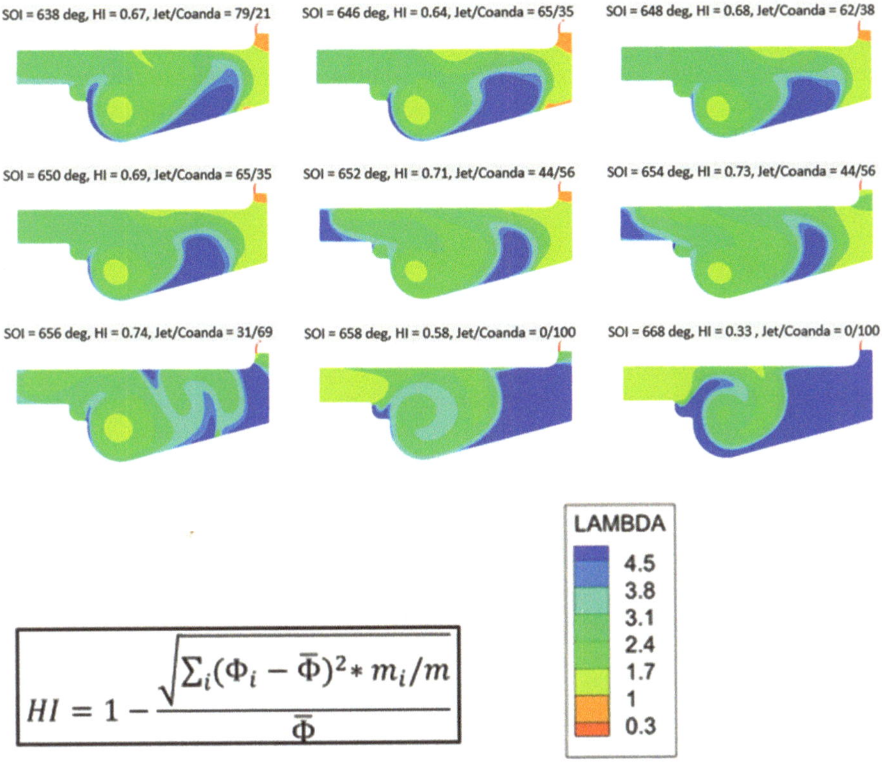

$$HI = 1 - \frac{\sqrt{\sum_i (\Phi_i - \bar{\Phi})^2 * m_i/m}}{\bar{\Phi}}$$

Fig. 10. Mixture formation influenced by different injection timing using omega piston shape, pictures show the left-hand-side cross section of a combustion chamber

The nine visible flow patterns start with a comparatively early start of injection (638 deg Crank) and end up with a late start of injection (668 deg crank). Parallel to the later SOI, the cylinder pressure is higher at later injection windows. This creates different shares of jet and Coandă, causing very different flows and Lambda distributions. The best HI can be detected using the SOI of 658 deg crank, with a HI of 0,74 and a jet/Coandă masse ratio of 31/69. Later SOIs than 658 deg crank eliminate the jet flow totally, the differences in Lambda distribution are just due to the different time difference between SOI and FDTC.

The calculations were made without any air charge motion induced into the combustion chamber. All kinetic energy (TKE) has been induced by the flow of the hydrogen. These figures just show a very first example of the investigations made. Further steps will be the optimization of the start of injection (SOI) considering the air intake flow and different engine conditions like rail pressure and boost pressure. In addition, combustion simulations will be done to understand the link between mixture formation and related knock sensitivity. It is expected that less inhomogeneous mixtures will reduce NO_X emissions.

The flow patterns change based on pressure ratios and injector needle lift. These different flow patterns can now be applied by different crank angles during the mixture formation process and it is therefore possible to reach out to extreme regions in the combustion chamber which cannot be reached by rigid elements like flow caps or fixed positions of injectors. This new approach combines variable injector positions and a controlled Coandă effect to improve the mixture formation process.

3 Durability

Due to the missing lubrication, the H_2 injectors are tremendously exposed to very difficult conditions regarding wear, both at the needle guide as well as the needle seat. The issue is even worse compared to natural gas injectors where in some cases residuals of lubricating oil are still existing. Fuel cells require a high purity of hydrogen used to avoid poisoning of the catalytic surfaces. As the combustion engine has to use the same purity of hydrogen by regulation, the challenge for a highly reliable injector is even more difficult.

In the past, Schaeffler has understood a lot of different wear processes under very difficult and different conditions (Fig. 11). The result is a set of methods to analyze and understand the wear process faced by different surfaces. In addition, coating recipes with the focus on wear, corrosion, friction as well as the assigned production recipes need to meet the necessary requirements (Fig. 12).

The development priority of durable injectors is at least as high as a good mixture formation process under all perceived engine conditions.

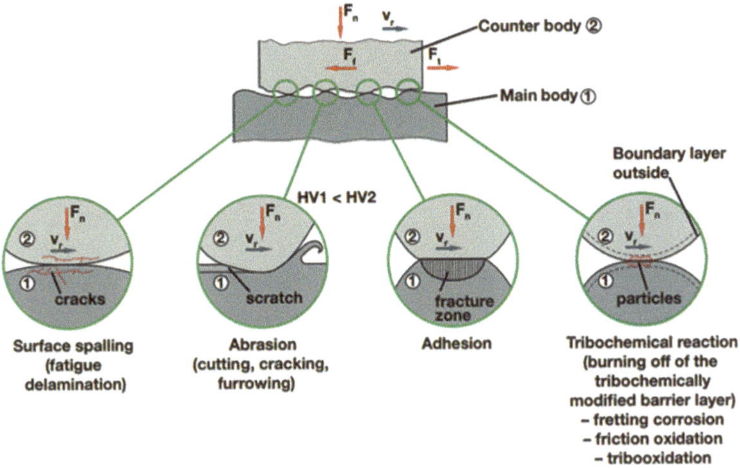

Fig. 11. Structure and mechanisms leading to wear

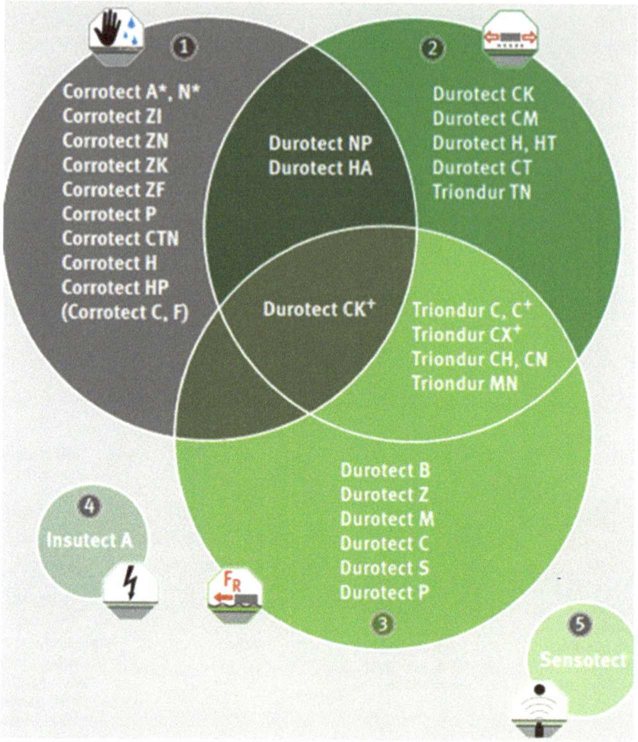

Fig. 12. Excerpt of schaeffler durability concepts for corrosion, wear, friction and insulation

References

1. Heinz Stutzenberger, Untersuchung des Betriebsverhaltens eines Wasserstoff-Wankelmotors, Dissertation RWTH Aachen, 1983
2. David Lejsek, Philippe Leick, Paul Jochmann, Peter Grabner, Eberhard Schutting: Analysis of Mixture Formation and Combustion in H2 Engines for Passenger Cars and Light Commercial Vehicles. 9th International Engine Congress, BadenBaden, 2022.
3. Dr.-Ing. Niklas Geiler, Dipl.-Ing. Klaus Moritz Springer, Dipl.-Ing. Thomas Lorenz, M.Sc. Michele Blomberg, M.Sc. Jens Achenbach, Dr.-Ing. Peter Bloch: H2 engine Hybrid Powertrain for future Light Commercial Vehicles, 10th International Engine Con-gress, BadenBaden, 2023.
4. David Lejsek, Dimitri Seboldt, Jan Niklas Geiler, Philippe Leick, Michael Frank, Karl Georg Stapf: Mixture formation and Combustion in a Passenger Car engine with Low Pressure H2 Direct Injection, 19th Symposium "sustainable Mobility, Transport and Power Generation, Graz, September 28–29, 2023
5. Dr. Maximilian Brauer, Lorenz von Römer, Michael Riess, Dr. Jochen Maaß, Anja Fink, Oliver Nett: Hydrogen DI Combustion System Development on a Retrofitted Medium Duty Diesel Engine, 19th Symposium for Sustainable Mobility, Transport and Power Generation, Graz, September 29th, 2023

6. https://de.wikipedia.org/wiki/Coand%C4%83-Effekt#/media/Datei:CoandaEffect3.png
7. Hans-Jürgen Berner, The investigation of hydrogen based vehicle propulsion systems at FKFS and IFS, The 7th Shanghai-Stuttgart-Symposium on Automotive and Powertrain Technology, Shanghai, December 2023
8. Anton Arnberger, Martin Wieser, AVL List GmbH, Andre Ferranese, Tupy S.A. Join-ville, Paul Christoforette, Peter Grabner, Helmut Eichelseder, TU Graz: Hydrogen En-gine with 50% BTE – Benefits of High Pressure Direct Injection – ITNA, September 29th 2023
9. M. Schumacher, P. Renninger, C. Barba, J. Lehmann: Hydrogen Engine – Robust Low Emission Powertrain for Heavy Duty Trucks, Wasserstoffmotorkonferenz 2023, September 12th.
10. Gavin Dober, Guy Hoffmann, Walter Piock, BorgWarner, Luxemburg; Laurent Dora-doux, Guillaume Meissonnier, Emmanuel Quali, Borgwarner, France; James Kewley, Christophe Cardon, Simon Coster, Jonathan Wray, BorgWarner, U.K.; Stefan Münz, Sascha Weiske, BorgWarner, Germany: Application of H2ICE Technology on Commercial Vehicles, 31st Colloquium for sustainable Mobility, Aachen, October 2022

Tailored Cooling and Charge Motion Concept for Spark Ignited Hydrogen Combustion Engine to Achieve Highest Efficiency and Power Density

Lukas Virnich[1(✉)], Avnish Dhongde[1], Björn Franzke[1], Ramund Tensing[1], and M. Can Kayacan[2]

[1] FEV, Aachen, Germany
virnich@fev.com
[2] Thermodynamics of Mobile Energy Conversion Systems, Aachen, Germany

1 Introduction

Several different combustion system options exist for H_2-ICEs as shown in Fig. 1 Suitability varies by application and depends also on the availability of en-abling technologies. Market intelligence shows that current hydrogen engine devel-opment mainly relies on the "Otto" combustion principle. The air/fuel mixture is ignit-ed by the external energy source provided by a spark plug as described in the left path.

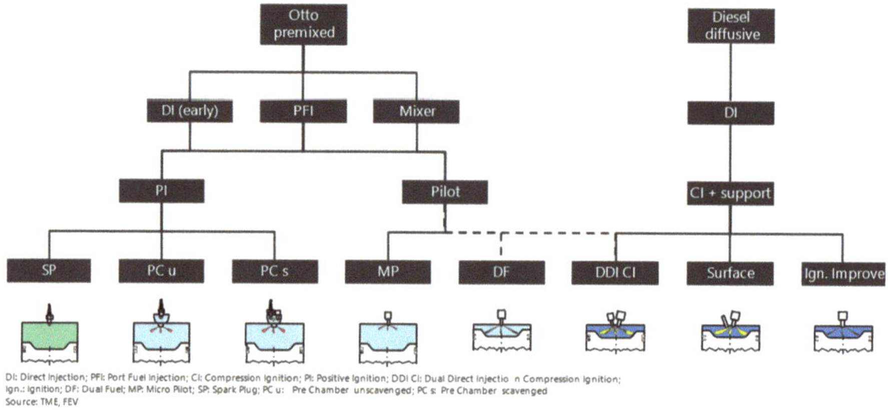

Fig. 1. Combustion system overview

A. Heintzel (Hrsg.): Int.Engine.Congr. 2024, Proceedings, S. 130–138, 2025.
https://doi.org/10.1007/978-3-658-47966-4_10

Combining hydrogen with a second fuel with a high cetane number provides additional options. One of them is to use a micro pilot injection directly into the cylinder while H_2 fuel injection is realized via port fuel injection (PFI). The second fuel is injected into a premixed air fuel mixture and the compression ignition of the second fuel ignites the hydrogen-air mixture. The hydrogen-air mixture is burning following the SI (spark ignition) principle. This principle is applied mainly in retrofit solutions but limits the engine's power density. Moreover, the achievable hydrogen share is rather low, especially in full load conditions.

With dual direct injection, diffusive combustion of H_2 can be enabled. A small amount of high cetane fuel is used to raise the temperature of the cylinder charge, and hydrogen is then directly injected with high pressure—around 300 bar—and combusted diffusively. Utilizing this combustion principle, the hardware of a diesel engine can be used almost entirely. Also, power density and efficiency of a conventional diesel engine can be achieved. On the other hand, the high injection pressure and high NOx raw emissions are two of the major drawbacks of diffusive combustion of hydrogen. Also, when using micro-pilot diesel fuel injection, CO_2 emissions are not zero according to Article 9 of Regulation (EU) 2017/2400.

2 Combustion Systems

In the following, only monofuel SI hydrogen combustion principles will be further explained. Port fuel injection (PFI) and direct injection (DI) of hydrogen will then be evaluated in detail.

2.1 Port Fuel Injection System

With PFI, hydrogen fuel is injected into the intake manifold at relatively low pressure levels (~10 bar). The relatively long time for mixture preparation and the turbulence created at the intake valves ensure a high level of mixture homogeneity. Hence, the occurrence of rich air/fuel areas is limited. This reduces the NOx formation as well as the tendency for abnormal combustion occurrence during the compression stroke. Whereas the presence of hydrogen within the combustion chamber right after the exhaust stroke creates the risk of pre-ignition during the intake stroke, which needs to be suppressed by proper valve and injection timing. Thanks to the low injection pressure, longest possible driving ranges are ensured, as the hydrogen in the tank can be extracted down to low pressure.

On the other hand, external mixture formation reduces the cylinder air charge. The reduction in volumetric efficiency is highly dependent on the air/fuel ratio and decreases with leaner mixtures. At a typical relative air/fuel ratio of $\lambda = 2$, the cylinder air charge is reduced by ~ 20% due to the presence of hydrogen. Both fresh air and hydrogen must be forced into the cylinder through the intake valve. Consequently, the boost pressure demand is much higher than with direct injection.

Only minor hardware modifications are required to run a PFI system. Indeed, injectors are easier to integrate into an existing engine and they will be becoming available on the market soon. Hence, a PFI configuration represents a very attractive solution for retrofitting lower power applications, particularly in the off-road market. If PFI is used for on-road applications, the drawback in volumetric efficiency needs to be compensated by a two-stage boosting system. For MD applications, this setup seems to be an alternative solution, as two-stage boosting systems are also known from former diesel engines (Fig. 2).

	External mixture preparation	Space requirement cylinder head		
	Low pressure PFI	**Low pressure DI**	**Mid pressure DI**	**High pressure DI**
Fuel Injection	Port fuel injection ~ 5-10 bar	Direct injection ~ 15-30 bar	Direct injection ~ 40-60 bar	Direct injection ~ 300 bar
Specific Power (HD) engine	< 25 kW/l	> 25 kW/l	> 25 kW/l	~ 30 kW/l
Peak BMEP (HD) engine	< 20 bar	> 20 bar	> 20 bar	> 25 bar
Combustion	Stoich/Lean Spark ignited	Lean Spark ignited	Lean Spark ignited	Lean Spark ignited -- Diffusive
Knock tendency	↓	0	0	↑(SI) -- Not existing (Dif.[1)])
Boost pressure demand	↑↑	0	0	0
Transient load response	↓↓	0	0	↑
Main benefit	• Easy to integrate, • Hardware available • Low failure risk	• Robust against back-fire • Power density • Transient response	• Same as LP DI • Smaller packaging compared to low pressure • Potentially better mixture preparation	• Same as MP DI • Diffusive combustion possible
Main challenges	• Boosting • Safety (Backfire)	• Integration DI Injector • Uniform mixture preparation	• Integration DI Injector • Range	• Integration DI Injector • High pressure generation • For Diffusive combustion high NOx raw emissions

1) Diffusive combustion

Fig. 2. Injection systems for a 13 L heavy duty engine with single stage turbo charging

Nevertheless, load response remains a drawback with a PFI layout, but assistant measures for dynamic operation, such as electrically assisted boosting systems (e-Charger or e-TC) or a 48V BSG, can substantially improve the transient response. Another option to improve the transient behaviour and increase the power density is to adapt the engine to implement a DI system.

2.2 Direct Injection System

The integration of a DI system into the cylinder head of an existing engine is a challenge. Due to the very low caloric volumetric density of gaseous hydrogen, the injector is very voluminous to guaranty a sufficient flow area for delivering the required fuel mass. With higher hydrogen pressure, the packaging gets easier. For the same mass flow rate, the injector can be made slimmer. In most of the configurations already studied, a spark plug in a central position is beneficial for the combustion. The mixture is thus ignited from a central position, allowing the flame front to propagate symmetrically in the combustion chamber. Therefore, in most cases, the DI injector must be installed in a lateral position.

By feeding the hydrogen directly into the combustion chamber, the risk of backfiring is significantly reduced. Only air must be forced from the intake side into the cylinder, which prevents the reduction in volumetric efficiency that occurs with PFI technology.

Compared to PFI, the time for mixture preparation is shortened, and hence a lower level of mixture homogeneity is achieved in principle. Consequently, the possibility for rich areas increases and with it the knock and abnormal combustion tendency. Thus, a major challenge for DI systems is to provide PFI-like mixture homogeneity to prevent any drawback in robustness for the DI system.

3 Operating Limits of Spark-ignited Hydrogen Combustion Engines

Spark ignition hydrogen combustion systems with port or direct fuel injection are an attractive zero-GHG emission option for commercial applications. While a diesel-like combustion system would generally offer higher efficiencies, low pressure injection systems in combination with spark ignition allow longer driving ranges by using more of the fuel tank content. On-board compression of the hydrogen is not yet feasible due to packaging constraints and efficiency deficits.

However, achieving power densities similar to those of state-of-the-art diesel engines is challenging because the operating limits of premixed combustions systems are generally limited by the occurrence of abnormal combustion.

To illustrate the operating limits of an SI hydrogen engine, several combustion anomalies, that may occur at high loads are displayed in Fig. 3.

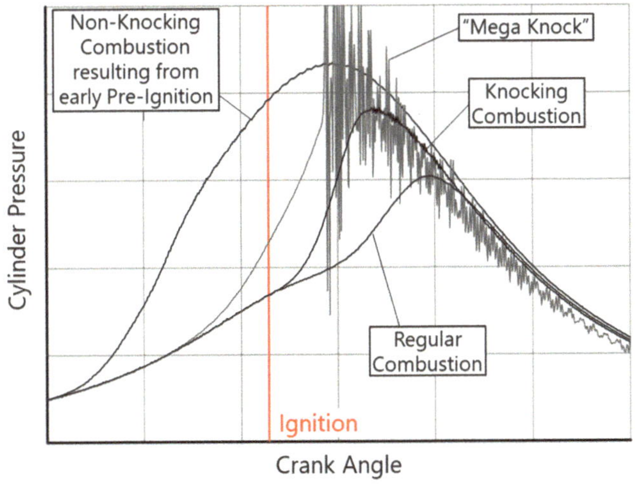

Fig. 3. Combustion anomalies at high loads in a hydrogen combustion engine

Knocking combustion can occur due to the natural fluctuation of the burn delay. Early combustion cycles promote auto-ignition in the unburnt zone, resulting in pressure oscillations that can potentially damage the engine.

Pre-ignition can be caused by ignition sources such as hotspots, glowing particles or burning oil droplets. In these cases, the combustion is much advanced, often resulting in knocking. In some cases, "Mega-Knock" can occur with pressure amplitudes exceeding 100 bar. Very early pre-ignitions may not result in knocking combustion because the fuel is completely burned before the boundary conditions for knock are reached. But even these cycles can reach severely high cylinder pressures.

3.1 Occurrence and Suppression Strategies of Abnormal Combustion

The main characteristics of an irregular combustion process are the self-ignition of unburned mixture in front of the flame front (knocking) or the ignition of the mixture before the actual spark ignition (pre-ignition). In both cases, excessive wall heat transfer and peak cylinder pressures, as well as pressure oscillations in the case of knocking, can lead to critical thermal and mechanical stress levels. Particularly high pressure amplitudes occur when large-volume self-ignition of the unburned mixture occurs during the compression ("mega-knocking") [2]. While the phenomenon of conventional knocking can be actively controlled by ignition timing retardation, safeguarding against pre-ignition is the central challenge in the combustion process development of a hydrogen engine. Figure 4 summarizes the two basic approaches to prevent pre-ignition.

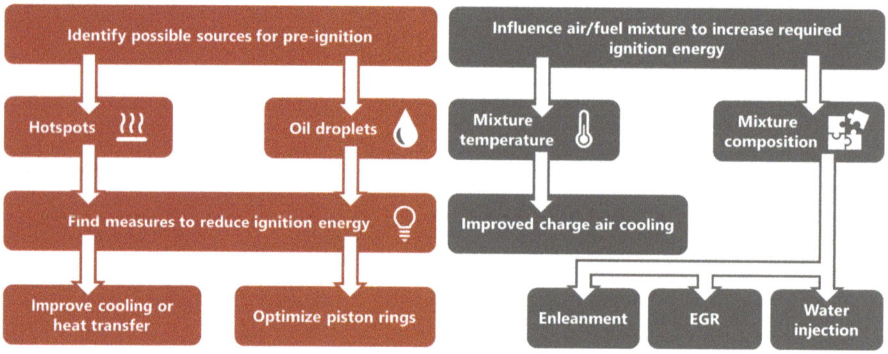

Fig. 4. Methods for the prevention of irregular combustion phenomena

The left-hand path pursues the approach of reducing the available ignition energy so that neither hotspots in the combustion chamber nor oil droplets act as ignition sources of the H_2-air mixture. While oil management at the cylinder wall can be influenced by the design of the pairing of piston rings and honing, hotspots can be minimized by optimized component design (especially with regards to cooling) as well as by appropriate material selection [3]. The right-hand path in Fig. 3 follows the approach of increasing the required ignition energy of the H_2-air mixture by charge dilution and improved mixture homogeneity. The improvement of mixture homogenization through the introduction of a strong tumble charge motion can significantly increase the robustness against abnormal combustion.

4 Cylinder Head Design for decreased pre ignition Tendency

4.1 Improved Mixture formation and Combustion with Tumble Motion

3D CFD RANS simulations were used to evaluate and optimize the mixture formation inside the combustion chamber at the rated power point. The details of the modeling and the calibration of the injection model to optical measurements and calibration of the combustion model to engine measurements can be found in [5].

A comparison of the base Diesel layout with the arterial cooling tumble layout is shown in Fig 5. The base layout with Diesel ports had no charge motion and relied solely on the cross-tumble created by the hydrogen injection to drive the mixture formation. Although this process was suitable for lower engine speeds and loads, the impulse of the injection was insufficient at the rated power point. The arterial cooling layout enabled the integration of dedicated tumble ports with eccentric bottom chamfers, which generated a significant amount of conventional tumble in the intake stroke. Moreover, the tumble motion is supported by the integrated jet forming machining in the cylinder head as well as the deep lens bowl aid as compared to the base pot like piston bowl. It must be noted that the tumble ports have lower flow coefficients and lead to a 4 % drop in air flow. However, the mixture formation was dramatically improved as indicated by the lambda distribution and has lesser rich zones ($\lambda < 2.0$) as compared to the base layout.

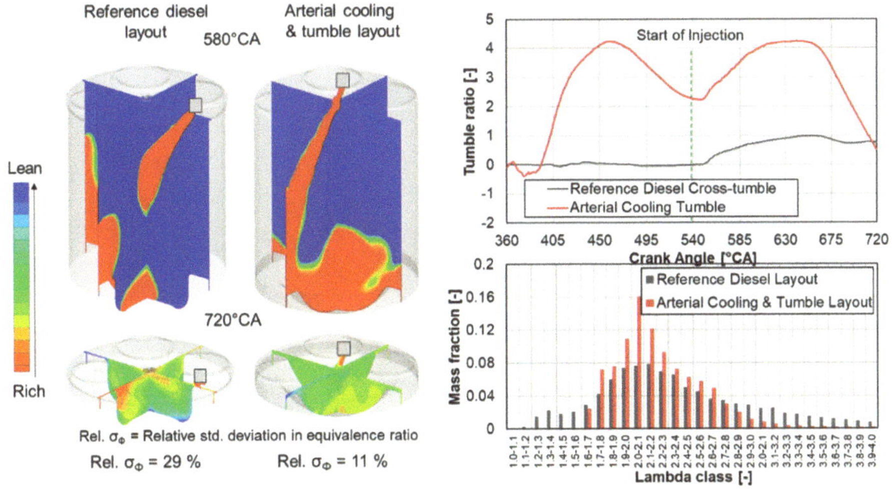

Fig. 5. CFD Cut-sections, Tumble and Lambda distribution. (© FEV)

In addition to the improvement in homogeneity, which decreases the NOx emissions by 60 %, the arterial cooling and tumble layout reduces the burn duration by 25 % because of increased turbulence and wrinkling of the flame front at the TDC via the destruction of the tumble motion. This layout is also expected to perform better with respect to abnormal combustion, which would shortly be confirmed with test bench investigations.

The simulation was run further to generate the 720 °C A cycle averaged heat transfer coefficients and reference gas temperatures, which were provided to the FEA simulation.

4.2 Temperature Reduction through Targeted Cooling

Due to the coolant supply from both intake and exhaust sides and the connections to the higher water plane, the conventional, large-volume Diesel cylinder head is subjected to unfavorable coolant flow and velocity distribution, particularly in the central area of the flame deck. However, the hydrogen injector and the spark plug require increased cooling, especially in this area.

In the optimized cylinder head with the arterial cooling ducts, the individual channels with defined cross sections flow around the valve seat inserts, the injector shaft and the spark plug in a targeted manner. In this way, the heat introduced by combustion is taken directly from the hot spots and transported away with high flow velocities and consequently high heat transfer coefficients, Fig. 5. As a positive side effect, the individual channels increase the rigidity of the combustion chamber plate compared with a large-volume conventional water jacket (Fig. 6).

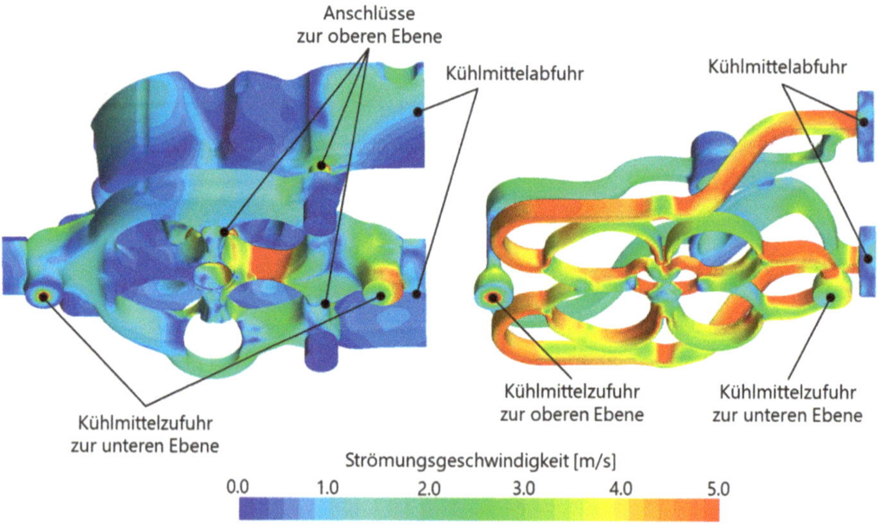

Fig. 6. Flow velocity of the coolant in the flame deck area with arterial cooling (right) and from the conventional cylinder head (left) at rated power. (© FEV)

The intensified charge motion does increase the average combustion side heat transfer coefficient in the middle region of the flame deck at a similar gas temperature by up to 45 % compared with the design without charge motion. Despite this, comparative thermal FE analyses show that the average material temperature on the flame deck of the cylinder head can be significantly lowered at rated power with constant volume flow because of higher heat transfer through arterial cooling (Fig. 7).

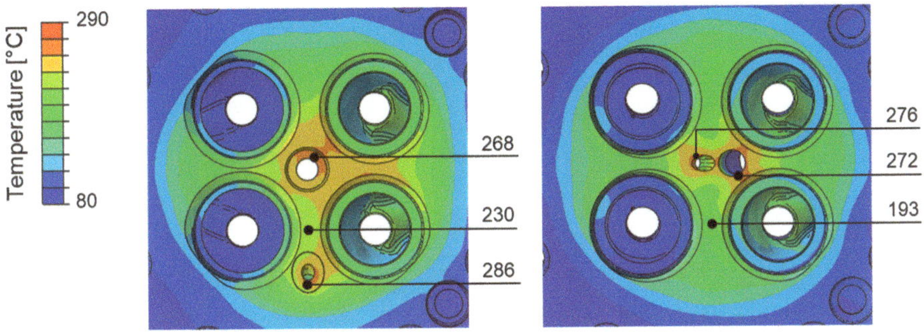

Fig. 7. Temperature distribution on the flame deck with arterial cooling and central injector (right) and on the conventional cylinder head of the reference engine (left) at rated power. (© FEV)

Arterial cooling is able not only to compensate for a temperature increase due to intensified charge movement, but also to reduce the maximum temperature by 10 °C.

5 Conclusions and Outlook

Latest developments in the manufacturing technology of sand cores for the casting process of large-volume engine components such as cylinder heads enable a novel design of arterial cooling channels with optimized flow cross-sections for the targeted cooling of hot zones of the combustion chamber flame deck. A centrally positioned injector and increased tumble motion improve the mixture homogeneity of a hydrogen engine, while decreased maximum temperatures reduce the tendency to uncontrolled pre-ignition of the hydrogen-air mixture. This enables a further increase in power density or, alternatively, in efficiency by raising the compression ratio above the current state-of-the-art level.

References

1. Virnich, L.,International Congress Engine Congress, Baden Baden Combustion system development for hydrogen fueled heavy duty internal combustion engines. 2021.
2. Boberic, A.: Measures to achieve high specific power with a heavy-duty H_2 internal combustion engine: A numerical and experimental analysis, 31. Aachen Kolloquium, Aachen, 2022
3. Franzke, B.: Leistungsstarke H_2 DI-Motoren für MD/HD-Anwendungen, HD ATZ/MTZ.
4. Klell, M.,Wasserstoff in der Fahrzeugtechnik. 4th Edition. Springer Vieweg Wiesbaden.
5. ISBN 978-3-658-20446-4, 2018.
6. Dhongde, A;.: Fortgeschrittene Simulationsmethoden für Wasserstoffverbrennungsmotoren. ATZheavyduty 14 (2021), Nr. 4, S. 46–51, 2021.

Direct Air Capture – Technology Status and Challenges in Technology Scaling

Marc-Simon Löffler(✉), Raphael Vollmer, Ulrich Zuberbühler, Dominik Roth, and Sebastian Thaler

Center for Solar Energy and Hydrogen Research Baden-Württemberg (ZSW), Stuttgart, Germany
{marc-simon.loeffler, raphael.vollmer,ulrich.zuberbuehler, dominik.roth,sebastian.thaler}@zsw-bw.de

1 Direct Air Capture

1.1 Technology Overview

The direct air capture technology which has been developed in recent years makes it possible to extract carbon dioxide from the ambient air in order to use it as a raw material or alternatively to remove it from the cycle by storing it

CO_2 is already used on an industrial scale as a solvent and cleaning agent, welding gas, fertilizer and as a raw material in the chemical industry. Large amounts of CO_2 will be required for future production of synthetic fuels (eFuels) and synthetic raw materials for the chemical industry such as methanol.

International interest in DAC has increased significantly the last couple of years. The advancing climate change necessitates the prevention of new CO_2 emissions and, consequently, the transformation of an economy based on fossil fuels towards one reliant on renewable energy sources. The carbon previously provided by fossil sources must, therefore, be extracted either from biogenic or unavoidable point sources or from the atmosphere as part of a CO_2 circular economy, known as Carbon Capture and Utilization (CCU). Since not all CO_2 emissions can be avoided in sectors such as agriculture, achieving a Net Zero balance requires the permanent removal of CO_2 through Carbon Capture and Storage (CCS). Various methods, including biochar, accelerated rock weathering, afforestation, and the restoration of peatlands, are employed in the field of CCS. In contrast, CCU—the extraction of CO_2 as a raw material—necessitates technologies like Direct Air Capture (DAC) or Bioenergy with Carbon Capture and Storage (BECCS).

Currently, DAC facilities with a capture capacity of around 0.1 Megatons per year (Mt/a) are operational. The required capture rates within the framework of the CCS approach to achieve climate protection goals and the Net Zero approach by 2050 vary

A. Heintzel (Hrsg.): Int.Engine.Congr. 2024, Proceedings, S. 139–149, 2025.
https://doi.org/10.1007/978-3-658-47966-4_11

according to different literature sources, ranging from approximately 980 to 3400 Mt/a (Smith et al. 2023, p. 10; IEA 2022, p. 8). The estimated CO_2 raw material needs for CCU by 2050 amount to about 6000 Mt (Galimova et al. 2022, p. 8). The proportion of DAC technology to this demand could range up to 63%, equivalent to 3830 Mt/a (Galimova et al. 2022, p. 10). Figure 1 compares the CO_2 demand and sources predicted by Galimova et al. (2022) for the year 2050.

An example of one of the currently largest DAC facilities is "Mammoth," currently under construction by Climeworks in Iceland with a capture capacity of 36,000 t_{CO2}/a. Given the aforementioned figures, approximately 167,000 facilities with Mammoth's capacity would need to be built by 2050 to meet the estimated demand of 3830 Mt_{CO2}/a for CCU and an average of 2200 Mt_{CO2}/a for CCS.

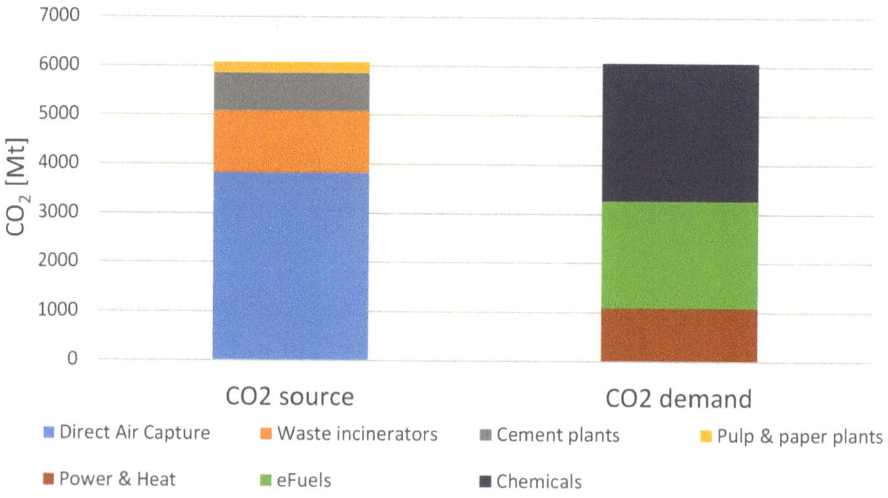

Fig. 1. Comparison of CO_2 demands in 2050 in the sectors power & heat, eFuels and chemical raw materials and expected CO_2 sources (Galimova et al. 2022)

The major challenge in extracting CO_2 from the air lies in the low concentration of the carbon dioxide of around 400 ppm, leading to high airflow rates in the absorption process and, by association, high levels of energy input. It is therefore essential when selecting possible methods of CO_2 capture to avoid changes in the airflow state as far as possible, such as changes in temperature or pressure. The low CO_2 concentration also means that there is a need for a sorbent with the very high selectivity in relation to CO_2 and one which is not carried out from the DAC system into the environment despite the high airflow rates.

As a general principle, most DAC methods consist of an absorption stage and a desorption stage. During the absorption step, CO_2 is stored in the sorbent until it is saturated. In the downstream desorption process, the CO_2 is usually separated from the sorbent again through changes in pressure and temperature whereupon absorption can be resumed. For the global use of the production of eFuels, especially at locations with good photovoltaic or wind conditions, the system-side (thermal technology) integration of DAC into the eFuel synthesis processes is of crucial importance. The aim is to use waste heat from electrolysis and fuel synthesis as efficiently as possible in order to cover most of the energy required to capture CO_2 from the air.

A distinction is currently drawn in DAC technologies, mainly differentiating between solid-bound sorbents and liquid sorbents and low-temperature and high-temperature processes (Fig. 2). High-temperature methods are mostly based on alkaline potassium hydroxide or sodium hydroxide solutions as sorbents and work with desorption temperatures of around 900 °C, whereas amine is used as the sorbent in low-temperature processes and the desorption temperature is around 100 °C. The high-temperature method can take the form of a continuous process due to the aqueous sorbent. In contrast, low-temperature processes predominantly feature the use of solid-bound amines and there is a general prevalence of batch operation because of their immobile nature. In general, when producing CO_2 from air, it is important to keep the specific consumption values for electricity and heat as low as possible or to cover the heat requirement from waste heat from associated processes to meet the economic targets

In almost all DAC absorption/adsorption processes, a large air volume flow (approx. 3000–5000 $m^3{}_{air}/m^3{}_{CO2}$) must be conveyed through the absorber (scrubber, fixed bed, filter) using a fan causing an electricity consumption of around 0.5–1 kWh/kg_{CO2}.

When CO_2 is absorbed from the air, washing processes with aqueous solutions result in a loss of water due to saturation of the air. In temperate latitudes the air has an average relative humidity of around 40–60% (except when it rains). After passing through the scrubber, the air is largely saturated. This water loss usually has to be replaced by deionized water and essentially depends on the relative humidity, the air and scrubber temperature.

In contrast, the ad- or absorption of CO_2 (deposition on the surface or transition through phase interface) on solid-bound amines is always accompanied by the co-adsorption of H_2O from the air. This fact can be used for the co-generation of the elementary raw materials CO_2 and H_2O from the air.

Various methods are used for desorption, such as a combination of increasing the temperature with a CO_2 partial pressure reduction (negative pressure) on the gas side or using stripping gas. The use of negative pressure requires gas-tight housings, which limits the module size (volume)[3]. If stripping gas is used (e.g. steam), it must then be separated from the produced CO_2. If water vapour is used as stripping gas, it can also serve as a heat transfer medium for heating the sorbent and can be relatively easily separated from the CO_2 by condensation. Using other stripping gases such as nitrogen or hydrogen makes the process more complex.

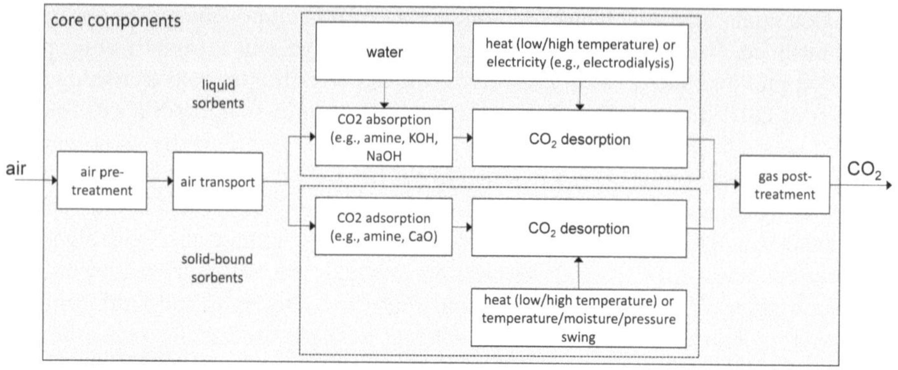

Figure 2. Overview of DAC process routes

Table 1 summarizes the characteristic process parameters of the two main DAC process routes today.

Table 1. Characteristic key values for solid (S-DAC) and liquid (L-DAC) direct air capture technologies [IEA, p. 28]

	Unit	S-DAC	L-DAC
Sorbents type	[-]	Solid	Liquid
Specific energy consumption	[kWh/tCO_2]	2000–2640	1530–2440
Share as heat consumption	[%]	75–80	80–100
Share as electricity consumption	[%]	20–25	0–20
Desorption temperature	[°C]	80–100	ca. 900
Net water requirement	[tH_2O/tCO_2]	-2 to none	0–50
Specific land requirement	[km^2/MtCO_2]	1.2–1.7	0.4
Levelised cost of CO_2	[\$/tCO_2]	Up to 540	Up to 340

With the companies Climeworks (Switzerland), Carbon Engineering (Canada) and Global Thermostat (USA), there are today 3 (pre)commercial competitors worldwide with existing plant references with production capacities >1000 t_{CO2}/a. Additional DAC processes are in development at R&D institutes and several dozen startups worldwide.

DAC systems in operation today have a total production capacity of <50,000 t_{CO2}/a.

The first plants with production capacities of several 10,000 t_{CO2}/a are currently under construction. For example, eFuels production plants require CO_2 quantities in the megaton range every year.

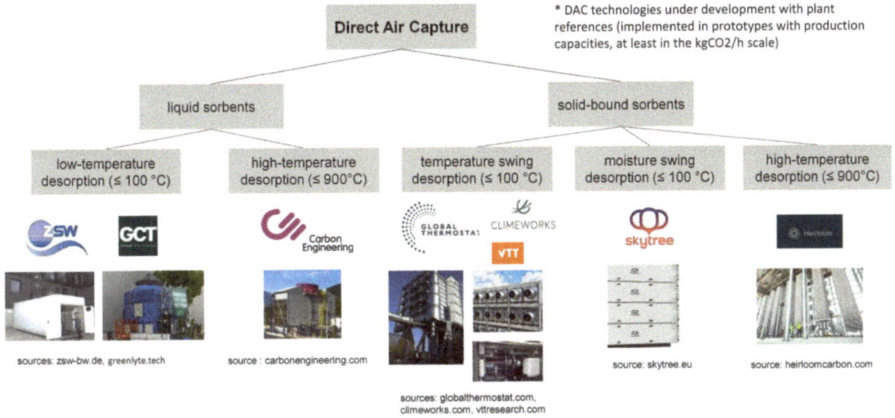

Fig. 3. DAC technologies under development with plant references (implementation in prototypes with production capacities at least on a kg_{CO2}/h scale)

2 Challenges in Technology Industrialization and Scaling

The main challenge for the coming years is the further industrialization of DAC technologies towards processes that can be operated robustly worldwide with regard to robustness/lifespan, scalability, stability and availability of the sorbents, automated process control, series suitability of the manufacturing processes. This particularly includes the following aspects:

- Cost efficiency must be improved including capital expenditures (CAPEX) and operational expenditures (OPEX), influenced by electrical and thermal efficiency and degradation behavior of the sorbent material.
- Processes must be improved or adjusted to be able to deal with different environmental influences (e.g. climatic conditions, salty air, water management issues) and to guarantee a stable operation at different locations worldwide.
- Possible environmental impacts of DAC plants (e.g. discharge of sorbents, climate impact) must be explored and be avoided as far as possible.
- Process integration e.g. use of waste heat or wastewater from adjacent process steps (electrolysis, synthesis) has to be optimized.
- The scaling of DAC technologies to plant sizes of up to >1 Mt_{CO2}/a will be crucial for the success of the technology to meet the cost targets < 100 €/t_{CO2}. This particularly includes the process scaling (scaling in module size vs. scaling in quantity), ensuring material availability, scaling production processes and dealing with resource-critical materials, water management as well as the implementation of recycling concepts.

3 ZSW DAC technology

The direct air capture technology developed by the ZSW the last couple of years (so called CORAL process) is a hybrid approach working with an aqueous solution of polyethyleneimine (PEI) as sorbent in a scrubber, thereby combining the advantages of a continuous processing and a low desorption temperature of around 100 °C

Using high molecular weight, branched polyethyleneimine (PEI) in an aqueous solution respectively CO_2 absorption in a scrubber is associated with the following advantages:

- For gases with a higher CO_2 concentration, the use of aqueous amine solutions for CO_2 separation is a well-known and proven technology. The CO_2 desorption takes place thermally (120–170°C). Compared to other known processes for CO_2 separation from air, the described washing process with aqueous PEI solution also enables thermal CO_2 desorption (100–120°C). This makes it possible, for example in combination with exothermic hydrocarbon synthesis, to use the waste heat for CO_2 desorption. This allows the power consumption of CO_2 desorption to be significantly reduced (<1 kWh/kg_{CO2}).
- High molecular weight branched PEI has a negligible vapor pressure of its own in aqueous solution and therefore cannot evaporate from the solution. That's why PEI is also suitable for CO_2 absorption from air.
- In aqueous solution the viscosity is sufficient for scrubber operation. In addition, the aqueous solution ensures good accessibility to the NH groups required for CO_2 bonds. What is important here is a large phase interface between gas and washing liquid, which can be achieved, for example, by fittings with a high specific surface area.
- High molecular weight branched PEI contains a variety of primary, secondary and tertiary amine groups, all of which are suitable for CO_2 binding and therefore has a very high CO_2 absorption capacity, which increases in the presence of H_2O.

The CORAL process is illustrated in Fig. 4. The liquid CO_2 sorbent, a PEI-containing washing solution, is continuously circulated in an absorber above a packing structure, in which a large phase boundary area results due to wetting. A fan ensures that air flows through the packing structure in the opposite direction, enabling CO_2 to be continuously absorbed into the scrubbing solution by the maximized phase interface. A pollen filter installed in front of the blower prevents impurities from entering the PEI solution. At the same time, scrubbing solution is drawn off and fed to the desorption process, where the CO_2 is separated from the scrubbing solution at an elevated temperature. The desorbed product gas is concentrated by condensation to CO_2 with a purity of approx. 99 vol.% and is directly available for subsequent synthesis processes. The CO_2-depleted PEI solution is pumped back out of the desorption unit into the absorber where the process starts again. The absorption process takes place close to ambient temperatures, while the temperature level in the desorption unit is at 100 °C. A heat exchanger is used between the absorber and desorber to preheat the fluid flowing from the absorber and to cool the fluid leaving the desorber. This reduces the amount of additional heat to be supplied to the desorption unit and, at the same time, lowers the amount of heat to be dissipated in the absorber. In contrast to PEI, water has a significant vapor pressure, leading to saturation of the air volume flow and withdrawing water out of the system. To prevent the process from running dry, water has to be added continuously.

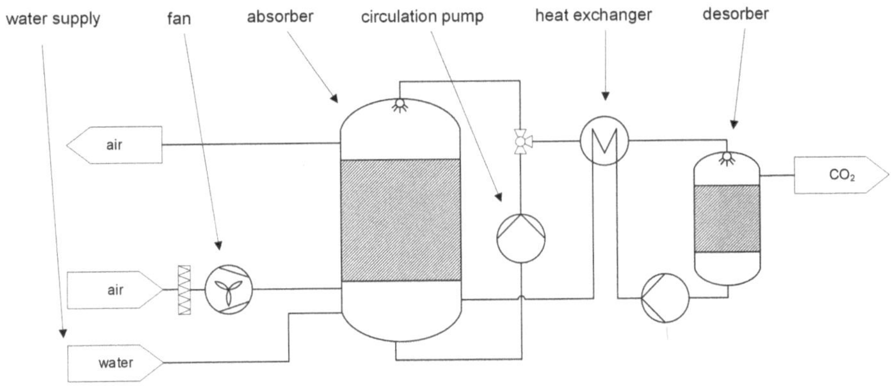

Fig. 4. Simplified flow diagram of the CORAL process at ZSW

A demonstrator of the previously described process with an absorption capacity of around 1 kg_{CO2}/h (DAC1) was commissioned in 2019. Since then, more than 13,000 research operating hours have been completed, of which over 6500 h were completed as part of an endurance test to demonstrate long-term stability. In total, around 5 tons of CO_2 were captured from the air during this period. The continuous process is beneficial in the avoidance of downtime, high material stress levels and fluctuating CO_2 qualities, and it is comparatively straightforward in terms of control engineering. The robust technology is another significant advantage, keeping maintenance costs low and scalability high—two fundamental requirements when it comes to achieving net zero emissions in 2050, especially in view of forecast separation rates of up to 10 gigatons per year. The high degree of CO_2 purity also enables not only the storage of CO_2 but also its further use as a raw material for such purposes as the production of eFuels

Fig. 5. left: CORAL demonstrator system DAC1 integrated in a standard 20 feet container with the desorber unit (grey pressure vessel in the open container door) and the absorber unit (black tower in the background) right: Setting the desorption temperature next to the desorber

The diagram in Fig. 6 shows the absorber-side CO_2 depletion of the DAC1 demonstration plant. At this operating day the air enters the absorber with a CO_2-concentration of approx. 440 ppm and leaves CO_2 reduced with a concentration of approx. 240 ppm. Even if the demonstration system only approximates a continuous CO_2 yield of 1 kg_{CO2}/h in the current operating mode, the robustness of the technology can be demonstrated by using commercially available systems with simple control technology and a consistently high CO_2 quality. The electrical energy requirement of the demonstration plant is around 1 kWh/kg_{CO2}, the thermal consumption for desorption averages 6 kWh/kg_{CO2}.

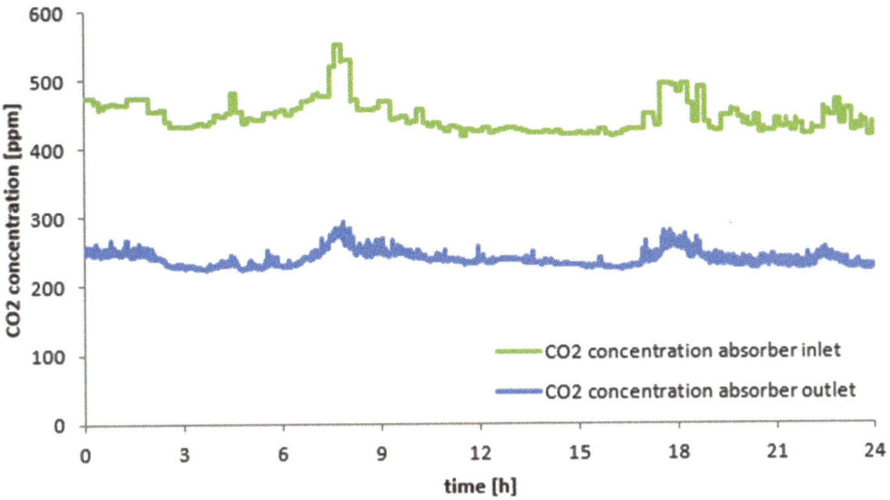

Fig. 6. Exemplary 24 h measurement data set of the DAC1 demonstrator with a constant CO_2 spread of approx. 200 ppm

The main challenges of the CORAL technology are water losses due to saturation of the air volume flow and a comparatively high thermal energy consumption during CO_2 desorption, which is subject of current research activities. While water consumption can be countered for example through optimized operational management, the specific thermal energy requirement can be reduced by optimizing the desorber unit and adjusting the used washing solution

As part of the "Direct Air Capture made in Baden-Württemberg" project funded by the Baden-Württemberg Ministries of Transport and Economic Affairs, the existing DAC system was scaled to a capacity of approximately 10 $kg_{CO2/}h$ (DAC10). The system is shown in Fig. 7 and is currently put into operation. The DAC10 system represents not only an initial advancement toward successful scaling but also a significant step towards achieving the aimed future target values of electrical energy consumption below 0.5 kWh/kg_{CO2} and thermal energy consumption below 2 kWh/kg_{CO2}.

Fig. 7. Front (left) and rear (right) of the DAC 10 system with absorption container and desorption container

4 Summary and outlook

DAC is a key technology for achieving the climate protection goals (CCS approach) and producing climate-neutral carbon-based fuels and basic chemicals (CCU approach). Currently, DAC facilities with in total a CO_2 capture capacity of around 0.1 Megatons per year (Mt/a) are operational with only a few (pre)commercial competitors worldwide with existing plant references with production capacities of some 1000 t_{CO2}/a. In contrast, an eFuels production plant will require CO_2 quantities in the range of approx. 1–2 Mt/a. The required capture rates by CCS to achieve the climate protection goals and the Net Zero approach by 2050 vary according to different literature sources, ranging from approximately 980 to 3400 Mt/a.

This results in the need to quickly industrialize and further scale the DAC technologies and to prepare them for operation at different locations with various environmental conditions (e.g. climatic conditions, salty air, water management issues) around the world. The scaling of DAC technologies to plant sizes of up to >1 Mt_{CO2}/a will be crucial for the success of the DAC technology to meet the long-term cost targets <100 €/t_{CO2}.

ZSW has successfully developed a DAC process using an easy scalable scrubber technology based on polyethyleneimine (PEI) in an aqueous solution and has implemented the technology in two demonstrators on the order of 1 kg_{CO2}/h or 10 t_{CO2}/a (DAC1) and 10 kg_{CO2}/h or 100 t_{CO2}/a (DAC10). The DAC1 demonstrator was successfully operated for a total of 13,000 operating hours. The DAC10 system is currently put into operation aiming to further improve the electrical and thermal energy consumption.

Economic considerations show that with the CORAL technology on a scaling level of 10,000 kg_{CO2}/h or 100,000 t_{CO2}/a and with electricity procurement costs of below 15 €/MWh and high system utilization (>6000 full operation hours per year), CO_2 supply costs in the range of 50–100 €/t_{CO2} can be achieved.

References

Galimova, Tansu; Ram, Manish; Bogdanov, Dmitrii; Fasihi, Mahdi; Khalili, Siavash; Gulagi, Ashish et al. (2022): Global demand analysis for carbon dioxide as raw material from key industrial sources and direct air capture to produce renewable electricity-based fuels and chemicals. In *Journal of Cleaner Production* 373. https://doi.org/10.1016/j.jclepro.2022.133920.

IEA (2022): Direct Air Capture. A key technology for net zero. Available online at www.iea.org/reports/direct-air-capture-2022.

Smith, S. M.; Geden, O.; Nemet, G.; Gidden, M.; Lamb, W. F.; Powis, C. et al. (2023): The State of Carbon Dioxide Removal. 1st Edition. Available online at https://www.stateofcdr.org.

Direct air capture: core technology for carbon negativity

Karl Dums[1(✉)], Arne Philipp Siemens[1], Fabian Ehrat[1], Klaus Behnke[2], Stephan Weber[2], and Marc Rueggeberg[3]

[1] Dr. Ing. h.c. F. Porsche AG, Stuttgart, Germany
{karl.dums,arne_philipp.siemens,fabian.ehrat}@porsche.de
[2] MAN Energy Solutions SE, Augsburg, Germany
{klaus.behnke,stephan.weber}@man-es.com
[3] Volkswagen AG, Wolfsburg, Germany
marc.rueggeberg@volkswagen.de

1 Introduction

Carbon dioxide is an essential molecule of our entire global atmosphere and ecosystem. As all life is mainly based among others on carbon, hydrogen, and oxygen the density of the global carbon dioxide layer is decisive for the actual global environmental conditions. As of now the average CO_2 concentration is at over 400 ppm and thereby more than 200 ppm above the preindustrial level causing global warming due to the increasing reflection of long wave sun radiation, that cannot leave the atmosphere back into space. Even after a successful climate change mitigation after mid of the century, there will still be several hundred ppm CO_2 left, because its degradation in the spheres takes up to 700 years. Thus, IPCC (Intergovernmental Panel on Climate Change) Studies [3] have shown already in 2018, that after reaching a carbon neutral economy worldwide mankind still has a huge second task to tackle: and that is the phase of carbon negativity that can last up to a century. [3] From 2050 on at the latest we must begin to extract CO_2 from the air by all available solutions at a scale of about 10–20+ megatons annually. That is ¼ or more of the total annual CO_2 emissions today.

This is necessary to neutralize the unavoidable overshoot of CO_2 Emissions in the 1,5 °C scenarios and even more in the 2 °C and above scenarios. This CO_2 extraction process then could literally lead to a cool-down-effect of ambient global temperature or a least to a no further increase of global warming.

A. Heintzel (Hrsg.): Int.Engine.Congr. 2024, Proceedings, S. 150–162, 2025.
https://doi.org/10.1007/978-3-658-47966-4_12

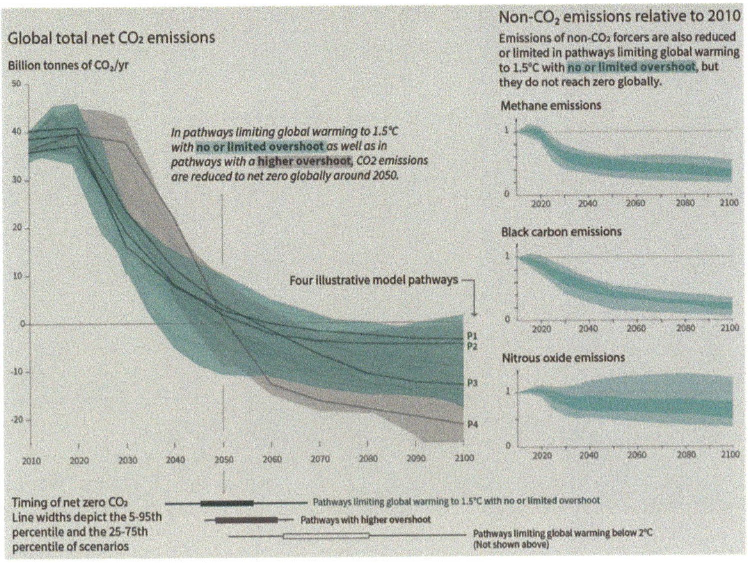

Fig. 12.1 [1] Source: IPCC-Report 1,5 °C, 2018

Hereby the simple relation accounts: the higher the overshoot the higher the in consequence required extraction rates in the decades to come.

That is why it´s vital to preserve the principal of technology neutrality and force all viable technologies into action. One of these technologies is the so-called **DAC or Direct Air Capture**. Based on a variety of different solutions in development all currently known processes have in common the active extraction of CO_2 from the ambient air based mainly on renewable electricity and heat.

2 Physical Process [4] in the Atmosphere

As solar radiation passes through the clear atmosphere some solar radiation is reflected. Most of the net incoming radiation is absorbed by the earth surface. It is then converted into heat causing the emission of longwave (infrared) radiation back to the atmosphere.

The greenhouse effect: Some of the infrared radiation is **absorbed and re-emitted** by the greenhouse gas molecules, chiefly CO_2. The direct effect is the additional warming of the earth´s surface and the Atmosphere. The surface gains more heat and infrared radiation is emitted again.

Some of the infrared radiation passes through the atmosphere and is lost in space.

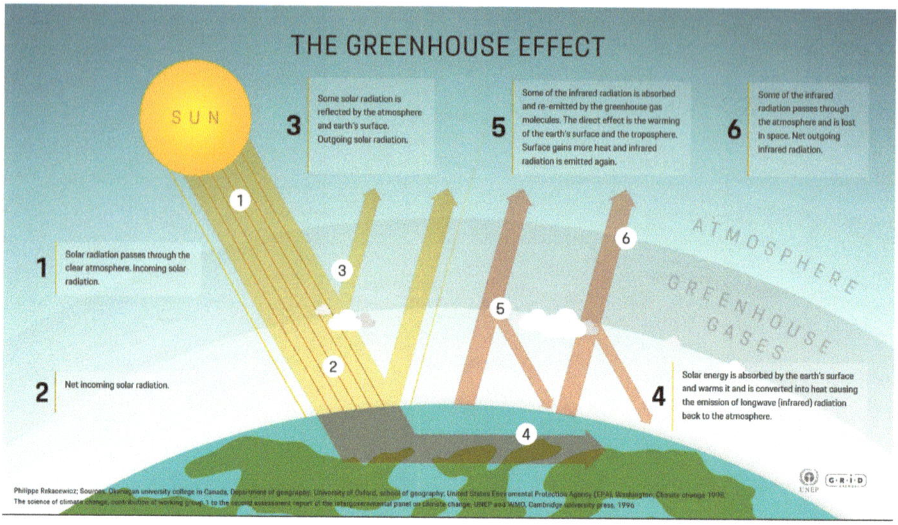

Fig. 12.2 [4] Source: 2. assessment report IPCC 1995, graphics adapted by STORMING AG

If we investigate a detailed analysis of the total amounts of incoming and outcoming solar radiation, we can deduct an additional small constant net energy absorption of the earth and stratosphere. In a non-anthropogenic global ecosystem with no or very slow global warming this absorption rate is at approximately 1 W/m^2.

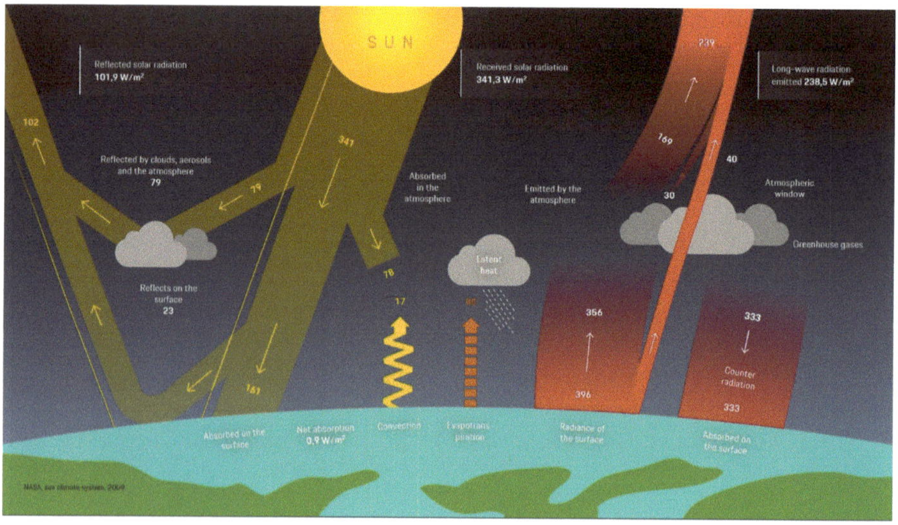

Fig. 12.3 [5] Source: NASA, sun climate system, alternative German, 2009, graphics adapted by STORMING AG

But through anthropogenic greenhouse gas emissions the "thickness" of the greenhouse gas layer increases when respectively the CO_2 concentration in the troposphere does. Then in consequence the absorption rate (radiative forcing label) increases too. The IPCC works with so called "representative concentration pathways" ("RCPs) to estimate future global warming scenarios. "The RCPs cover a wider range than the scenarios from the Special Report on Emissions Scenarios (SRES) used in previous assessments, as they also represent scenarios with climate policy." [6] This shows that former scenarios from 2014 correlate with newer simulations from 2021. The RCP value is the core physical link between CO_2 concentration [ppm] and global warming [°C].

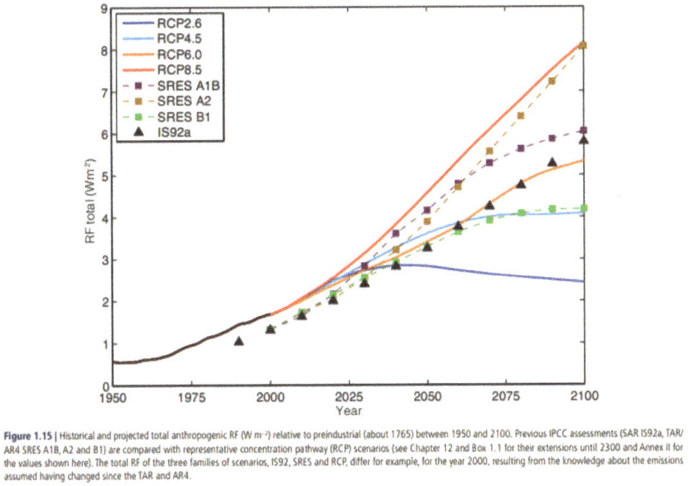

Fig. 12.4 [6] Source: Climate Change, The Physical Science Basis, IPCC 2021

The IPCC Report of 2021 correlates the global warming with the so-called representative concentration pathways measuring the different scenarios in the amount of the absorption rate [watt per square meter] from 2 W/m² up to 8,5 W/m² also defined as the radiative forcing level.

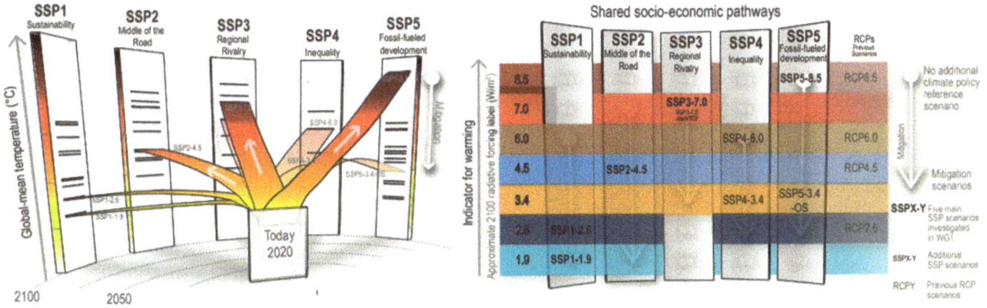

Fig. 12.5 [7] Source: Climate Change, The Physical Science Basis, IPCC 2021

This is the reason why the global CO_2 concentration in the atmosphere is decisive for climate change mitigation. Because carbon removal and carbon negativity bring down the CO_2 concentration in the troposphere to a stable ambient temperature level. In consequence carbon capture and storage technologies like Direct Air Capture are crucial and should be developed and deployed as soon as possible.

3 The Global Carbon Cycle and the DAC Functional Technology Scheme

CO_2 Can Be Obtained From Different Sources Some technologies count as removal and others simply use existing CO_2 coming out of fossil processes, many of them which are on the long term avoidable, like coal power plants. It is important to also point out a special type of sources: unavoidable point sources like cement production. Here we need technological long-term solutions.

Carbon dioxide point sources from fossil fuels and industrial processes are today all around the world broadly available but will be phased out step by step obviously by the mid of the century. Still, they remain the most relevant carbon capture process for the next 2 decades. When it comes to policy: On the US Side the IRA (Inflation Reduction Act), on the EU side the Green Deal and in China the 5 Year plans make the use of point sources in example to produce renewable energy carriers possible in this and the next decade, at least. But the capacity will decrease due to the availability and upcoming regulatory restrictions.

With **biological plant-based processes** there is a second most relevant path for CO_2 removal. On a global basis it is by far the biggest turnover in the atmospheric and ecological system. As example [7]Science Direct states for the global carbon cycle that it is made up of carbon reservoirs (stocks) and the dynamic transfer of carbon between them (fluxes): on basis of pure carbon only by photosynthesis the turnover is at 120,4 gigatons carbon trough terrestrial vegetation and 92 gigatons carbon through ocean absorption and photosynthesis per year (not measured in CO_2 but in carbon atoms per year from the year 1990). For comparison the carbon emission through fossil fuel was in the year 1990 6.3 gigatons and is up to today at approximately 13 to 15 gigatons. This is a factor of ~15.

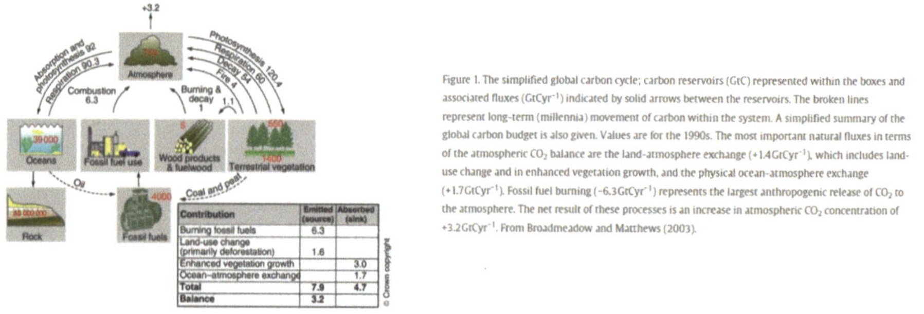

Fig. 12.6 [8] Source: Biomass: Impact on Carbon Cycle and Greenhouse Gas Emissions, Carly Green, Kenneth A. Byrne, in Encyclopedia of Energy, 2004; Global Carbon Cycle—an overview | ScienceDirect Topics

Meaning that the biological turnover in the balanced ecosystem is 10–15× bigger, then the anthropogenic fossil fuel emissions. It becomes clear, that the global potential for biobased solutions is necessary for long term carbon capture. Reforestation, regreening and many others are thereby key to a sustainable future. But on the long-term absolute capacities will be somewhat limited due to land scarcity and competition with other use cases like agriculture. In addition, the global community of states can only unlock a certain potential of bio-CCS, because we are still witnessing a strong deforestation and land use change in many countries.

Nevertheless, the third major pillar is **technological Direct (Carbon) Air Capture with renewable energies**. Compared to the other two, DAC poses the only technological approach, which is as such scalable, independent from the availability of industrial CO_2 sources or from the usage of biomass. To achieve a stabilized ecosystem and limit global warming to 1,5 °C DAC is a key technology. DAC will increase as technology; it is potentially available at large-scale and offers both carbon as resource and the possibility for negative emissions. The process itself can easily be described by several phases: The Environment air is taken in by ventilation and then filtered. By a wide range of available processes, a sorbent material absorbs the CO_2 from the air and desorbs it into a separate storage.

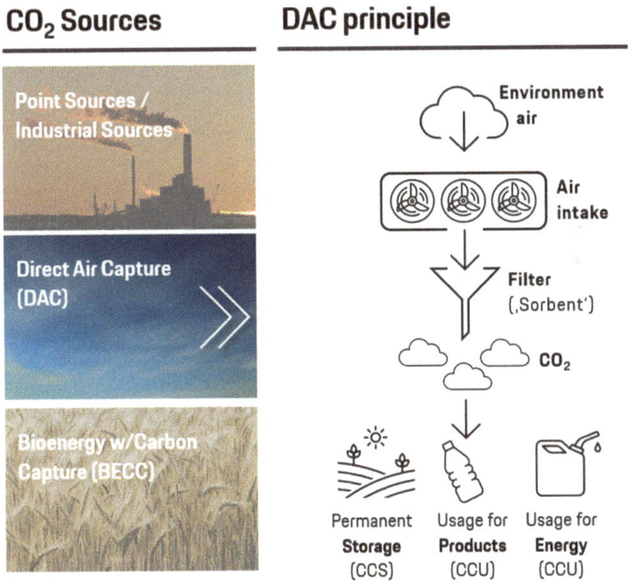

Fig. 12.7 [9] Source: DAC Project Team, 2022, climate.ec.europa.eu

[11] One technological approach is the adsorption principle. The ambient air is ventilated through a housing in which the CO_2 adsorbent material is located. Once the adsorbent material has reached its adsorption capacity, the CO_2 is discharged and extracted in a defined manner. The concentrated CO_2 extracted from the air can then

be stored or reused. Through temperature- and pressure-swing adsorption can be realized with certain materials. By changing from temperature T_{ads} to T_{des} and from pressure P_{ads} to P_{des} the process unlocks a working capacity Q_{ads} that can be measured in [mg/g].

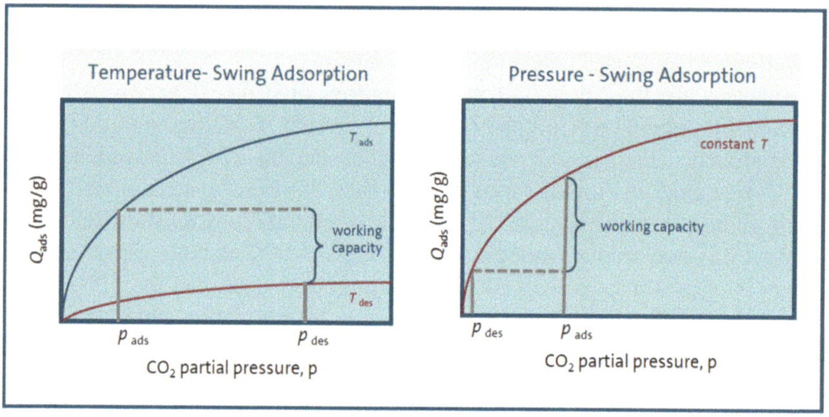

Fig. 12.8 [10] Source: Volkswagen AG, 2023

[10] The CO_2 extraction in a 2-step process (adsorption/desorption) from the ambient air can be summarized as follows in several steps

Air intake  Air filtering (particles, sand etc.) Lowering temperature (better adsorption)  Air drying (better adsorption)  CO_2 adsorption (deposition on sorbent material)  Air outlet (low CO_2 exhaust air)

[10] The CO_2 adsorption/desorption as core process is again detailed in the following steps

Closure adsorption chamber  Vacuum formation (CO_2 still bound in the sorbent)  Increase in temperature (among other things, with waste heat from electrolysis etc.)  CO_2 desorption (CO_2 dissolves and is in gas form)  Suction of the CO_2 and, if necessary, temporary storage  CO_2 use, e.g. for eFuels, via methanol synthesis

4 Plant Setup

There are several plant setup solutions for DAC to be considered. As part of P-t-X ("Power-to-X") technologies the DAC can be either realized as a "stand alone" or "integrated" solution. While considering "on- and off-grid" for the supply of the renewable electricity and heat it makes much sense to integrate DAC plants into other eFuels plants because of the wide availability of excessive heat, electricity and many other components like operating personnel. This enhances the efficiency of all processes: combining chemical and process engineering solutions is state of the art. On the other side eFuel plants need CO_2 and the ideal solution is to use it from DAC or biogenic sources. This is where the Haru Oni Pilot Plant in Punta Arenas, Chile, comes in:

The eFuels pilot plant is the starting point for a huge variety of global opportunities.

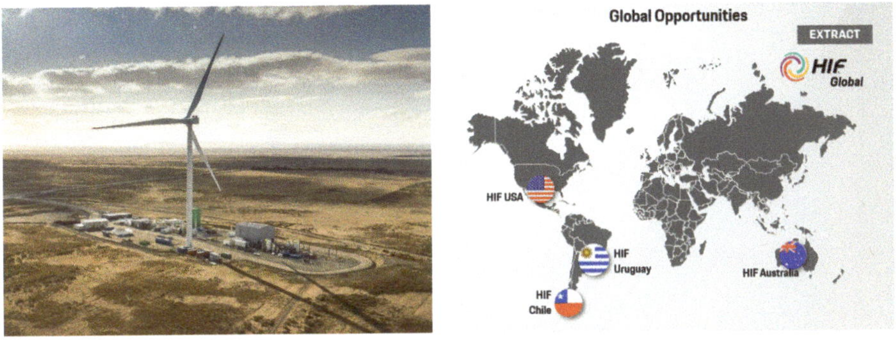

Fig. 12.9 [11] Source: HIF Global, Porsche AG, 2023

The DAC plant can achieve the extraction of CO_2 directly from the air—greenhouse gas can become raw material (e.g. for eFuel production), or can be stored safely in various underground storage reservoirs for negative CO_2 emissions.

With the existing wind in Haru Oni the DAC plant can be operated completely on renewable energy.

The DAC plant can possibly filter/purify the air (~950.000.000 m^3/a) and thus can remove CO_2. The output quantity is aimed for ~ 600 t CO_2 per year. Hence the average amount of air filtered for one ton of CO_2 is ~ 1.600.000 m^3.

In comparison the empire state building has volume of 1.000.000 m^3 and the Olympic stadium in Munich has a volume of approximately 2.500.000 m^3. So, for 630 kg of CO_2 the air volume of the empire state building and for 1.6 tons of CO_2 the air volume of the Olympic stadium in Munich must be filtered.

Proof -of-concept Plant Project Partners:

PORSCHE | VOLKSWAGEN AKTIENGESELLSCHAFT | MAN Energy Solutions | HIF

Fig. 12.10 [11] Source: MAN Energy Solutions SE, DAC project team, 2023

“Front” view
“Rear” view

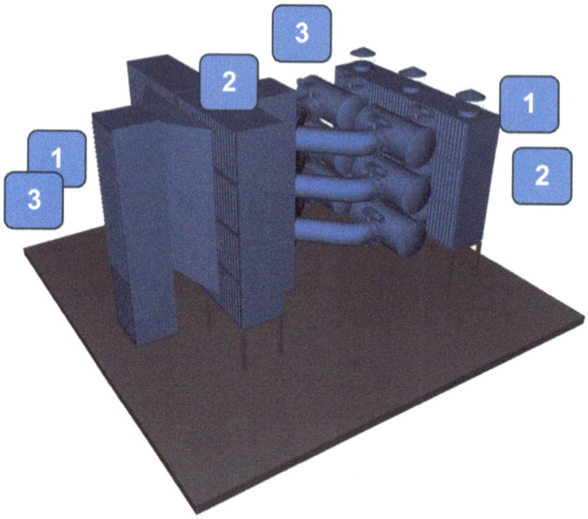

Fig. 12.11 [11] Source: MAN Energy Solutions SE, DAC project team, 2023

1. Air intake and prefilter
2. CO_2 filter and humidity management
3. Air exhaust treatment

The following periphery components are important to mention: Air fans, vacuum pumps, heat pumps, air ventilation shafts, heat exchanger. The overall technical design is aiming at several advantages: It is very robust; the sorbent has low costs and is bulky available. It´s designed for long lifetime and the plant has as on-top output: Water. This is a possible technological breakthrough because the water could be used for eFuels production or other water use cases like hydrogen or irrigation. The sorbent is of non-toxic composure and enables an easy disposal. Anyhow there are also technological challenges to be tackled:

This design in the **first of a kind (FOAK)** plant. Space requirements and energy demand will decrease. The current setup intends to enable technological experiments and variations. The next step in industrialization is refinement of surfaces, materials, and costs. But as shown by Boston Consulting [12] from a FOAK to best case the costs can be cut down by approximately—90% in average.

The DAC POC Project Is Combining Strengths of the Volkswagen AG, MAN Energy Solutions SE, Porsche AG and Partners

Porsche AG: Project Leader and networking, “Lighthouse” customer
Volkswagen AG: Expertise in DAC, CO_2, sorbent materials, expertise in industrial mass production

MAN Energy Solutions SE: Expertise in plant projects, producer of large decarbonization components, like H_2-electrolyser, industrial heat pumps and CO_2 compressors

HIF Global LLC., Highly Innovative Fuels: Operator, integration into eFuels plant (CO_2 utilization)

5 Market Perspectives

If we compare different carbon capture methods, we have to acknowledge that there is a huge variety of more cost efficient solutions starting with forestry covering a range from ~20 to ~80 $/ton CO_2. Coal point sources reach up to ~100 $/ton, cement CCS up to 150 and bio energy CCS up to 200 $/ton CO_2. The current range for DAC target costs is between 200 and 100 $/ton, with a long-term technology goal to go below 100$/ton.

As DAC is not depending on point sources it might be the technology of choice in the time when they are depleted.

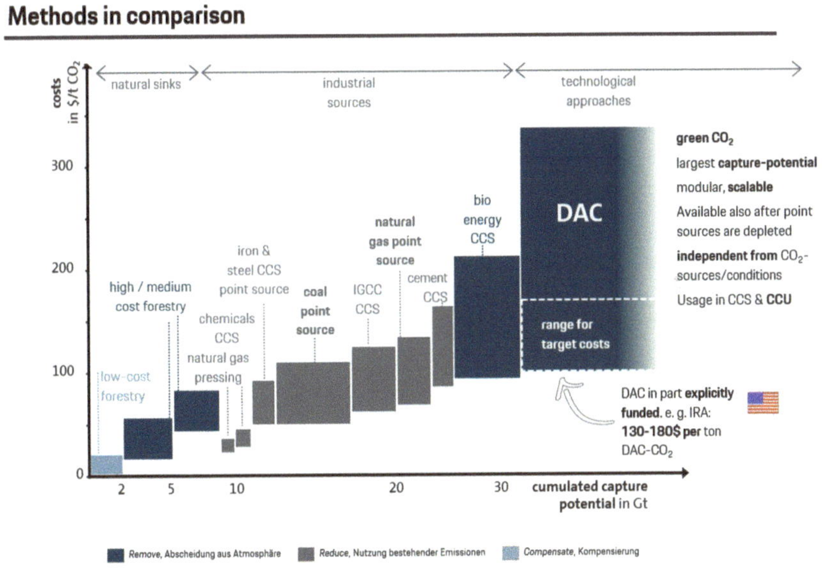

Fig. 12.12 [14] Source: Volkswagen AG, 2022, DAC project team

DAC is intended for variety of solutions:

Direct Utilization
CO_2 used without further processing for e.g., food preservation and the carbonation of drinks. This market might increase due to population growth.

Energy Carriers
Through energy input and methanol synthesis CO_2 can serve as a raw material for energy carriers e.g., hydro carbonates. This market might increase due to the decarbonization of the transport sectors.

Material Conversion
CO_2 used for urea production (fertilizer, skin care) and building materials This market might increase due to the decarbonization of building materials and consumer sectors.

Storage
Long-term and safe CO_2 storage proven with certificates that can be traded in CO_2 certificate markets: increase due to climate change mitigation commitments.

Based on required demand for eFuels in the transportation sector, reduction targets in construction and necessary abatements for high emission industries experts predict that already by 2030 the global CO_2 market might reach a respectable size in 2030 by possibly 1500 megatons from which a DAC share of ~6% at 85 megatons might be possible.

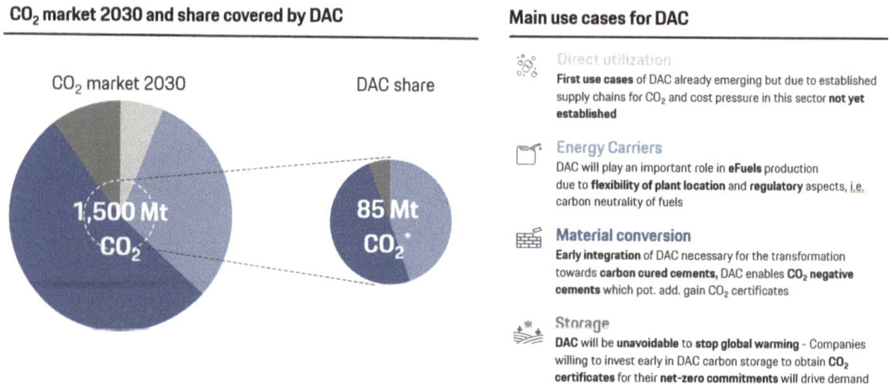

Fig. 12.13 [15] Source: DENA (2020), IEA (2021), IEA. (2020), IEA (2019), ETC (2022)

[14] Source: DENA (2020), IEA (2021), IEA. (2020), IEA (2019), ETC (2022)

Studies [15] show a 27 bn. USD* market that needs to be established to reach the climate goals in 2030. Assumption: 85 Mt CO_2 capturing by DAC at a price of approx. USD 300 in 2030

Analysis led to a possible huge potential for carbon capture and DAC until 2040 by 4000 megatons and until 2050 by potentially 7000 megatons.

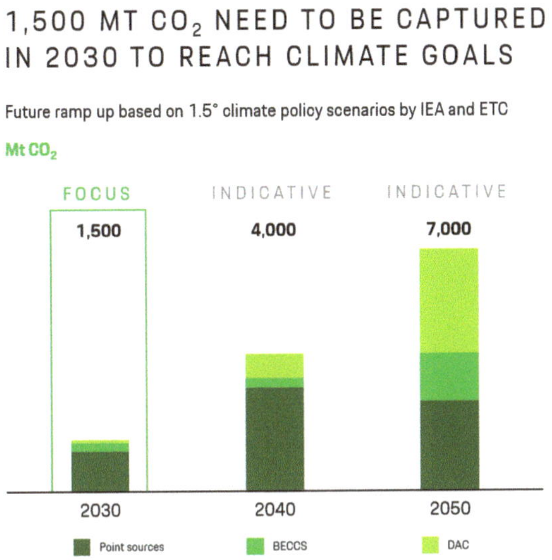

Fig. 12.14 [17] Source: According to IPCC (2022), 2 average values of IEA and ETC scenario, Sources: IEA (2021); IEA. (2020); ETC (2022), DAC project team

6 Conclusion

Overall, we can deduct that due to the physical circumstances of the greenhouse effect there is currently no alternative to the mass deployment of carbon capture across this century. Otherwise, climate change cannot be mitigated and the 1,5 or 2 °C scenarios cannot be reached. Hence next to point sources and biological solutions DAC is the technology of choice with several standalone advantages making it a mandatory third pillar in carbon capture. Porsche AG and partners are currently working on a proof-of-concept plant with the motivation to push this technology forward.

References

1. https://en.wikipedia.org/wiki/Radiative_forcing
2. https://earthobservatory.nasa.gov/features/EnergyBalance/page7.php
3. IPCC-Report 1,5°C, 2018
4. Assessment report IPCC 1995
5. NASA, sun climate system, alternative German, 2009
6. Climate Change, The Physical Science Basis, IPCC 2021
7. Biomass: Impact on Carbon Cycle and Greenhouse Gas Emissions, Carly Green, Kenneth A. Byrne, in Encyclopedia of Energy, 2004; Global Carbon Cycle - an overview | ScienceDirect Topics
8. DAC project team; climate.ec.europa.eu
9. Volkswagen AG, 2023

10. HIF Global LLC., Porsche AG, 2023
11. MAN Energy Solution SE, DAC project team, 2023
12. Shifting the Direct Air Capture Paradigm, Boston Consulting, 2023
13. Volkswagen AG, 2022, DAC project team
14. DENA (2020), IEA (2021), IEA. (2020), IEA (2019), ETC (2022)
15. IEA (2022), ETC (2022), Global CCS Institute (2019)
16. According to IPCC (2022), 2 average values of IEA and ETC scenario, Sources: IEA (2021); IEA (2020); ETC (2022), DAC project team

Liebherr Hydrogen System and Control Development

Richard Pirkl[1], François Masson[2](✉), and Guenther Neuhaus[1]

[1] Liebherr-Components Deggendorf GmbH, Deggendorf, Germany
{richard.pirkl,guenther.neuhaus}@liebherr.com
[2] Liebherr Machines Bulles SA, Bulles, Switzerland
francois.masson@liebherr.com

1 Liebherr CCR on the way to alternative Fuels

Not only since the European "Green Deal" Liebherr as engine manufacturer is forced to deal with current and future changes in legislative and market changes. Liebherr engines are used in Liebherr products, but also in several applications of external customers around the globe.

Since a few years there is a strongly growing interest in alternatives for diesel fuel. At the same time the combustion engine as power-source is well known and appreciated for its robustness in rough conditions. Furthermore, a diesel engine is—with minor modifications—a good basis also to be used with other fuels [1]. Be-neath activities regarding hydrogen Liebherr CCR currently develops a multi-fuel injection system for liquid fuels including methanol, ethanol, ammonia and others. As most of the alternative fuels show a lower calorific value, higher flow rates through the injection equipment are requested to maintain the engine power on the existing level.

As the pressure levels are way below diesel systems (injector upstream <80 bar), the basis for the injectors is a direct driven concept (compared to servo hydraulic in diesel applications). This allows a design for high flow rates in a small package.

It´s clear that there is a big potential for CO_2 reduction by using alternative fuels which can be additively manufactured under use of CO_2 out of the atmosphere.

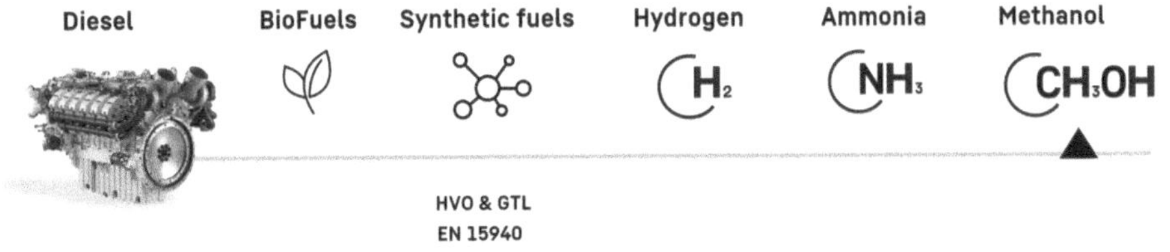

Anyway, the only available fuels to power an ICE are hydrogen and ammonia. As these fuels are the only ones to not contain carbon and therefore to not react to CO_2 among other exhaust gases.

A. Heintzel (Hrsg.): Int.Engine.Congr. 2024, Proceedings, S. 163–177, 2025.
https://doi.org/10.1007/978-3-658-47966-4_13

2 Hydrogen as Fuel, Comparison to Diesel

2.1 Hydrogen Characteristic

When used in a combustion engine hydrogen is in gaseous state with a density approximately thousand time lower than diesel (*reference at 80 degC 10 bar for H_2/1000 bar for diesel*). Despite of having a higher energy per unit mass the volumetric flow is significantly greater than diesel resulting in specific constraint in system dimensioning. Special attention is required for managing pressure drops through the system.

In addition to flow challenges, the sensitivity to pressure and temperature is much higher in comparison to diesel, creating significant obstacle in controlling injected mass quantity. This sensitivity is visible in Fig. 1. especially when the variation of density is displayed in percentage [2].

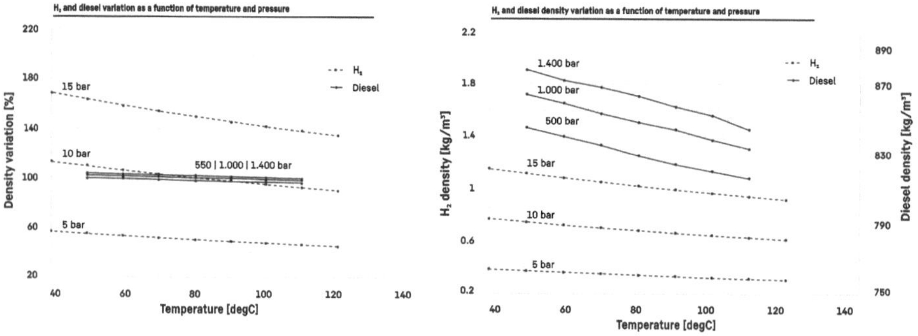

Fig. 1. H_2 and diesel density variation as a function of pressure and temperature

2.2 Hydrogen Storage

A general challenge related to the usage of H_2 concerns the gas storage. The energy per volume is more than 10 times lower between compressed H_2 and diesel leading to tank integration challenges in the vehicle. The Fig. 2. compares the characteristic of different fuels [3].

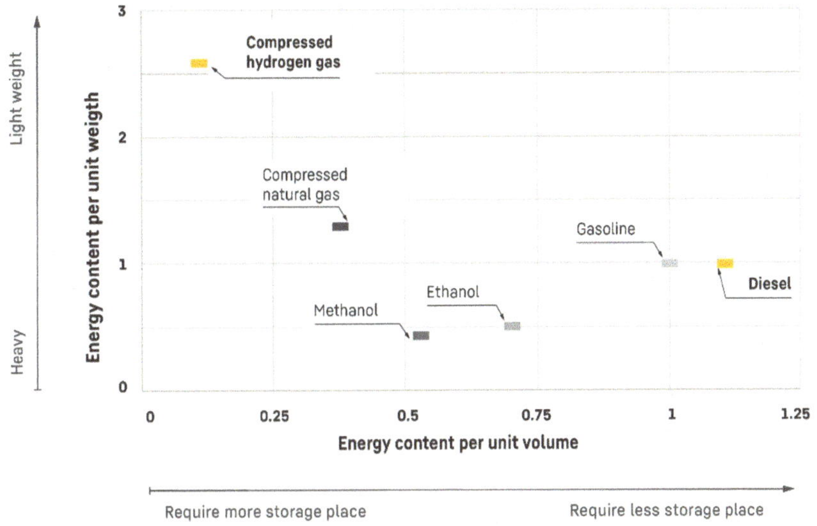

Fig. 2. Energy density comparison of several fuel (indexed to gasoline = 1) Source: U.S Energy information Administration

2.3 Effect of Hydrogen on Component Design, System Design and Control

The usage of hydrogen in comparison to diesel significantly influences the design and functionality of the components, system- and control strategies. As illustrated in Fig. 3. the actuators demand high magnetic forces due to the high cross section necessary to ensure the hydrogen flow and limit the pressure drop through the system. The **g**as **m**etering **(GMV)** valve and injector used in Liebherr H_2 system are directly actuated valves.

	Diesel System 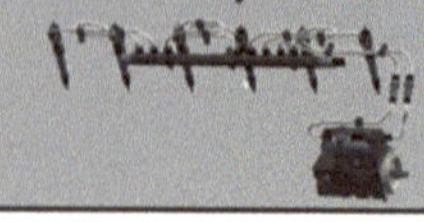	Hydrogen System 
Material	**Martensitic & Bainitic Steels** high pressure resistance, low sulfur, high purity class, high strength	**Austenitic Steels** resistance against hydrogen embrittlement weldable, low carbon content
Sealing	**Metal to Metal** + DLC coating, +cone seat	**Elastomer to Metal** + flat seat
Guidance	**Metal to Metal** + Fuel/Oil lubrication + DLC coating	**PTFE to Metal** + Dry friction + DLC coating
Fuel density & heat value	**Small Cross Section** hydraulic actuated valve (servo)	**High Cross Section → High Power Magnet** direct actuated valve
Solenoid	**Wet** operating in oil	**Dry** hermetical separated from oil

Fig. 3. Comparison between diesel and hydrogen injection system

The system tightness requires the usage of specific material such as elastomer leading to potential sensitivity in terms of control of low quantity, part to part and ageing. This requires the development of specific control structure and strategies.

3 Liebherr Hydrogen System

3.1 System Overview

The H_2 system guarantees a precise delivery of hydrogen to the engine within a specified timeframe, typically the intake valve opening, and at a given pressure. Pressurized H_2 is stored in a tank with a maximum pressure of 700 bar. Prior to entering the Liebherr H_2 system, a coalescing filter eliminates any potential liquid residuals like water and the pressure is reduced to an acceptable level via a mechanical pressure regulator (Fig. 4 and 5).

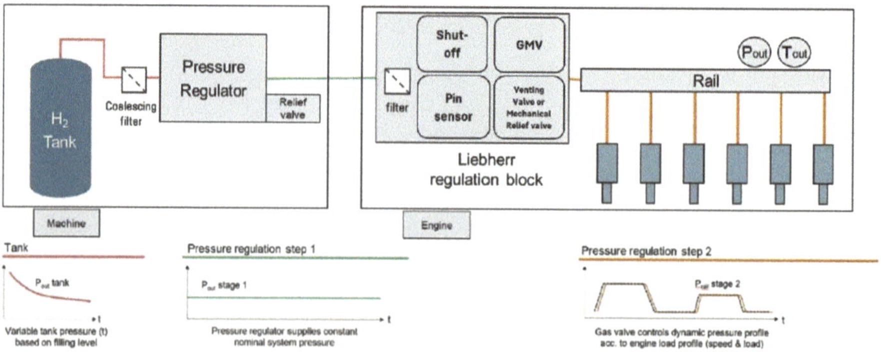

Fig. 4. H_2 system schematic view: machine and engine related part

Fig. 5. H_2 system view on engine

The Liebherr H_2 system is easily adaptable for engine integration and can be conveniently packaged. Its modular design offers the advantage of requiring minimal adjustments to be mounted in any kind of customer engine [4].

The Liebherr H_2 platform system consists of the following elements: 6 H_2 fuel injectors, 1 hydrogen regulator block, 1 rail including pressure and temperature sensor, H_2 control modules. (Fig. 6).

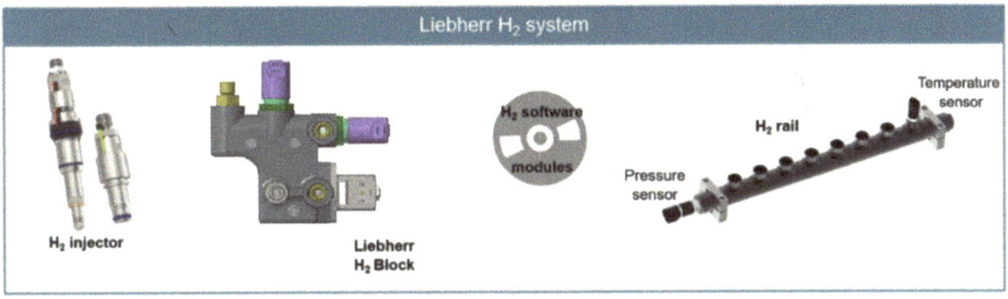

Fig. 6. Liebherr H_2 system content

3.2 Liebherr Hydrogen Injectors

The Liebherr hydrogen injector is available in side feed and top feed version. The injector has a gas-tight design, dry-run compatible and delivers an H_2 flow enabling a maximum power of 60 Kw per cylinder on the engine (Fig. 7).

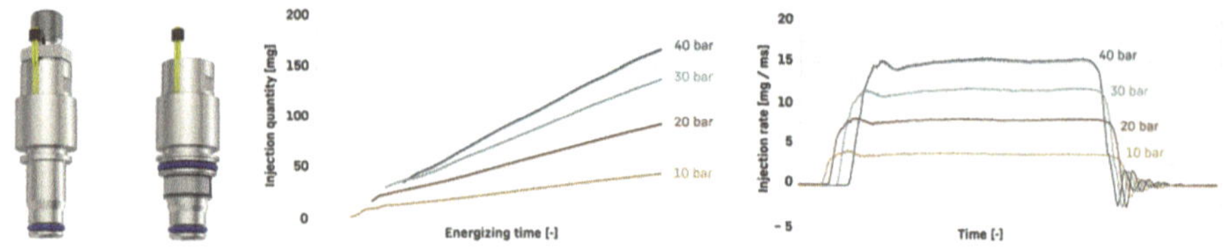

Fig. 7. Injector side feed and top feed version—example of injection quantity curve and injection rate

3.3 Liebherr H_2 Regulation Block

The Liebherr H_2 regulation block ensures the system pressure control and safety functionality. It includes a proportional valve, the **G**as **M**etering **V**alve (**GMV**) which allows the control of H_2 flow entering in the system and consequently the regulation of the H_2 pressure from atmospheric to typically 15 bar of max pressure. The pressure upstream the GMV is also measured in the block allowing a fine tuning of the pressure control.

There are two naturally closed valves at the inlet and outlet of the block, which allow the H2 filling line to be closed and the remaining hydrogen to be removed from the system when the engine is stopped. Specific control sequences are defined to ensure a safe engine start and engine stop and to monitor the system and component functionality (Fig. 8).

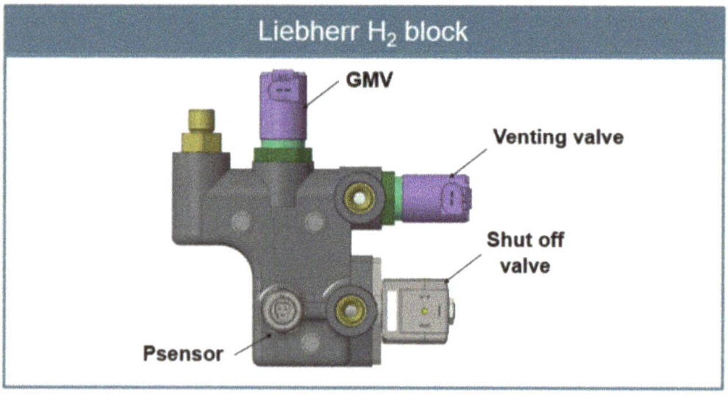

Fig. 8. Content of Liebherr H2 block

3.4 Liebherr H_2 Gas Metering Valve

As shown in Fig. 4. the Liebherr H_2-injection system uses variable rail pressure to ensure proper injection quantity over the whole power range of the system. As this pressure is controlled by the GMV, this valve is one of the key components in the regulation block. It is a normal closed, proportional driven valve. The design features a pressure supported opening needle with elastomer sealing. The elastomer sealing is needed to ensure a negligible leakage rate in closed condition.

Typically, this type of design comes with an inlet pressure threshold at which the valve opens. In combination with a mechanical PRV this can be used as overpressure protection (Fig. 9).

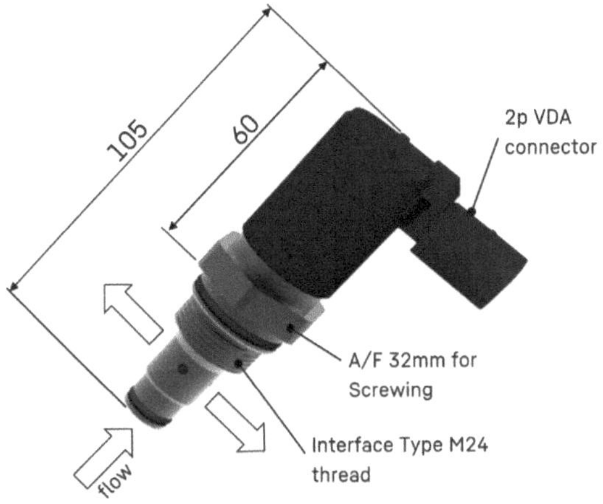

Fig. 9. LHV-ICE (Liebherr Hydrogen Valve—internal combustion engine)

Regarding DI-/PFI-systems there are two separate valves available which are based on the same platform but with adapted internals: different sealing diameter, sealing pin and actor configuration to manage their respective requirements.

The valve uses a dry running concept with specialized coatings, and current samples can be full adjusted in terms of valve lift and spring preload. It uses a PWM-signal with up to 2A of mean current.

4 System Test

4.1 Overview of Test Conditions

The system development is supported by a pneumatic test bench using compressed air as fluid. The system interfaces can be modified to represent the different test conditions encountered on engine. The measurement of pressure, temperature and flow in different localization of the system helps to fully characterize the component and system behavior. The measured mass flow of compressed air is transferred into H_2 equivalent mass flow by using a transfer factor. This factor (f) is equal to the root square of the density ratio between air (ρ_{air}) and hydrogen (ρ_{H2}) which is approximatively equal to the molar mass ratio of air (M_{air}) and H_2 (M_{H2}) assuming both gases are respecting the perfect equation gas law (Fig. 10).

$$f = \sqrt{\frac{\rho_{air}}{\rho_{H2}}} = \sqrt{\frac{M_{air}}{M_{H2}}} = 3.79$$

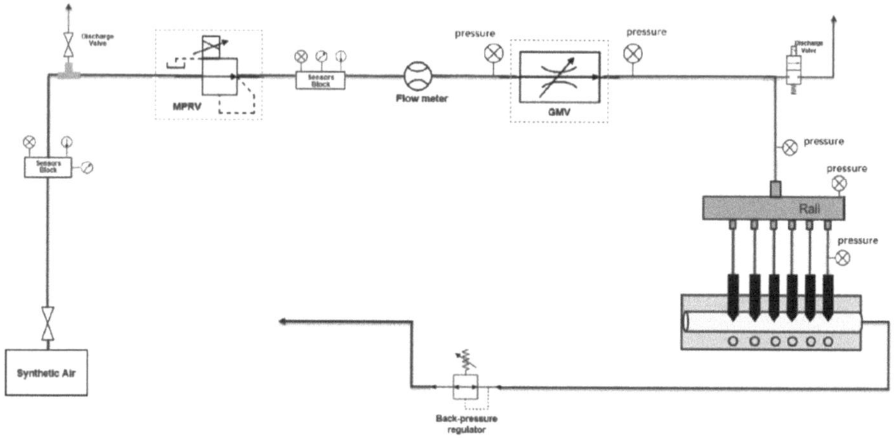

Fig. 10. Pneumatic bench schematic

The tests were performed on different cruise points that are representative of the real engine operating points of a 6 cylinder off-road engine with a maximum power of 60 kW/cylinder.

4.2 System test results

The system should be able to deliver the requested mass of H_2 during a defined timeframe at a requested pressure.

The GMV inlet pressure is an important parameter. The variation depending on the operating point as shown in Fig. 11. is relatively low ensuring good condition to have stable pressure control. The injector mass flow is requested based on injector characterization realized on a component test bench with injection analyzer and nitrogen. The system results displayed in Fig. 11. show very similar measurements in comparison to the request.

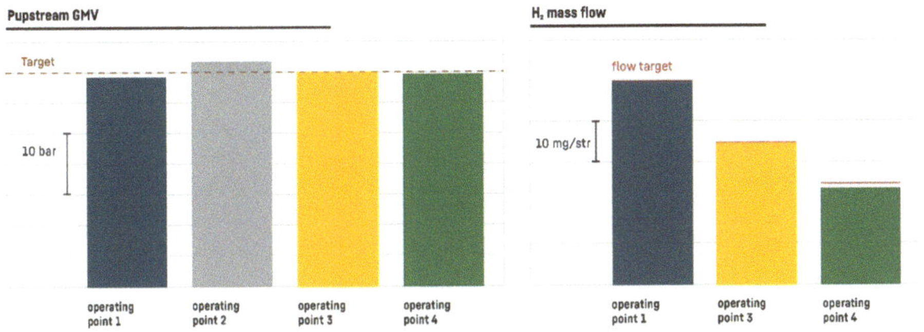

Fig. 11. Pupstream GMV/injected mass quantity measured during H_2 system tests

The rail pressure is adjusted depending on typical calibrated value on engine and can be easily followed by the system (Fig. 12). The injection time should be short enough to inject the full requested quantity while the intakes valves are opened as displayed in Fig. 12.

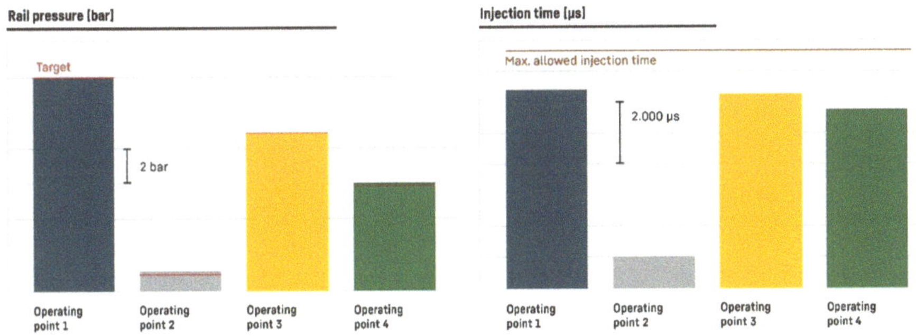

Fig. 12. Rail pressure and injection time measurement on H_2 pneumatic system bench

Finally, the pressure along the system plays an important role. The measurement used to control the system is performed on the rail while the injector behavior is impacted by the pressure at the injector inlet (Fig. 13).

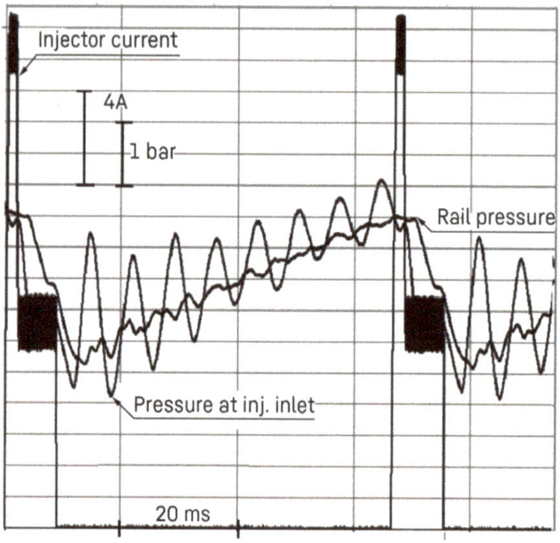

Fig. 13. Pressures measurement on H_2 system bench—@1800 erpm/110 mg/str of H_2 injected quantity

The injector closing creates pressure waves which are significantly damped in the rail. This phenomenon may influence the injector behavior and therefore the fueling accuracy. These crucial system measurements serve as input for the control team to develop the necessary strategies to improve the system performance.

5 System Control Strategies

5.1 Control Strategies Overview

The H_2 system delivered by Liebherr includes control strategies divided into three main areas: pressure control, injection control and diagnosis (Fig. 14).

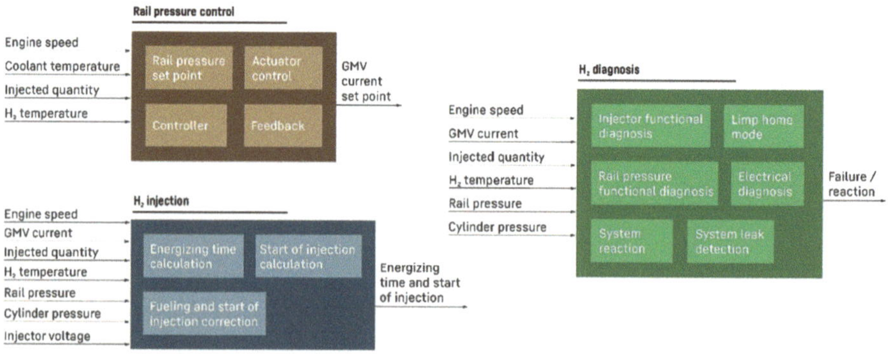

Fig. 14. Overview of H_2 system control structure

5.2 Rail Pressure Control Overview and Results

The development of the rail pressure control implies challenges related to the valve's high flow, sensitivity to inlet pressure and gas temperature, and the specific design needed to ensure tightness against H_2 when the actuator is closed. Therefore, the control design relies on the actuator characteristics and the gas flow model through an orifice (Fig. 15 and 16).

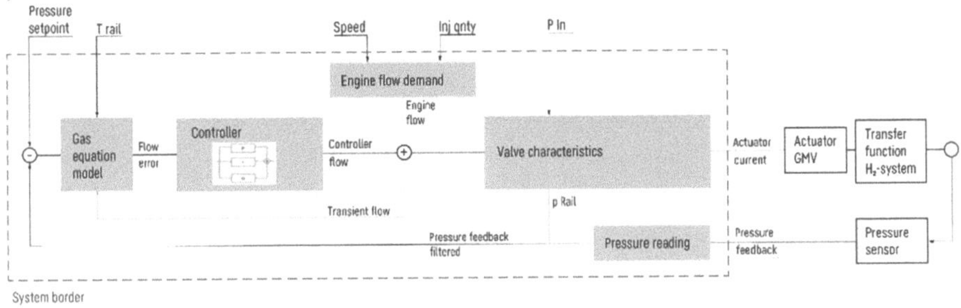

Fig. 15. Overview of pressure control structure

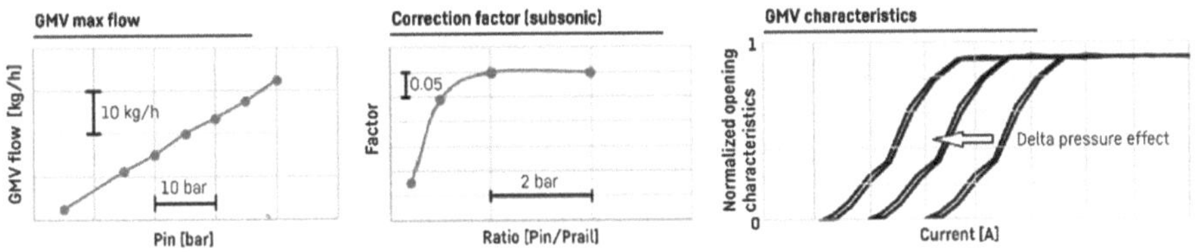

Fig. 16. GMV characteristics used in the control structure

The GMV characteristic is described thanks to the max flow of the valve and the normalized flow which depends on the current applied to the GMV and the delta of pressure across the valve. The flow is also corrected by a factor until the condition for choked flow is reached. This enables an excellent pre-control under various test conditions and limits the closed loop correction of the controller to a negligible part. Adopting this approach ensures optimal control quality in transient and requires minor calibration activities to fulfill the control requirements under varying inlet conditions (Fig. 17).

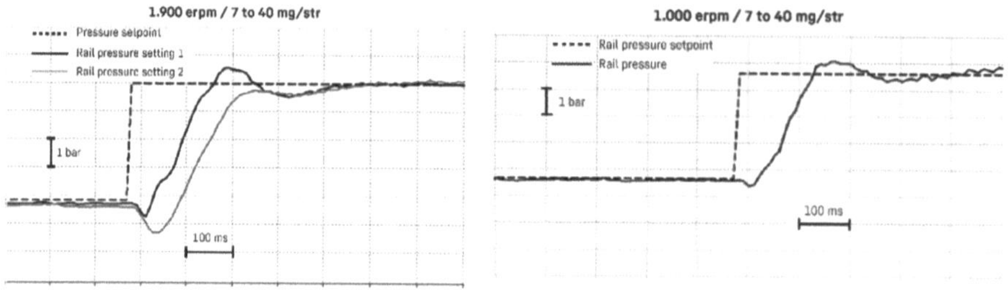

Fig. 17. Rail pressure control performance on pneumatic bench

The calibration optimization guarantees a system response time below 100 ms and a reasonable overshoot. The pressure slope, typically around 40 bar/second ensures an excellent fuel delivery, meeting the engine performance requirement.

5.3 Injection Control

Injection control is facing challenges related to the sensitivity to injector back-pressure and driving current. The driving current directly affects the injector opening, requiring optimization and calibration as a function of the operating point.

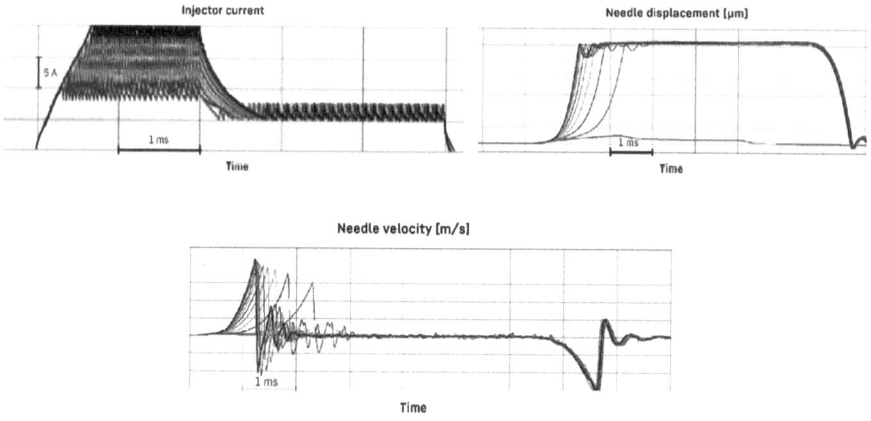

Fig. 18. Effect of driving on current on needle displacement and velocity

The Fig. 18. shows the significant effect of the driving current value on the needle speed and velocity. The driving current becomes therefore a tuning parameter to optimize the injector performance and durability. The fueling control stratregy is adapted accordingly to be able to calibrate the driving current in an optimized way as a function of the operating point.

The injector opening is influenced by various forces acting on the needle: the force of the spring (Fspring), the force of the magnet (Fmagnet), the force of the rail pressure (FPrail) and the force of the manifold pressure (FMan pressure) as illustrated in the Fig. 19. The fueling is characterized under reference conditions. When the conditions change, such as a variation in the manifold pressure, the force balance is different impacting the injector opening. The back pressure compensation function developed by Liebherr corrects the effect on injection timing and quantity.

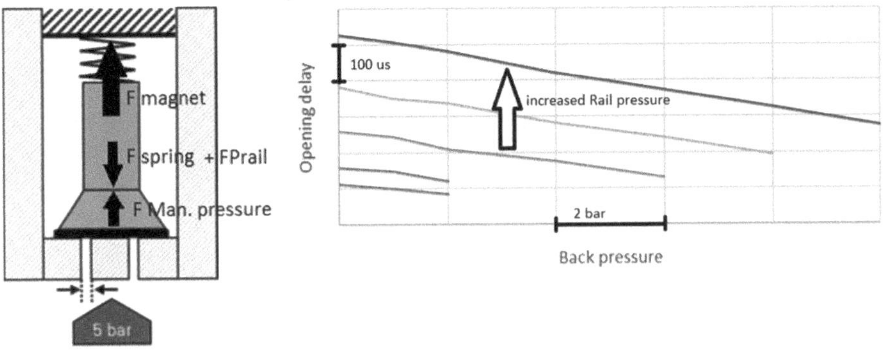

Fig. 19. Forces applied to the needle/effect of back pressure on opening delay

6 Engine Test Results

Liebherr's business unit fuel injection equipment is cooperating closely with the engine team to develop the H_2 injection system. The engine tests are used to finalize the validation of the system including the control function development. The final objective is to reach the necessary dynamic, accuracy and durability of the injection system to achieve the desired engine performance and reliability. Figure 20 shows as example the pressure control dynamic reached on Liebherr's H_2 engine on NRTC cycle.

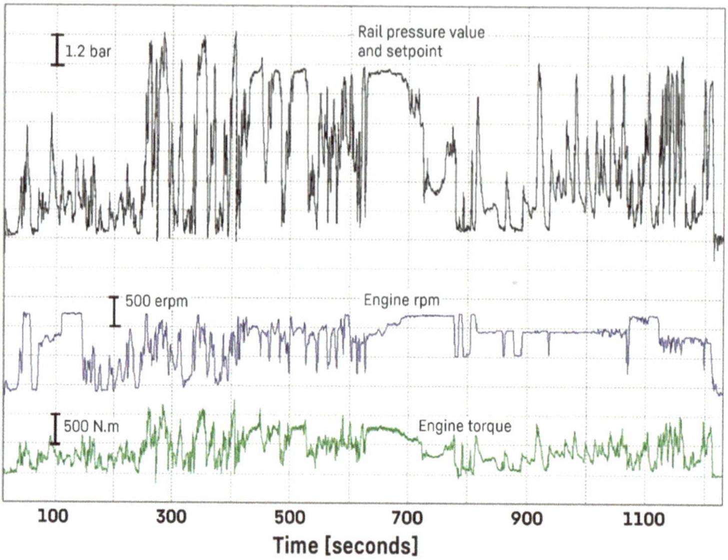

Fig. 20. H_2engine NRTC cycle

7 Conclusion

The challenge of reducing CO_2 emissions in internal combustion engines necessitates the development of a new injection system to use alternative fuels such as methanol, ammonia and hydrogen. Liebherr leverages its experience in diesel common rail system to develop a hydrogen injection system. The characteristics of H_2 as a fuel in comparison to diesel but also the specific component design and functionality are considered to develop a system layout and control functions allowing excellent performance on engine test bench. The next challenge is to operate the system in a machine environment including diverse test conditions such as starts, stops, cold and hot temperatures. The stability of performance across multiple systems (part to part) and after aging will need to be demonstrated. The Liebherr team is ready to overcome these challenges.

References

1. Weiss, U., The off-highway sector in the field of tension of future power train concepts – Which chances has the internal combustion engine (ICE) in this industry?, International engine congress Baden-Baden 2022 (2022).

2. National Institute of Standards and Technology Chemistry WebBook, https://webbook.nist.gov/chemistry/
3. U.S. Energy Information Administration, https://www.eia.gov/
4. Pirkl, R., D'Onofrio, M., Kapusta, L. : Liebherr's approach to hydrogen fuel injection systems. In: 17th International MTZ Conference on Heavy-Duty Engines, Donaueschingen (2022).

Effects of Advanced Truck Fleet Monitoring on Engine Efficiency, Reliability and Sustainability

Luis Serrano[1,2](✉), João Fonseca Pereira[1,2], João Braguez[1,2], and Ricardo Almeida[1,2]

[1] Association for the Development of Industrial Aerodynamics, ADAI, Coimbra, Portugal
{luis.serrano,fonseca.pereira,joao.braguez, ricardo.almeida}@ipleiria.pt
[2] ESTG - School of Technology and Management, Polytechnic of Leiria, Rua General Norton de Matos, Leiria, Portugal

1 Introduction

While in use, engine oils naturally deteriorate due to continuous chemical and physical changes that lower performance in comparison to their original specifications. The assessment of oil aging is specifically related to fuel dilution, exhaust gas entrainment, and soot loading [1].

The use of different fuels, considering the incorporation of different amounts of biofuels significantly affects engine oil condition, since oil performance is most affected by fuel dilution, which cause changes in the oil characteristics, but also in the soot and exhaust emissions production and mixed in oil. It's common to associate engine failures as being caused by fuel-related problems, and the use of biofuels and e-fuels increases the uncertainty associated to the oil aging, once these fuels affect the blow-by, the exhaust gas composition, the soot formation, increasing the unpredictability of the oil characteristics to assure engine lubrication and protection [2].

To evaluate whether fuel may have some influence on engine degradation by lubricant contamination, a set of 12 vehicles fueled with a blend of 7% biodiesel in fossil diesel (B7), from two heavy transportation companies, was monitored, in which periodic analyses of the lubricating oil in the engine of these vehicles were performed.

Monitoring the condition of lubricating oil in internal combustion engines, through a program of periodic analysis, presents several advantages in terms of the useful life of monitored vehicles, providing increased reliability. At the same time, it allows for an improvement in the definition of maintenance programs and the reduction of maintenance costs over time. Carrying out periodic lubricant analysis allows the engine's condition to be evaluated, by analyzing the indicators of condition, contamination, and wear revealed by the lubricant properties and characteristics.

A. Heintzel (Hrsg.): Int.Engine.Congr. 2024, Proceedings, S. 178–193, 2025.
https://doi.org/10.1007/978-3-658-47966-4_14

It was defined a planning that involved collecting engine lubricant oil after the first half of its useful life and at the end of it, according to the maintenance plan of each vehicle.

2 Used Engine Oil Analysis

2.1 Engine Oil Indicators

Usually, the effect of the contamination and wear indicators causes a change in the condition of the lubricant, so it is necessary to evaluate the need for changing the lubricant replacement interval based on the results of periodic analysis. Defining the correct sampling period allows the lubricant to be used until it reaches the indicators limit values.

Table 1 shows the limits of lubricant condition indicators, viscosity and TBN, according to three degradation degrees, "Caution", "Alarm" and "Critical", which allow a progressive characterization of lubricant contamination, to evaluate the lubricant useful life. The lubricant contamination directly affects the viscosity and TBN values, so it's necessary to manage the changing of these properties in relation to the reference values for the "new" oil.

Table 1. Viscosity and TBN limits [3]

	Units	Caution	Alarm	Critical
Viscosity	% (of new oil)	15	20	25
TBN	mg KOH/g	50% reduction	3	2.5

Viscosity and TBN may increase or decrease from the standard values of the "new" oil during its useful life, depending on the type of indicators to which it's subjected. Table 2 shows the effects os some indicators/faults on these properties.

Table 2. Possible causes for lubricant viscosity and TBN changing [4]

Oil property	Possible causes
Viscosity	**Increase:** • Aging/oxidation • Soot contamination • Blown engine head gasket (oil contamination by engine coolant) **Decrease:** • Fuel dilution,
TBN	**Increase:** • Addition of incorrect oil to the system **Decrease:** • Engine overheating • Extension of lubricant use; • Lubricant contamination

Table 3 and Table 4 shows the limit values of contamination and wear indicators in the lubricant, respectively, according three degrees of contamination, “Caution”, “Alarm” and “Critical”.

Table 3. Contamination indicator limits in lubricant [3]

	Units	Caution	Alarm	Critical
Water	%	0.1	0.2	0.5
Soot	%	2.5	3.5	5
Fuel	%	2	4	5

Table 4. Wear indicator limits in lubricant [3]

	Units	Caution	Alarm	Critical
Aluminium (Al)	mg/kg	12	19	24
Iron (Fe)	mg/kg	75	86	100
Copper (Cu)	mg/kg	35	45	60

The most found wear metals and contaminants through the analysis of used lubricants, as well as their possible origin, are shown in Tables 5 and 6. The relationship between the elements in the oil and their possible origin makes it possible, in cases where high concentrations are found, to predict the occurrence of engine faults caused by excessive wear of internal engine components and/or the presence of contaminants in the lubricant that can damage them.

Table 5. Common wear metals in internal combustion engines and their possible origins [4]

Element	Origin
Aluminium (Al)	• Piston skirts • Turbocharger • Main/rod bearings • Dirt (with Silicon) • Oil cooler
Copper (Cu)	• Main/rod bearings (with Lead and Tin) • Oil cooler (first 2000 h of use) • Wrist pin and valve train bushings • Thrust washers
Iron (Fe)	• Liners • Valve train • Crankshaft • Piston rods • Oil pump • Wriston Pins • Pistons

Table 6. Common contamination substances in internal combustion engines and their possible origins [4]

Element	Origin
Water	• Blown engine head gasket • Faulty oil cooler
Fuel	• Excessive idling • Low engine temperature • Injector fault • Incomplete combustion • Incorrect air-fuel ratio
Soot	• Poor fuel quality • Low engine compression • Worn/badly piston rings • Defective injector spraying • Incorrect air-fuel ratio

As previously mentioned, the three degrees of oil contamination allow for a gradual characterization of the lubricant, with the "Caution" degree meaning that the condition of the lubricant must be monitored, the "Alarm" degree indicating the need to reduce the period until the next collection, and the "Critical" degree indicating the need for immediate engine shutdown. To define the action to be taken according to the results obtained, the information presented in Table 7 was considered.

Table 7. Actions to take according to the indicators

Contamination degree	Intermediate sample (Action depending on the indicator)	Pre-service sample (Action depending on the indicator)
Caution	Check possible origins Track oil properties evolution	Track oil properties evolution in next samples
Alarm	Check possible origins Anticipation of the next collection	Check possible origins Evaluation of the maintenance schedule
Critical	Immediate vehicles stop until causes are identified Change oil if vehicle starts running again after evaluation	Check possible origins Consider stopping the vehicle for repairs

3 Experimental

3.1 Fleet Monitoring

In this study, the distance traveled weekly by each vehicle was monitored using the dedicated software platform, through the log of the vehicles' weekly kilometers, obtained from the odometer via CAN communication, during the entire period of the study.

This information was organized in a database created for this purpose, called "Fleet Monitoring Tool", which allowed to register the information taken from a proper software application, considering weekly periods, and based on this, to calculate the weekly distance traveled, as well as to estimate the number of kilometers traveled in the following weeks. Each vehicle has a "sheet" identical to the one shown in Fig. 1, which allows an individual analysis.

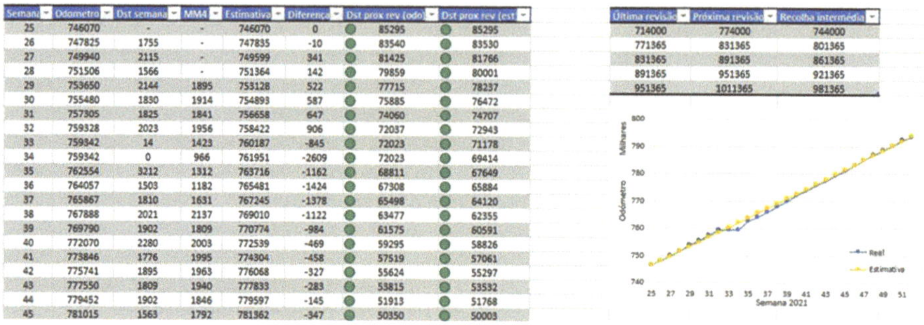

Semana	Odometro	Dst semanal	MM4	Estimativa	Diferença	Dst prox rev (odo)	Dst prox rev (est)
25	746070	-	-	746070	0	85295	85295
26	747825	1755	-	747835	-10	83540	83530
27	749940	2115	-	749599	341	81425	81766
28	751506	1566	-	751364	142	79859	80001
29	753650	2144	1895	753128	522	77715	78237
30	755480	1830	1914	754893	587	75885	76472
31	757305	1825	1841	756658	647	74060	74707
32	759328	2023	1956	758422	906	72037	72943
33	759342	14	1423	760187	-845	72023	71178
34	759342	0	966	761951	-2609	72023	69414
35	762554	3212	1312	763716	-1162	68811	67649
36	764057	1503	1182	765481	-1424	67308	65884
37	765867	1810	1631	767245	-1378	65498	64120
38	767888	2021	2137	769010	-1122	63477	62355
39	769790	1902	1809	770774	-984	61575	60591
40	772070	2280	2003	772539	-469	59295	58826
41	773846	1776	1995	774304	-458	57519	57061
42	775741	1895	1963	776068	-327	55624	55297
43	777550	1809	1940	777833	-283	53815	53532
44	779452	1902	1846	779597	-145	51913	51768
45	781015	1563	1792	781362	-347	50350	50003

Última revisão	Próxima revisão	Recolha intermédia
714000	774000	744000
771365	831365	801365
831365	891365	861365
891365	951365	921365
951365	1011365	981365

Fig. 1. Fleet Monitoring Tool

The weekly distance traveled by each vehicle was calculated by subtracting the total number of kilometers recorded between two consecutive weeks. This information was the starting point for the method developed to estimate the total number of kilometers of the vehicles in each following week.

The estimated mileage of the vehicles in each week was calculated by adding the average of the 4-week moving averages, detailed in Fig. 2, which allowed the effect of deviations in the average weekly distance traveled that occurred in certain weeks, in which the vehicles were immobilized or traveled an atypical number of kilometers, to be amortized and thus allowing to obtain an accurate calculation method.

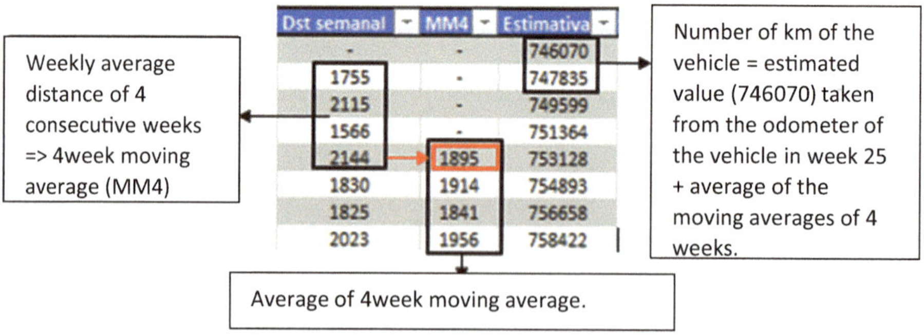

Dst semanal	MM4	Estimativa
-	-	746070
1755	-	747835
2115	-	749599
1566	-	751364
2144	1895	753128
1830	1914	754893
1825	1841	756658
2023	1956	758422

Fig. 2. Calculation of the total mileage of the vehicle each week according to MM4

The fleet monitoring tool also allows the calculation of the "distance" that the vehicle is from the next revision, according to the actual and calculated number of kilometers. The moments of completion of each type of collection are indicated through a "semaphore" designed for this purpose, whose yellow color indicates the realization of the intermediate oil sampling and red color the pre-review oil sampling, with a margin of 3000 km from the values defined in the table presented on the right in Fig. 3.

Dst prox rev (odo)	Dst prox rev (est
85295	85295
83540	83530
81425	81766
79859	80001
77715	78237

Última revisão	Recolha Pré-revisão	Recolha intermédia
714000	774000	744000
771365	831365	801365
831365	891365	861365
891365	951365	921365
951365	1011365	981365

Fig. 3. Distance to next service according to real and estimated mileage

Oil sampling was planned at two different points in its life, designated as "intermediate" and "pre-service":

- Intermediate sampling—carried out when the lubricant is halfway through its useful life, in accordance with the maintenance plan for each vehicle and organization. Analysis of the lubricant's properties at this stage makes it possible to identify significant changes in its properties and/or the abnormal presence of contaminants/wear metals in it and, as a result, assess the need to reduce the period until the next oil change and the viability of using the lubricant until the end of its expected useful life.
- Pre-service oil sampling—carried out at the end of the lubricant's useful life before the oil filters and the corresponding lubricant are replaced. The analysis of the lubricant properties in this final stage allows to observe whether the change in its properties and the evolution in the presence of contaminants and wear metals follows the trend seen in the first half of its useful life and, above all, whether the lubricant is within the recommended parameters for engine lubrication, as well as to verify that the frequency of oil changes is appropriate.

The tests performed on the used lubricants included analysis of the following parameters:

- TBN
- Viscosity at 100 °C
- Wear metal content (Aluminum, copper, and iron)
- Water, fuel, and soot concentration

3.2 Engine Oil Sampling Procedure

The lube oil sampling was performed using a vacuum pump that collects the fluid into a 250 ml bottle, by inserting the collection pipe through the oil level dipstick tube. Table 8 lists the equipment and consumables needed to perform the sample collection.

Table 8. Oil Sampling Equipment and Consumables

Equipment	• 250 ml sampling bottles • Collection pipe with 6mm outer diameter and pipe cutter • Kevlar gloves (for high temperature) • Safety glasses • Sample transport box • Vacuum pump and collection pipe transport box • Labels for sample identification and zipper bags

The oil collection procedure was divided into 3 phases, and follows the most common recommendations [5, 6]:

- Instrumentation preparation—Locate the oil dipstick and cut the collection pipe according to the length of the dipstick (sampling collected near, but not in the bottom of the engine sump).
- Engine preparation—Run the engine for a time that allows the engine to be in normal temperature condition and the oil to be adequately mixed, allowing to collect a sample that represents as well as possible the condition of the oil and the engine.
- Sample collection—Collect two samples of engine oil, in the minimum quantity of 100 ml, through the dipstick housing.

4 Results

4.1 Engine Oil Analysis

Tables 9 and 10 summarize the results obtained from the analysis of oil samples from the vehicles of fleets A and B, respectively. The condition and contamination indicators, as well as the corresponding analysis of those results for the vehicles with more significant effects, are thoroughly disclosed in the parts that follow, considering the circumstances in which certain problems or doubts were found.

Table 9. Results of oil samples from fleet A vehicles

	1st oil sampling			2nd oil sampling	
Fleet A	Intermediate sample	Pre-service sample		Intermediate sample	Pre-service sample
A1	OK	OK		–	–
A2	–	TBN; Fuel		OK	–
A3	–	OK		–	–
A4	–	OK		–	–
A5	–	OK		OK	–

Table 10. Results of oil samples from fleet B vehicles

	1st oil sampling		2nd oil sampling	
Fleet B	Intermediate sample	Pre-service sample	Intermediate sample	Pre-service sample
B1	–	Fuel	–	–
B2	Fuel	–	OK	–
B3	OK	–	–	–
B4	OK	OK	–	–
B5	OK	OK	OK	–
B6	–	Water; Fuel		
B7	–	Fuel		

Vehicle A2. Figure 4 and Table 11 present the lubricant condition indicators of vehicle A2, TBN and viscosity, for the pre-service and intermediate samples. For the vehicle in question, the samples do not match the same oil, which may influence the evolution of the indicators.

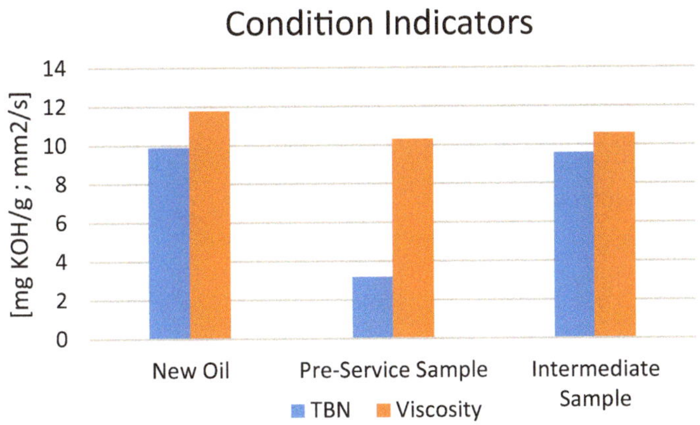

Fig. 4. Lubricant condition indicators of vehicle A2

Table 11. Comparing the lubricant condition indicators of vehicle A2

Sampling sequence	Viscosity (1 cSt = 1 mm^2/s)	TBN (mg KOH/g)
New oil	11,78	9,9
Pre-service sample	10,30—OK	3,2—Caution
Intermediate sample	10,59—OK	9,6—OK

Figure 5 and Table 12 show the lubricant contamination indicators of vehicle A2 for the pre-service and intermediate samples.

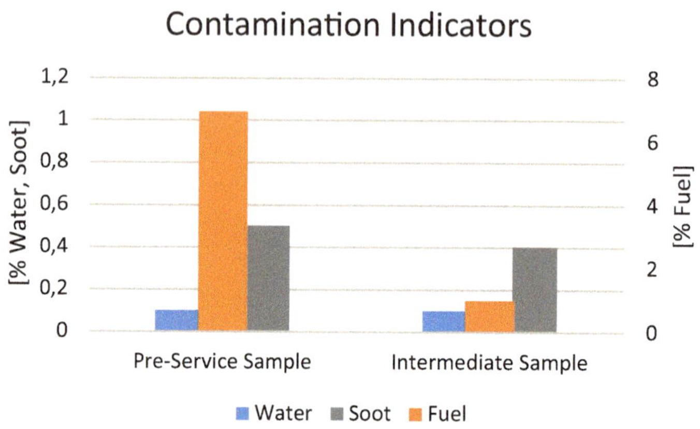

Fig. 5. Lubricant contamination indicators of vehicle A2

Table 12. Comparing the lubricant contamination indicators of vehicle A2

Sampling sequence	Water (%)	Fuel (%)	Soot (%)
Pre-service sample	0,1—OK	6,9—Critical	0,5—OK
Intermediate sample	0,1—OK	1,0 —OK	0,4— OK

Figure 6 and Table 13 show the concentration of wear indicators in the lubricant of vehicle A2, for the intermediate and pre-service samples.

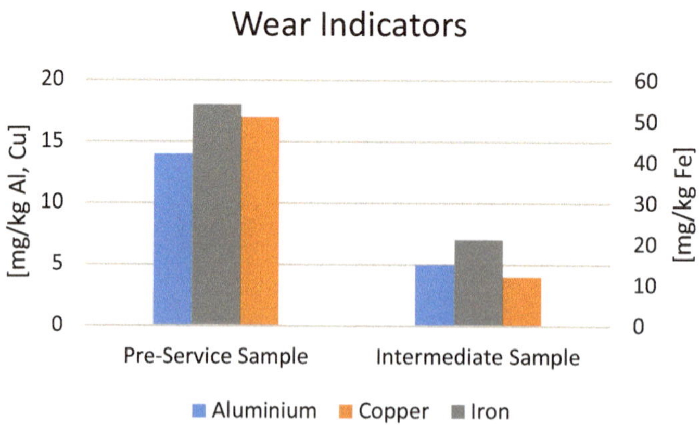

Fig. 6. Lubricant wear indicators of vehicle A2

Table 13. Comparing the lubricant wear indicators of vehicle A2

Sampling sequence	Aluminum (mg/kg)	Copper (mg/kg)	Iron mg/kg)
Pre-srvice sample	147— OK	17—OK	547—OK
Intermediate sample	57— OK	47—OK	217—OK

The pre-service sample contains a critical fuel concentration of 6.9%. This increase could not be predicted by analyzing the intermediate sample because the pre-service sample was taken before the intermediate sample, which corresponds to different oils cycle. The high fuel concentration is followed by a decrease in the viscosity value of 12.6% compared to the original value and 2.5% compared to the intermediate collection, and an increase in the concentration of wear metals, which, however, are in acceptable parameters, but reveal a level of wear that should be carefully analyzed. The TBN shows a worrying value of 3.2 mg KOH/g, corresponding to a decrease of 67.7% compared to the original value, which can be explained by the presence of the contaminants mentioned above.

An atypical decrease in viscosity value is observed for the intermediate sample, of 10.1% compared to the original value, without an abnormal concentration of fuel, contaminants, and wear metals, observing only a soot concentration of 0.4% that theoretically would cause an increase in this value. The TBN value reveals an insignificant decrease compared to the original value for this sample.

The high fuel concentration value for the pre-service sample, followed by an intermediate sample (after oil change) that showed normal fuel concentration, suggests that lubricant fuel contamination occurred due to engine operating conditions rather than engine problems. The high concentration of fuel in the lubricant may be associated with the vehicle's type of use (excessive idling, cold starting, excessive load). The results indicate engine operation outside the critical limits defined for lubricant condition for a significant period of the last half of its useful life, due to the high concentration of fuel.

Vehicle B1. Figure 7 and Table 14 show the lubricant condition indicators of vehicle B1, TBN and viscosity, for the pre-service sample.

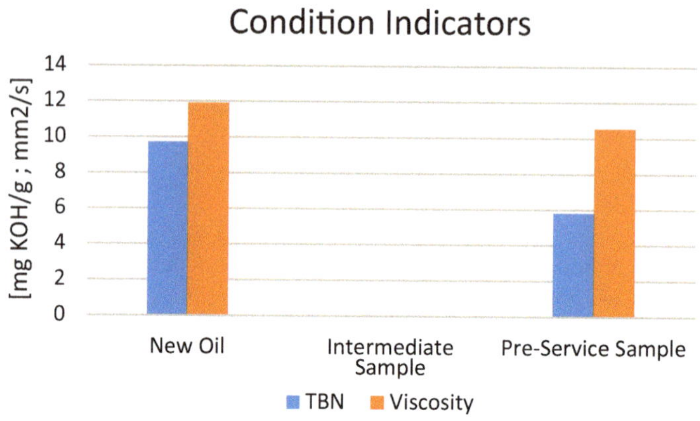

Fig. 7. Lubricant condition indicators of vehicle B1

Table 14. Comparing the lubricant condition indicators of vehicle B1

Sampling sequence	Viscosity (1 cSt = 1 mm^2/s)	TBN (mg KOH/g)
New oil	11,88	9,7
Intermediate sample	–	–
Pre-service sample	10,51—OK	5,8—OK

Figure 8 and Table 15 show the lubricant contamination indicators for vehicle B1, for the pre-service sample.

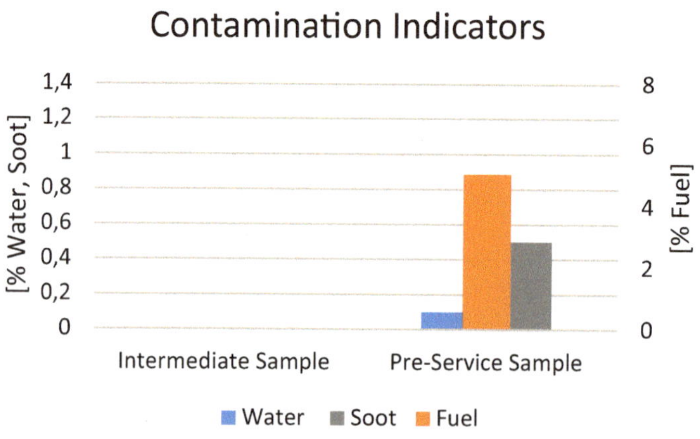

Fig. 8. Lubricant contamination indicators of vehicle B1

Table 15. Comparing the lubricant contamination indicators of vehicle B1

Sampling sequence	Water (%)	Fuel (%)	Soot (%)
Intermediate sample	–	–	–
Pre-service sample	0,1—OK	5,1—Critical	0,5 —OK

The pre-service sample indicates a critical fuel concentration of 5.1%, associated with a decrease in viscosity value of 11.5% and TBN value of 40.2% compared to the respective original values, which, however, disclose normal values. The high presence of fuel in the lubricant may be associated with problems in the engine injection system or the vehicle's type of use (cold start, idle time, excessive load). There is a normal concentration of the other indicators. The results obtained indicate that the engine operates outside the critical limits defined for the condition of the lubricant, due to the high concentration of fuel in it, but the lack of an intermediate collection does not ensure the period of its operation under these conditions.

Vehicle B2. Figure 9 and Table 16 presents the lubricant condition indicators of vehicle B2, TBN and viscosity, for the intermediate sample.

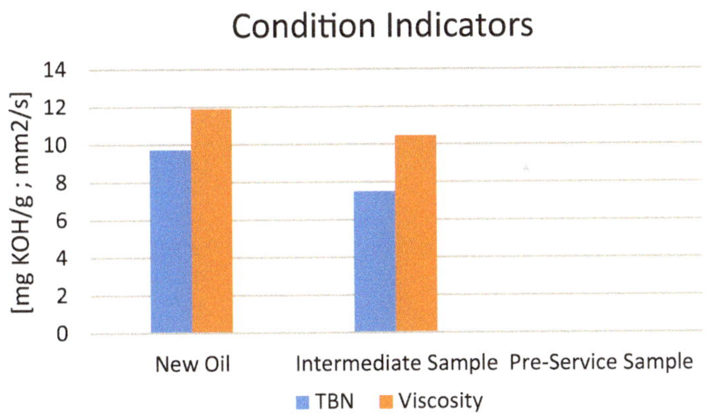

Fig. 9. Lubricant condition indicators of vehicle B2

Table 16. Comparing the lubricant condition indicators of vehicle B2

	Viscosity (1 cSt = 1 mm^2/s)	TBN (mg KOH/g)
New oil	11,88	9,7
Intermediate sample	10,45—OK	7,5—OK
Pre-service sample	–	–

Figure 10 and Table 17 reveal the lubricant contamination indicators for vehicle B2, for the intermediate sample.

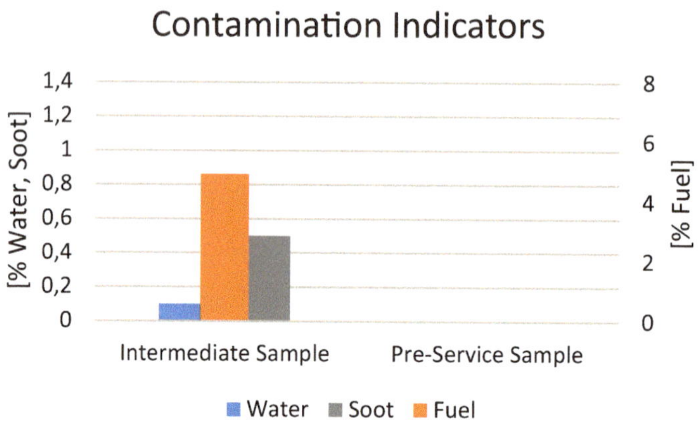

Fig. 10. Lubricant contamination indicators of vehicle B2

Table 17. Comparing the lubricant contamination indicators of vehicle B2

Sampling sequence	Water (%)	Fuel (%)	Soot (%)
Intermediate Sample	0,1—OK	4,9—Alarm	0,5—OK
Pre-Service Sample	–	–	–

An alarm fuel concentration of 4.9% is observed in the intermediate sample analyzed. The viscosity value decreased by 12.0% and the TBN value by 22.7% compared to the original values, while the remaining indicators are within the normal limits established. Since this is an intermediate sample, it's recommended that the vehicle be stopped until the causes of the high presence of fuel in the oil in the middle of its useful life were identified, which may be associated with the type of use of the vehicle, particularly the high idle time. In summary, the lubricant presents only an alarm level of fuel concentration, but nothing can be concluded about the operation of the engine within the recommended parameters for the condition of the lubricant throughout its useful life, given the lack of a pre-service sample.

Vehicle B4. Figure 11 and Table 18 show the lubricant contamination indicators for vehicle B4, for the intermediate and pre-review samples.

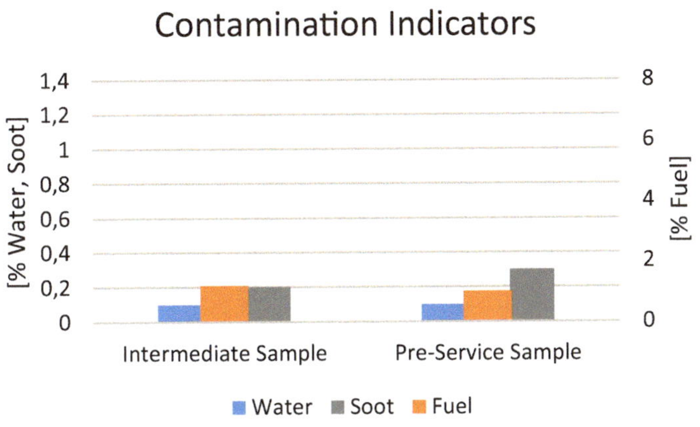

Fig. 11. Lubricant contamination indicators of vehicle B4

Table 18. Comparing the lubricant contamination indicators of vehicle B4

Sampling sequence	Water (%)	Fuel (%)	Soot (%)
Intermediate sample	0,1—OK	1,2—OK	0,2—OK
Pre-service sample	0,1—K	1,0—K	0,3—OK

An abnormal decrease in fuel concentration is observed between the intermediate and pre-service samples, within normal limits for this indicator. At the time of the pre-service collection, an oil leak was identified in the engine valve cover area, and the driver indicated the recurrent need to replace the engine oil level with "new" oil, which may explain the atypical behavior of this parameter.

Vehicle B6. Fig. 12 and Table 19 show the lubricant contamination indicators for vehicle B6, for the pre-service sample.

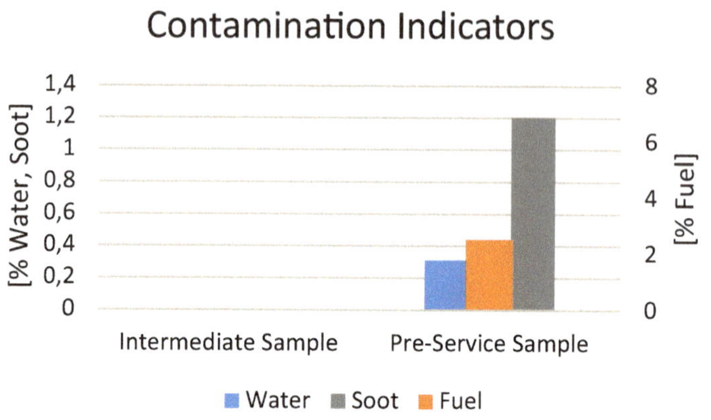

Fig. 12. Lubricant contamination indicators of vehicle B6

Table 19. Comparing the lubricant contamination indicators of vehicle B6

Sampling sequence	Water (%)	Fuel (%)	Soot (%)
Intermediate Sample	–	–	–
Pre-Service Sample	0,3—Alarm	2,5—Caution	1,2 – OK

An alarming water concentration of 0.3% was observed in the lubricant, indicating the presence of coolant in it. It was reported by the vehicle's maintenance supervisor the existence of an overpressure problem in the vehicle's cooling circuit, which may be associated with a faulty engine head gasket and, consequently, cause lubricant contamination by coolant.

There is also a worrying concentration of 2.5% of fuel, which may be associated with the fact that the vehicle in question has been out of service for several weeks, which may compromise the operating condition of the engine injection system, particularly the proper spraying of the injectors, and causing the presence of a considerable soot concentration of 1.2%, which is, however, within the normal limits established. The water and fuel concentration in the lubricant show an alarming and worrying value, respectively, which indicates that the lubricant conditions may not assure the proper lubrication of the engine, as well as the lack of an intermediate collection does not allow us to assess its operation period under these conditions.

5 Conclusions

The engine oil analysis made it possible to draw some conclusions regarding the lubricant degradation, and the condition and type of engine use. However, the time window in which the study was carried out was still reduced, what would be necessary for a more accurate and objective analysis, since it represented about 60.000 km of circulation of each vehicle, i.e., only one maintenance cycle.

The defined oil sampling methodology allowed for representative samples of the engine's condition to be obtained, which validates the procedures and equipment used, since in the cases in which faults were detected and/or reported in the vehicles, they influenced the condition of the engine's lubricating oil.

The developed vehicle monitoring process proved to be an advantage in the fleet management, providing an adjustment of the oil drain interval, the assessment of the condition of the lubricant and its engine, as well as the detection of engine failures.

This procedure reveals that an appropriate maintenance program would include a more frequent assessment of the lubricating oil conditions, enabling the oil service to be adjusted in accordance with the fuel type and engine operating conditions. Additionally, this enables the possibility of preventative maintenance and the prevention of malfunctions before severe engine deterioration may happen. Furthermore, it makes it possible to control the frequency of oil changes more precisely, which increases engine longevity. It also provides a tool that can lengthen the time between oil changes, if that is possible, and reduce waste because a large amount of oil was only squandered when it did not effectively present the necessary qualities to lubricate the engine. All of these benefits contribute to an environmentally friendly process. Knowing better the engine's condition, makes it feasible to maintain it more efficiently, which lowers fuel consumption, CO_2 emissions, and travel expenses.

Additionally, no anomalies were identified in the oil clearly related to the type of biofuel used (commercial B7). It is intended to use this methodology in a subsequent investigation to examine and evaluate the effects of employing various fuels with higher proportions of biofuels or e-fuels.

References

1. J. A. Shenker, "The compatibility of semi-synthetic engine oil with conventional diesel and biodiesel fuels", PhD Thesis, University of Birmingham, 2015.
2. W. Meyring, "Influence of alternative fuels on oil aging." Development Fuels. Springer Professional – Automotive (www.springerprofessional.com/automotive)
3. Oil Analysis, L. General Guidelines to Alarm Limits (2021).
4. Oil Analysis, L. Source Element Guide - contamination, wear metal, additives, and multi-origin elements (2018).
5. Apex Oil Lab, L. Used Oil Sampling Techniques. In: https://apexoillab.com/used-oil-sampling-techniques, last accessed 2021/11/20.
6. Caterpillar Energy Solutions GmbH. Specification for lube oil. Carl-Benz-Str. 1, Mannheim, Germany (2019).

Development of a Passive Pre-Chamber Spark Plug for Passenger Cars Feasible for Mass Production

Alexander Eichhorn[1](✉), Matthias Blankmeister[2], Corinna Vonau[2], Michael Frank[1], und Fabian Lemke[2]

[1] Robert Bosch GmbH—Power Solutions, Schwieberdingen, Germany
{Alexander.Eichhorn2,michael.frank}@de.bosch.com
[2] Robert Bosch GmbH—Mobility Aftermarket, Stuttgart, Germany
{matthias.blankmeister,corinna.vonau, fabian.lemke}@de.bosch.com

1 Concept of Combustion Systems with PCSP

Future ICE powertrains will have to strike a balance between being extremely efficient in terms of fuel consumption and cost, while at the same time complying with future emission standards. Even under today's conditions, considerable efforts are being made in engine and component development to achieve these goals.

One component currently under development that can significantly improve engine efficiency by reducing engine knock is the "Passive Pre-Chamber Spark Plug" (PCSP). Base idea is to fit a cap to the front of the spark plug to form a cavity. Holes in the cap connect the cavity of the PCSP to the main combustion chamber of the engine. During compression, the PCSP volume is filled with fresh mixture to ensure a robust ignition within the PCSP volume. Flame jets then enter the main combustion chamber and ignite the cylinder charge at multiple points. As a result, this type of spark plug provides improved ignition of the cylinder charge, resulting in very robust combustion with increased combustion speed. Finally, this improved combustion will reduce knock sensitivity which is used under high-load conditions to advance ignition timing and therefore improve the center of combustion.

The industrialization of PCSP technology requires major efforts in component development, but also in combustion system engineering and calibration. This paper aims to provide an insight into the detailed development process from both a component and system development perspective (Fig. 1).

A. Heintzel (Hrsg.): Int.Engine.Congr. 2024, Proceedings, S. 194–218, 2025.
https://doi.org/10.1007/978-3-658-47966-4_15

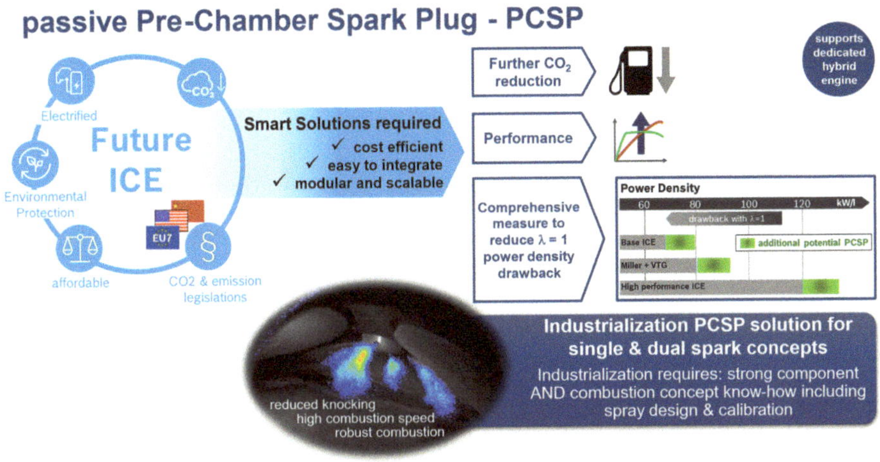

Fig. 1. Overview passive Pre-Chamber Spark Plug (PCSP) technology

During the development process it became increasingly clear that with improving the PCSP design towards high-load performance, the behavior under low-load conditions is getting worse [1]. This leads to two feasible applications for a PCSP concept:

1. "High-efficiency approach" in a hybrid powertrain.
   Focus on PCSP design trade-offs between high-load performance and low-load feasibility, while eliminating the lowest engine loads by exploiting the capabilities of an adapted hybrid operation strategy. An approach to the development of a highly efficient powertrain concept was presented last year [2]
2. "High-performance approach" in a high-power ICE-only powertrain.
   Focus on maximum high-load performance PCSP design and use a second conventional J-gap spark plug. In a so called "Dual spark concept" this spark plug is installed additionally into the cylinder head to cover the most demanding operation modes for PCSP, such as idling and catalyst heating.

2 Variant Frame and Criteria for PCSP Development

A variant frame was defined during the design process of the passive M12 prechamber spark plug. This variant frame still offers a high degree of flexibility to adapt the PCSP to specific engine/combustion system concepts, while scaling the variant diversity to a level suitable for efficient mass production.

There are applications for a PCSP that lead to different approaches to the design of the pre-chamber spark plug. Firstly, there is a distinction between so-called "high-efficiency" variants with small chamber volumes and "high-performance" variants with larger chamber volumes (factor 1.5 to 2), see Fig. 2. In general, smaller chamber volumes result in better low-load performance and are therefore more suitable for engines in hybrid powertrains with the PCSP as the sole ignition source.

For dual spark engines the design of the pre chamber spark plug can be more aggressively optimized towards knock reduction at high load. This is usually achieved by increasing the chamber volume as the main dominant factor.

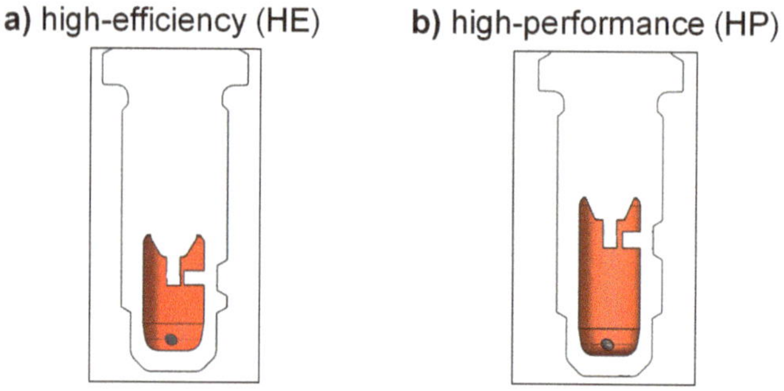

Fig. 2. High-efficiency (a) vs. high-performance (b) version of the PCSP

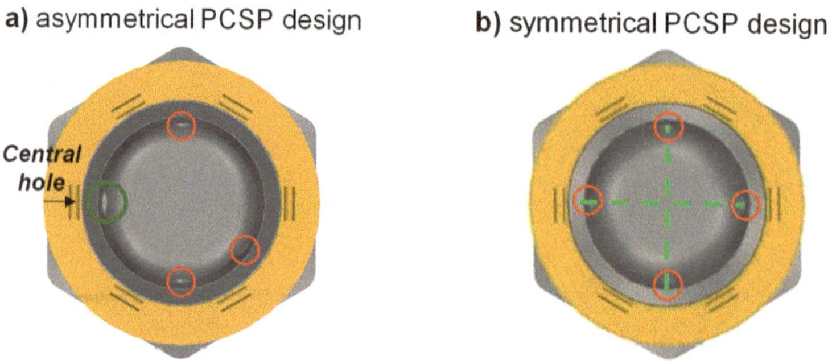

Fig. 3. Asymmetrical (a) vs. symmetrical (b) PCSP design

The second categorization within the defined variant frame is the distinction between symmetrical and asymmetrical distribution of holes around the circumference of the PCSP cap. The asymmetrical design offers advantages in the gas exchange process of the PCSP, resulting in improved part-load and low-load performance, as well as greater tolerance to external and internal EGR. However, this design type requires a defined thread gate on the cylinder head thread as an additional engine design requirement. Alternatively, a symmetrical hole design can be chosen, which is free from orientation constraints during spark plug installation. Such spark plugs offer very similar performance at high loads but lag at partial loads.

Bosch therefore offers a range of PCSP variants so that the best option can be selected for the particular application.

3 Development Method

The adaptation of PCSP technology to different engine topologies typically follows a three-step approach. In the first step, the CAD geometry of the combustion chamber is used to distribute the holes and thus the flame jet orientation in the most favorable way for the given geometry. In a second step, the PCSP design variant is tested in transient 3D-CFD calculations that include combustion. Once the most promising designs have been selected and manufactured, the samples are evaluated on the engine test bench.

3.1 Static Design: "Flame Targeting"

The initial step to configure a PCSP design involves adapting the PCSP to the cylinder geometry of the customer's engine. This adaptation is accomplished by utilizing the CAD engine geometry of the dedicated engine, determining the number of holes, the angle at which they are placed, and their distribution around the periphery of the PCSP cap. The challenge during the design phase is finding the optimal balance between a design that provides the best fit for the main combustion chamber and a design that provides the best performance for the pre-chamber itself.

Symmetrical PCSP designs are characterized by evenly arranged holes—typically four holes—around the circumference, see Figure 4.

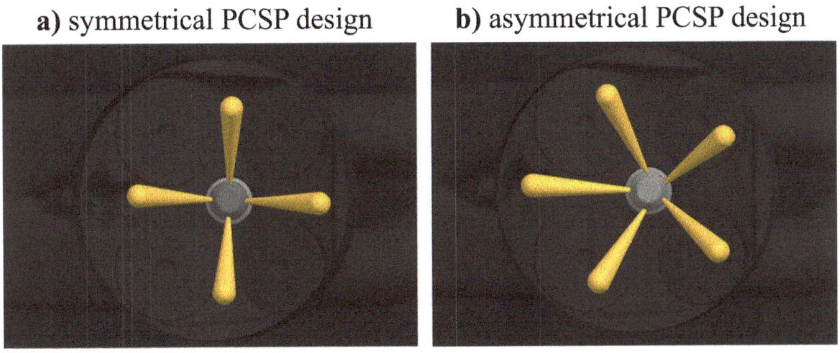

Fig. 4. Static Flame Targeting: (a) symmetrical vs. (b) asymmetrical PCSP design

This ensures an even flame jet distribution in the main combustion chamber, resulting in an evenly distributed ignition. Although this design results in high turbulence levels within the pre-chamber volume itself, it also leads to relatively high residual gas content inside the pre-chamber because it cannot be purged in an optimal way.

Asymmetrical PCSP designs typically feature a four or five-hole design. The largest diameter hole is positioned perpendicular to the crankshaft axis, toward the intake side of the engine. This "central" hole (Fig. 3) enhances purging of the PCSP volume. The remaining holes are distributed toward the four gas exchange valves, as shown in Fig. 4. Although this design offers more flexibility, it increases the installation requirements of the PCSP in the cylinder head.

3.2 Dynamic Evaluation: CFD

The CAD design of the PCSP is transferred to a transient CFD model for performance evaluation. Normally, both part-load and high-load operation points must be calculated to provide an overview of the performance across a wide range of engine operation points.

Due to the complexity of the PCSP behavior, it is necessary to simulate a complete cycle including combustion.

There is a stark contrast in the behavior of a pre-chamber spark plug between part and full load operation. Also, there are opposing effects to consider for the inflow and the filling phase as well as the ignition and the combustion in the main combustion chamber. For symmetrical pre-chamber spark plugs, the important phases in the cycle are the compression phase (filling), combustion in the pre-chamber and combustion in the main combustion chamber.

Regarding asymmetrical pre-chamber spark plugs, the intake stroke needs to be considered additionally: To ensure an ignitable mixture in the pre-chamber at ignition timing, the residual gas must be purged out of the pre-chamber volume during this phase (Fig. 5).

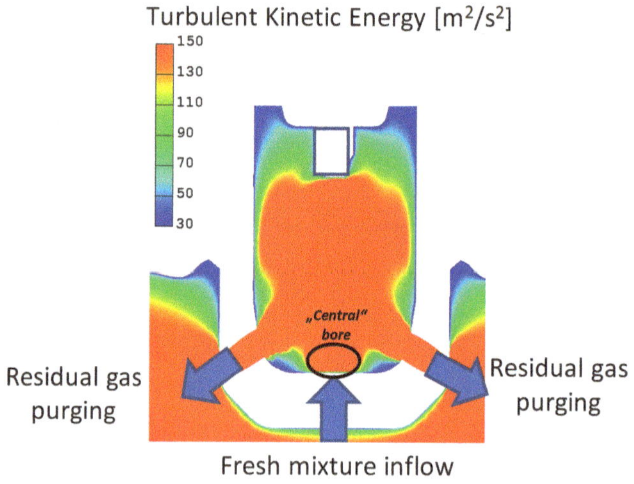

Fig. 5. Purging process of asymmetrical PCSP

Hence to allow the fresh mixture to enter for and the flow of residual gas to exit the pre-chamber spark, the design of the hole angles and hole diameters is essential. If designed properly, the fresh mixture is pushed inside the pre-chamber through the central hole which is oriented towards the intake side and the residual gas is leaving the pre-chamber through the other holes (compare Fig. 6, positive values correlate to flow direction into the pre-chamber).

The holes must be positioned on the cap in such a way that they can catch the tumble flow in the main combustion chamber through the central hole (see Fig. 5). On the other hand, low-pressure areas develop at the PCSP surface areas perpendicular to the central hole since the tumble flow is accelerated in these areas causing a local pressure drop. These low-pressure zones can be used to draw residual gas out of the pre-chamber. Consequently, more fresh mixture can be transported into the pre-chamber during the compression phase which is especially important for ignitability at low loads and low engine speeds. Here the asymmetrical PCSP achieves a significantly improved part-load capability compared to the symmetrical PCSP.

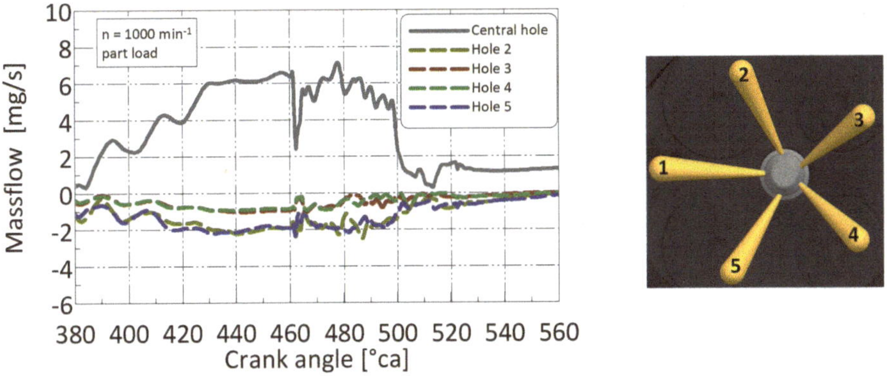

Fig. 6. Massflow through individual holes of PCSP at part-load operation point

In the compression phase, the fresh mixture is forced into the pre-chamber. Mixture homogeneity and mixing with remaining residual gas are an indication of the combustion quality in the pre-chamber. The flow design inside the pre-chamber ensures that the mixture is transported to the ignition location very quickly. The spark ignites the mixture and the high turbulent kinetic energy (TKE) ensures that combustion is highly accelerated. Figure 7 presents the CFD results for a part-load operation point, comparing symmetrical and asymmetrical pre-chamber spark plug designs.

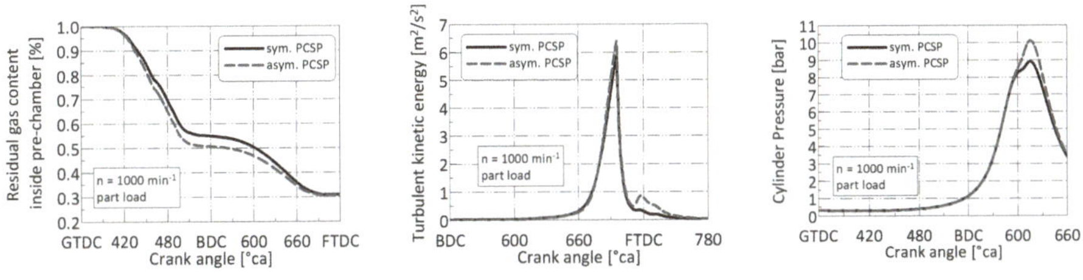

Fig. 7. CFD results part-load operation point

The asymmetrical design shows a significantly lower residual gas content than the symmetrical design due the improved purging of the pre-chamber volume. The turbulence inside the pre-chamber is comparable between the two designs. Finally, examining the resulting cylinder pressure trace of the main combustion reveals that the asymmetrical pre-chamber design results in faster combustion due to more stable and rapid ignition inside the PCSP volume based on the lower residual gas content. This ultimately leads to more robust flame jets and improved ignition in the main combustion chamber.

Figure 8 shows the CFD results for an operation point at rated power. At high loads, there is no difference in the purging process between the two types of pre-chamber spark plugs. The symmetrical design results in higher turbulence levels than the asymmetric design, which directly affects the combustion inside the pre-chamber and the ignition and combustion of the main chamber. The higher turbulence value leads to faster combustion. This can be seen in the higher pressure in the symmetric pre-chamber and subsequently in the higher resulting cylinder pressure in the main combustion chamber.

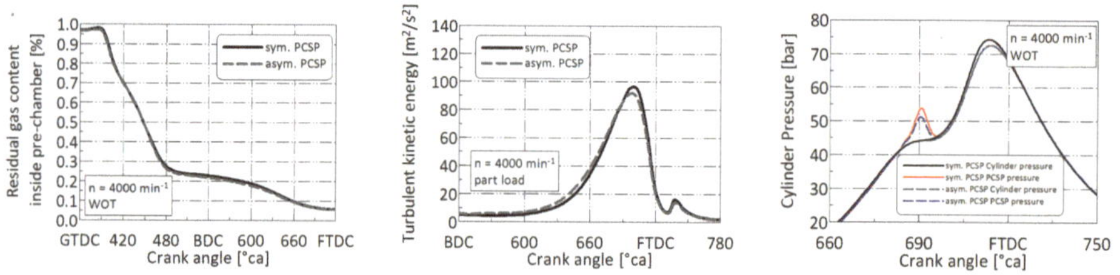

Fig. 8. CFD results WOT operation point

Figure 9 and 10 show the start of the flames jets emerging from the holes and their development 2°ca later. Another design criterion is the penetration depth of the flame jets into the main combustion chamber. The penetration depth is determined on the one hand by the ratio of the pre-chamber volume to the cross-sectional area and on the other hand by the length of the holes. Longer holes in the pre-chamber cap increase the penetration depth.

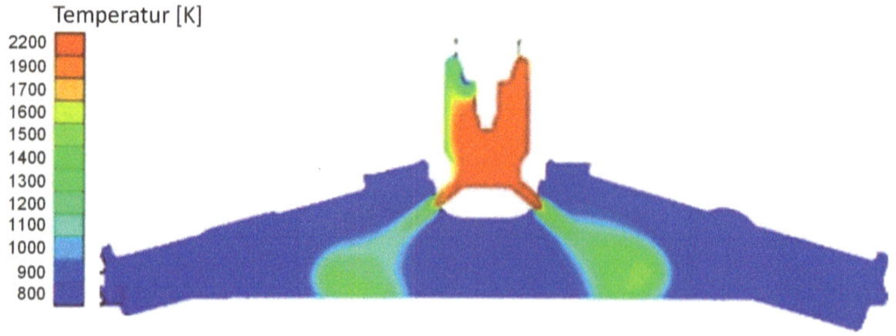

Fig. 9. Gas temperature distribution inside combustion chamber

In summary, the parameter set consisting of the number of holes, diameter and length of the holes and their angle can be used to describe the design space of a pre-chamber spark plug.

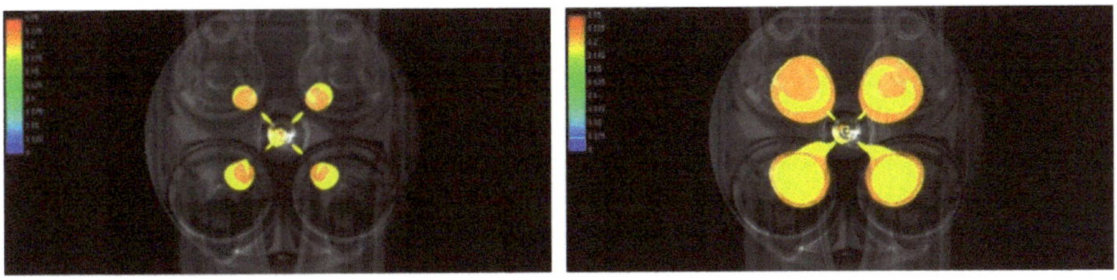

Fig. 10. Progress of the flame front inside the main combustion chamber (2°ca time difference) using a symmetrical PCSP

3.3 Engine Tests

The engine tests performed to evaluate the behavior of the PCSP should cover all relevant engine operation ranges so that the benefits of the PCSP can be quantified, but also to identify the challenging areas of engine operation with PCSP. Engine operation with PCSP will always be a trade-off between maximizing high load benefits and adequate low/part-load performance. The areas that should be considered during engine testing are:

Benefits of PCSP

- High-load performance: MFB50%-shift by knock reduction.
- $\lambda = 1$ power increase by knock reduction and increased combustion speed
- If external EGR is used: Tolerance of PCSP to external EGR

The results of the high-load operation can be used to assess whether an additional compression ratio increase of the engine is enabled by the PCSP. If so, this would be a measure to improve engine efficiency across the engine map, while preventing excessive knock at high-load.

Challenging Operating Modes for PCSP

- Part-load fuel consumption
- Idle performance
- Minimum achievable torque
- Catalyst heating performance
- Cold start at lowest required temperatures

The results of the demanding operating modes can be used to assess whether design changes to the PCSP are required to improve low-load performance, or whether even a dual spark concept would be a valuable option.

4 Knock Detection and PCSP

The flame jets emerging from the PCSP during ignition induce a pressure fluctuation in the main combustion chamber in a very similar frequency range to the regular knock-induced pressure fluctuations. The most dominant frequency is the first mode of the combustion chamber, but higher modes are also excited by the flame jets. A new challenge is presented because the effects on the cylinder pressure signal are quite similar between a non-knocking engine operating with PCSP and engine knock. As the pressure oscillations are transmitted as structure-borne noise, the knock sensor is also affected by these effects. A method for correcting these phenomena was developed for the test bench engine measurements and an adapted ECU knock control software was generated. In addition, the calibration strategy of the knock sensor was developed. This is essential, as only high-quality knock detection and knock control will allow the engine to operate at the knock limit and thus allow the benefits of the PCSP to be translated into a real increase in engine efficiency.

4.1 Correction Method for Test Bench Indication System

Using pressure indication systems, the standard method to judge the knock intensity is to evaluate the maximum pressure peak in the high pass filtered pressure trace, see eq. 1

$$KPPK = \max(abs\left(p_{cyl_{highpass_{filtered}}}\right)$$

The induced cylinder pressure oscillations caused by the PCSP flame jets even in a non-knocking operating cycle are clearly visible in Fig. 11.

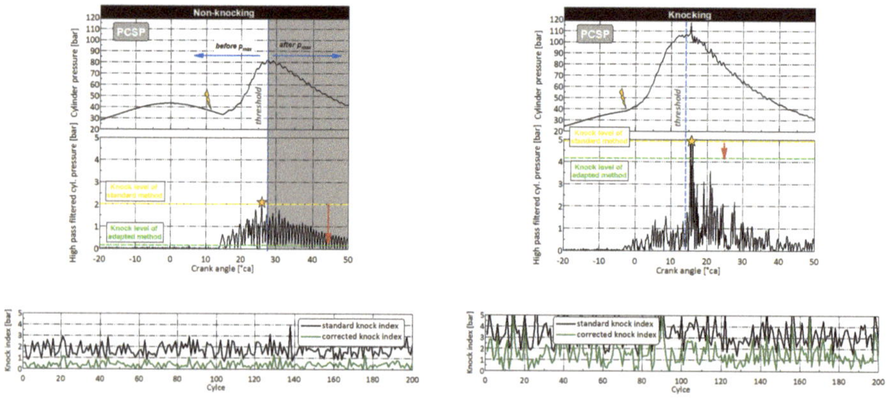

Fig. 11. Pressure trace and high pass filtered pressure trace for a non-knocking & knocking working cycle with conventional PCSP

This oscillation has an amplitude that is often in the range usually attributed to the onset of engine knock. The bell-shaped high pass filtered pressure curve showing its maximum value usually before the maximum of the unfiltered cylinder pressure is reached. This is the key indicator that no knocking has occurred.

The conventional spark plug usually does not show such a bell-shaped oscillation in the high pass filtered pressure curve, see Fig. 12.

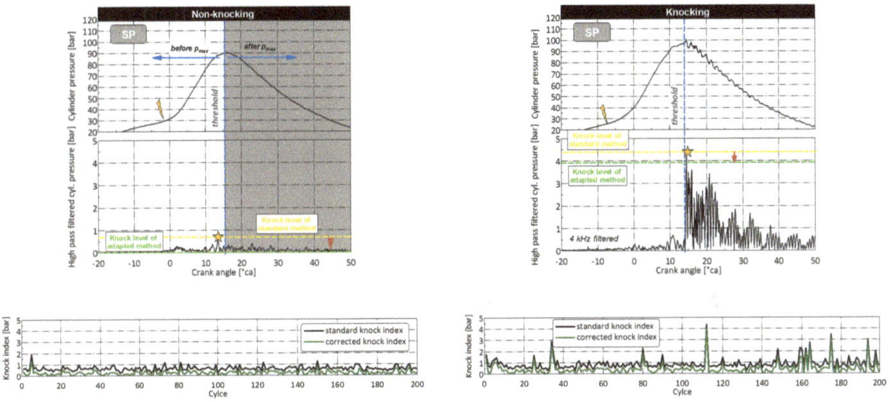

Fig. 12. Pressure trace and high pass filtered pressure trace for a non-knocking & knocking working cycle with conventional SP

For analysis purposes it is therefore essential to correct for the non-knock related pressure fluctuations in the pressure signal to achieve good comparability between the knock limits of different PCSP samples. The calculation rule can be found in Eq. 2 and 3.

$$KPPK_{PCSP} = KPPK_{after\ threshhold} - KPPK_{before\ threshold}$$

$$Threshold = AP_{max} + 5^{\circ}ca$$

The high pass filtered pressure trace is divided into two sectors. In each sector the maximum value in the signal is determined. The corrected knock index is the difference between the maximum value before and after the maximum cylinder pressure.

To compare the quality of the knock limit of different PCSP samples, it is useful to evaluate a statistical value of the new knock index, namely the 97th percentile of at least 1000 measured working cycles for the engine.

4.2 Knock Detection via Knock Sensor and Knock Control

Model Based Knock Detection (MBKD) was developed by Bosch as knock detection method with the best detection performance and is used for engines with PCSP.

This method is based on filtering the structure-borne noise signal in the frequency domain (see Fig. 13).

A normalization of the knock characteristic to a reference level (base noise characteristic of the engine) is used to calculate the knock intensity of each combustion. A comparison with an adjustable knock detection threshold is then used to decide whether this combustion is knocking or not.

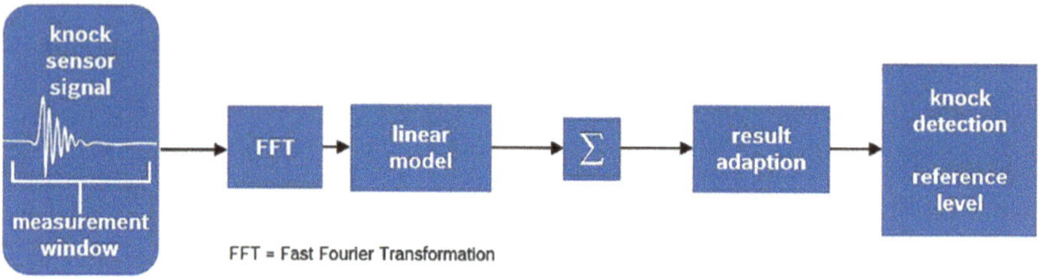

Fig. 13. Flow chart of the Bosch Model Based Knock Detection

The MBKD results in improved knock detection quality, reduced knock control intervention, increased generalizability, and increased robustness against electrical and mechanical interference. Therefore, this knock detection method provides several benefits for the detection of knock events in a PCSP engine.

The interfering noise due to the flame jets caused by a pre-chamber ignition is challenging the knock detection. In Fig. 14, the comparison between a non-knocking combustion with interfering noise and a knocking combustion with interfering noise is given for the cylinder pressure, as well as the knock sensor signal. The time-frequency-analysis is given for each signal below the respective one. The pressure signal is shown unprocessed (blue) and filtered (gray) with a 4 to 45 [kHz] bandpass.

The time-frequency analysis of the non-knocking combustion shows the stimulation of the resonance frequency of the combustion chamber by the pre-chamber jets in the cylinder pressure and knock sensor signal. This resonance frequency and its modes, which are typically used for knock detection, are contaminated by disturbing noise and require advanced signal processing to separate knocking from non-knocking combustion.

Additional functional enhancements, such as a wider evaluated frequency range and more accurate modeling of basic noise dependencies and noise frequency characteristics, improve detection quality for PCSP engines.

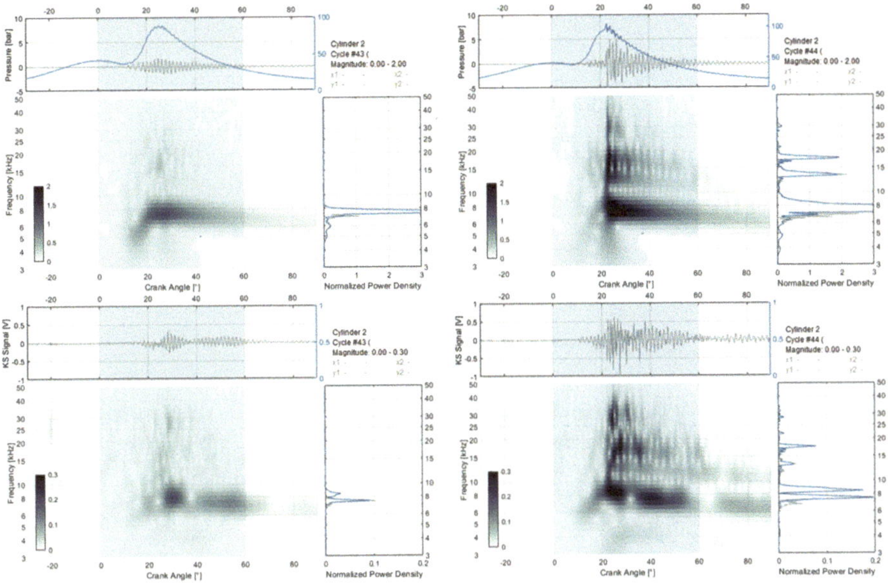

Fig. 14. Comparison of non-knocking (left) and knocking combustion (right) with PCSP

5 Application of PCSP: Benefits

Passive pre-chamber spark plug technology is primarily a knock reduction measure, so the benefits of using PCSP are seen at those engine operating points where knock is a limiting factor.

5.1 Knock Reduction Potential in Knock limited Operating Points

Figure 15 shows the result of an ignition angle sweep at an operating point of $n = 4000\ min^{-1}$ & IMEP = 16 bar. Four spark plug samples are compared:

- conventional J-gap spark plug as a reference,
- symmetrical PCSP design
- asymmetrical PCSP design
- asymmetrical PCSP with increased chamber volume.

Comparison of the knock index for all four spark plug samples shows that, as expected, the largest PCSP-chamber volume results in the most advanced MFB50% shift compared to the conventional J-gap spark plug. The largest PCSP-chamber volume correlates with the highest ignition energy and therefore the largest flame jets exiting the PCSP. This is also reflected in the short combustion duration and high cylinder pressure for a given MFB50%.

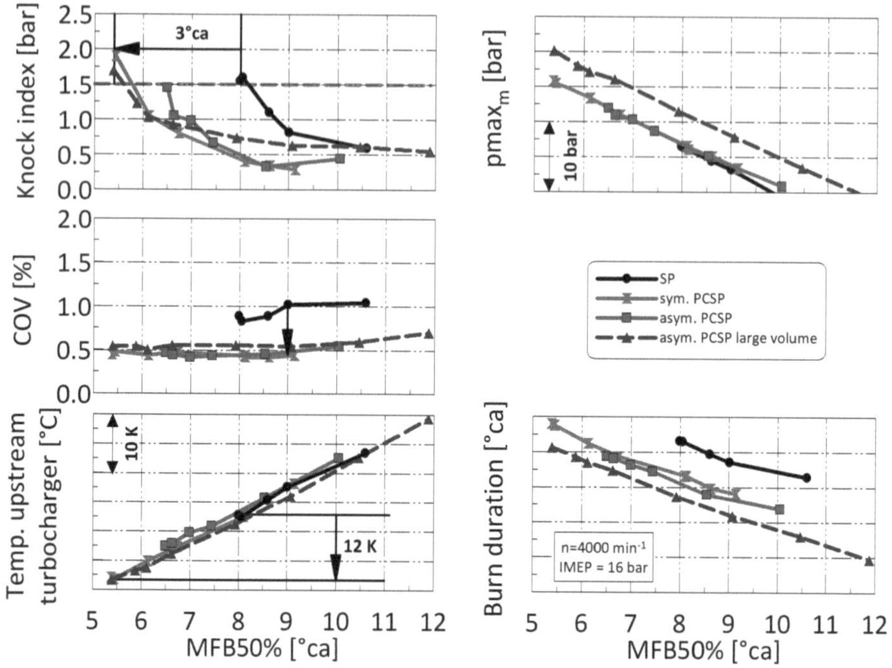

Fig. 15. Comparison of different PCSP design at a knock limited operating point

A comparison of the two PCSP samples with smaller chamber volume shows a slight advantage for the symmetrical design over the asymmetrical design. In this case, this is due to the fact that the "flame targeting" of the symmetrical design is more beneficial than the asymmetrical design. The asymmetrical hole design is mainly focused on improving the purging process of the PCSP. This slightly reduces the knock reduction capability, but will be beneficial in PCSP critical areas of operation.

A major benefit common to all PCSP samples is improved combustion stability, with the COV of the torque output significantly reduced by a factor of two.

Summary of results: PCSP offers advantages in knock limited operating points by reducing engine knock. In addition, combustion speed is increased, and combustion stability is significantly improved. Increased PCSP-chamber volumes are advantageous at these operating points and a symmetrical PCSP design has no disadvantage compared to an asymmetrical design. All variants are designed to avoid abnormal ignition events such as pre-ignition occurring inside the pre-chamber.

5.2 $\lambda=1$ Power Increase in Operating Points Limited by Exhaust Temperatures

With fuel enrichment for component protection no longer applicable under the forthcoming emission regulations in the European, US and Chinese markets, one of the key features of a pre-chamber spark plug is its ability to increase maximum engine power under stoichiometric operation.

With a PCSP, the exhaust gas temperature is affected by three different factors. Firstly, the combustion phase can be shifted earlier by the knock suppression effect of the PCSP. Second, combustion stability is increased. There is less variation in the combustion phase, which reduces the proportion of work cycles with retarded MFB50% values, resulting in lower exhaust temperatures. Finally, the third effect is the increased combustion speed and hence increased cylinder pressure, which leads to higher wall heat transfer and exhaust gas temperature reduction due to increased heat flow to the coolant system instead of the exhaust system.

PCSP technology offers a λ=1 power increase of 8 to 15 kW/l depending on the functional goals of the target engine and on the respective PCSP design.

The PCSP samples presented to date were tested on engines with a specific power output > 130 kW/l. Despite the high specific power outputs, no abnormal combustion events occurred due to the "cold" design of the pre-chamber spark plugs.

5.3 Part-load Engine Operation Using PCSP

As discussed in the previous section, the benefit of a PCSP concentrates in the knock limited operating points. Therefore, target of PCSP component development is to achieve the same part-load performance like a J-gap spark plug. Also, fuel consumption and combustion limits (lowest possible air charge and latest possible ignition angle) of the conventional spark plug are benchmark for engine operation with PCSP. Figure 16 shows a variation of intake valve timing at an operating point of n = 2000 min^{-1} and IMEP =4 bar under warm engine conditions.

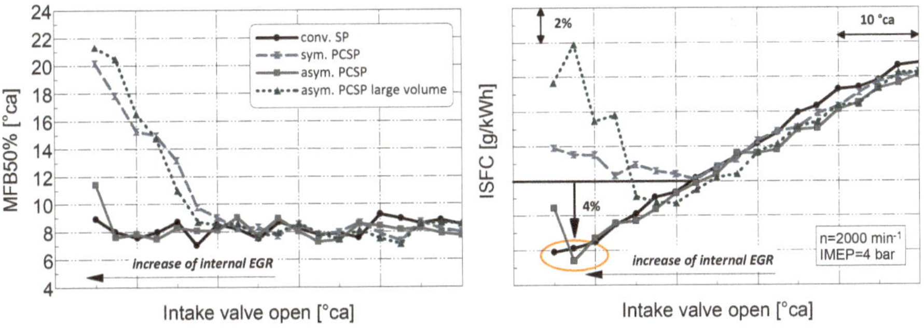

Fig. 16. Intake valve timing variation

With advanced valve timings, the internal residual gas fraction increases, which subsequently leads to higher required manifold pressures and thus de-throttling of the engine. Therefore, earlier intake valve timings reduce specific fuel consumption at part load. Figure 17 shows significant differences in the behavior of different PCSP variants. The asymmetrical PCSP can operate at the same valve timings as the conventional spark plug and therefore has no fuel consumption penalty. The symmetrical PCSP sample and the sample with increased PCSP-chamber volume reach the stability limit (MFB50% = 8°ca) at a less advanced intake valve timing and therefore show an ISFC disadvantage of ~4% at this operating point.

The symmetrical PCSP variant is not designed to purge the residual gas. The high PCSP-chamber volume design suffers from the much larger volume to be purged. Therefore, the residual gas content inside the pre-chamber is always higher for these two designs than for the low volume asymmetrical PCSP. This results in less EGR tolerance at part-load conditions.

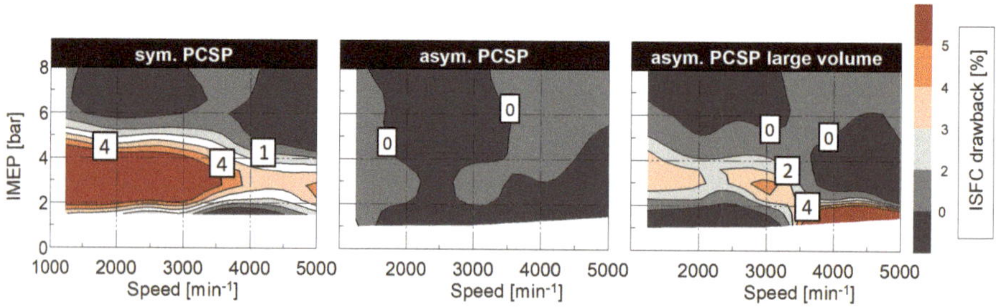

Fig. 17. ISFC drawback PCSP to conventional SP at part-load

In summary, Fig. 17 shows the difference in ISFC of each pre-chamber spark plug compared to the conventional J-gap spark plug. The advantages of the asymmetrical PCSP design, with its focus on chamber volume purging, are evident as there is no ISFC drawback over the entire operating map. The symmetrical PCSP shows ISFC losses of up to 4% at loads lower than IMEP = 4 bar.

5.4 Behavior of PCSP in Combination with Cooled External EGR

As already presented in [2], one measure to further improve the overall efficiency of an engine concept is the use of cooled external EGR. Figure 18 shows a sweep of the external EGR rate at a BSFC optimum operating point. At lower EGR rates, earlier combustion phasing is possible due to the knock reduction effect of the PCSP. Therefore, at lower EGR rates, fuel consumption with PCSP is lower than with a conventional spark plug. The minimum fuel consumption is reached at the same EGR rate and both types of spark plug still provide stable ignition. Subsequently, the EGR rate can be increased, but without additional optimization no further efficiency can be achieved due to the disproportionate increase in back pressure caused by the turbocharger. With an optimized valve timing strategy, the EGR rate can be further increased and the BSFC further improved. This is only possible with the asymmetrical PCSP, as it provides a more robust ignition than the conventional spark plug, which reaches its operating limit earlier.

Figure 18 also shows the behavior of the symmetrical PCSP. Since no residual gas is purged out of the PCSP by design, the external EGR tolerance is lower, very similar to the behavior at part-load.

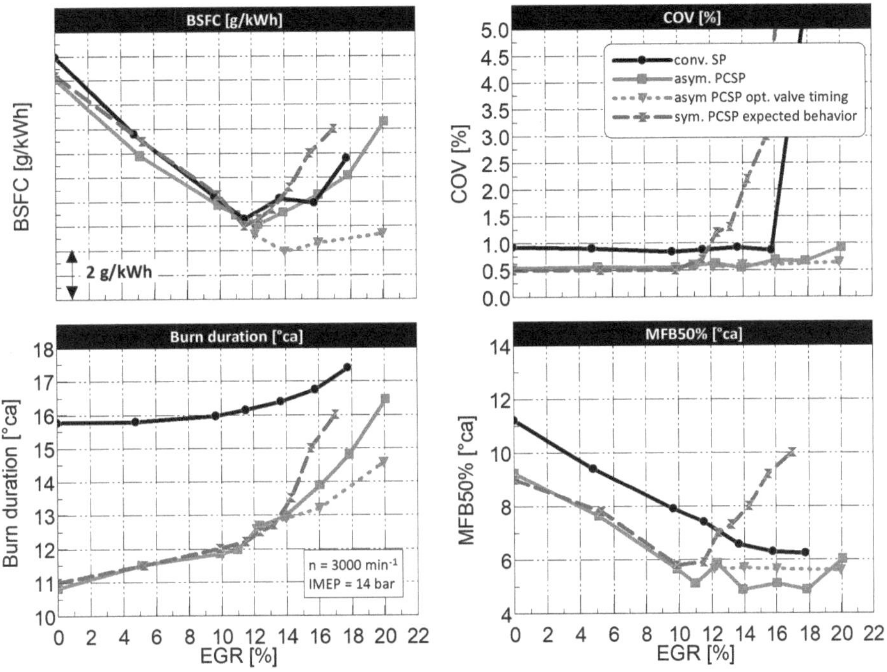

Fig. 18. EGR-Tolerance of PCSP

6 Application of PCSP: Challenges

In addition to the benefits outlined in the previous section, there are also challenges to be considered. For each engine and powertrain concept, there are three operating modes that need to be carefully considered to determine if the use of a PCSP as a single ignition source is feasible. These are idling or negative engine torque, catalyst heating and engine start at cold temperatures. All three operating modes can be solved by implementing a dual spark concept. Hybrid powertrains with high degree of electrification offer many options for handling these operating modes with a single PCSP combustion system, but must be evaluated individually for each engine/powertrain concept. At this stage, the use of PCSP as a single ignition source in conventional ICE powertrains does not appear to be practical.

6.1 Catalyst Heating

The combination of low engine load and retarded ignition timing is challenging for a PCSP concept. During catalyst heating, the ignition angle is retarded in order to increase the exhaust gas temperature and thus the enthalpy flux to the catalyst. The later the ignition angle with sufficient combustion stability, the faster the catalyst heats up. Potent catalyst heating performance is essential, especially for future, even stricter emission regulations. Figure 19 shows a first evaluation of the four different spark plug samples at a typical catalyst heating operating point of n = 1200 min^{-1} and IMEP = 3 bar at stationary engine operation and $T_{coolant}$ = 30 °C.

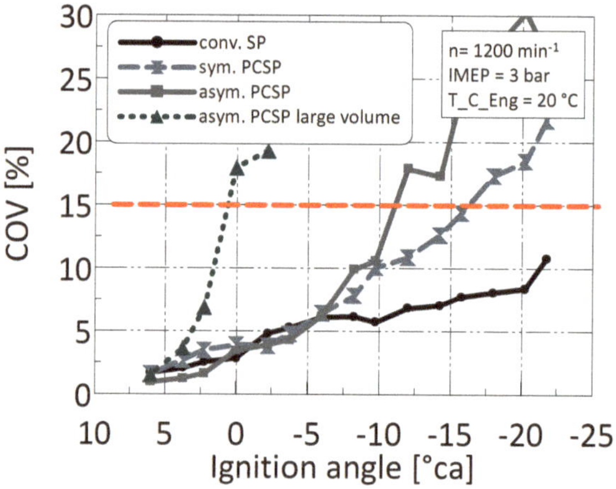

Fig. 19. Assessment of catalyst heating: Ignition angle sweeps

With a conventional J-gap spark plug ignition angle can be retarded to ign = −22 °C until the temperature limit of the turbocharger is reached, but combustion is still sufficiently stable. In comparison none of the investigated PCSP samples is able to achieve a comparable robust combustion stability with COV below 15%. Symmetrical PCSP samples offer slight advantages over asymmetrical samples as the symmetrical design inherently leads to higher turbulence levels within the chamber volume at late ignition angles. This slightly improves combustion stability at late ignition angles but is not sufficient to achieve comparable performance to conventional spark plugs.

Samples with increased chamber volume show even less tolerance to ignition angle retardation.

To improve the behavior in catalyst heating, further adaptation of the PCSP design could be one way to go. Keeping in mind that improvements at low load conditions already led to a compromise regarding maximum high-load performance, other possibilities beside component design changes have to be considered also. In a two-step approach first the calibration data mainly of injection parameters are optimized to the requirements of the PCSP ignited combustion. This is done not only in the base idling catalyst heating operation point but in a wider range of the engine map. Idea is to check if an increase in engine speed and/or engine load may help to stabilize the combustion with PCSP. If a heat-up and emission behavior comparable to J-gap spark plug could be achieved by changing the operation point, this might be an option for hybrid powertrains.

Figure 20 shows the results of the catalyst heating optimization using PCSP. Transient catalyst heating tests were performed on the engine dynamometer. Before starting the engine, it was completely cooled down to ambient temperature. The transient catalyst heating was measured while the engine was running at different speed/load combinations. The best calibration data at each operating point was first evaluated with steady-state engine tests and then used for the transient heat-up tests. Both engine out and post-catalyst emissions were measured to determine the cumulative emissions of the heat-up process. Figure 20 shows that a shift in the operating point to slightly higher engine speed and load allows a PCSP catalyst heating performance comparable to a J-gap spark plug. Beside identical heat-up process also the engineering target for NO_x and CO emissions can be maintained.

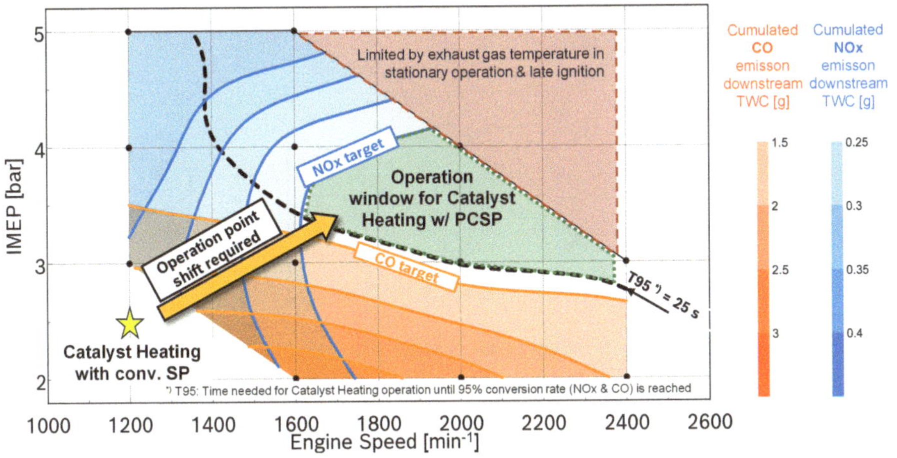

Fig. 20. Optimization of catalyst heating performance with PCSP

Generally, this operating point shift seems to be possible in dedicated hybrid powertrains. However, careful evaluation is required on a case-by-case basis.

Summary of results: Catalyst heating with symmetrical or asymmetrical PCSP design is feasible in Hybrid powertrains with high degree of electrification if the catalyst heating operating point can be shifted. In this case, comparable emission performance between conventional and pre-chamber spark plugs can be achieved at this new operating point. If this operating point shift is not applicable, a dual spark concept remains a possible solution. This is also the case for pre-chamber spark plugs with strongly increased chamber volume.

6.2 Engine Cold Start with PCSP

Another challenge for the PCSP is the engine start at cold ambient conditions. Even though the PCSP approach is connected to a Hybrid powertrain or a dual spark concept, the target should be to achieve a comparable engine startup performance with PCSP as for J-gap spark plug.

Optimization of Engine Start with PCSP

The behavior of an engine is always an interaction of hardware and calibration, so when changing the hardware (e.g. the spark plug), the calibration has to be adapted accordingly. Calibration parameters which need to be checked and optimized for the engine start with PCSP are ignition angles, mixture (fuel quantity in startup) and injection strategy (number of injections, injection angles, fuel mass split). To evaluate the sensitivity of the calibration parameters and to detect an improved calibration setting, engine starts are performed on engine test bench with engine start temperature of ~20 °C. The first and quite simple evaluation criteria for the engine start behavior is the time the engine needs to achieve 1000 min^{-1}. For the standard spark plug, a value of 0.6 s is measured as the baseline. Based on the standard calibration for J-gap spark plug consisting of a triple injection strategy with a late end of injection in the compression stroke, test results show quite fast that this strategy leads to very poor behavior for PCSP, see Fig. 21.

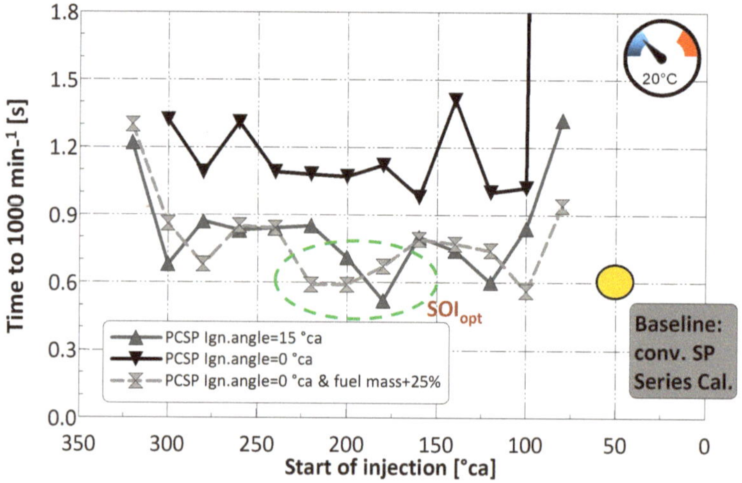

Fig. 21. Optimization of start calibration with PCSP

The injection strategy needs to be adjusted to reduce the amount of fuel injected late in the compression stroke. In addition, the ignition angle should be shifted towards earlier ignition. Changing the calibration parameters resulted in an engine starting time comparable to that of a conventional J-gap spark plug with its corresponding start calibration.

Cold Start Repeatability Test

To assess the robustness of a PCSP under cold start conditions, a cold start repeatability test is executed at the engine test bench.

The test profile is shown in Fig. 22—the engine and ambient conditions are set to $T_{amb} = -10$ °C. After engine start, acceleration to v = 50 km/h with gear changes up to third gear is simulated before the engine is switched off and cooled down for the next repetition. The test is passed if 25 repetitions can be performed without engine start performance loss. All tests are executed with the corresponding calibration setting, i.e. series calibration for the J-Gap spark plug and adapted calibration for the PCSP.

In Fig. 22 the engine speed profiles of all 25 engine start repetitions for the conventional J-gap spark plug and the pre-chamber spark plug are displayed, showing that both types of spark plug passed the test.

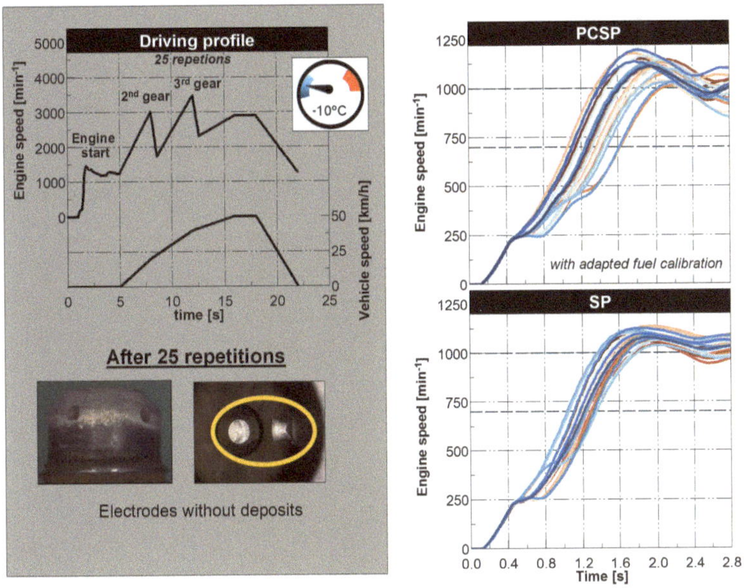

Fig. 22. Cold start repeatability test

Although the variation in engine rev-up is less for the conventional J-gap spark plug, the PCSP still achieves sufficiently robust behavior for this first test.

Optical evaluation of the PCSP sample and PCSP electrodes after the tests shows that there are no deposits, confirming positive the test results.

Low Temperature Cold Engine Start

The most challenging task is engine start at temperatures below $T_{amb} = -30$ °C. In order to evaluate the behavior of Bosch pre-chamber spark plug samples under these conditions, vehicle roller tests were performed in a cold box without any changes to the engine or engine calibration. A low vapor pressure fuel was used for these tests to cover worst case conditions.

Although knowing that parameter optimization does have a major impact to the engine start behavior with PCSP, sufficient time for adaption of the calibration at vehicle at −30 °C could not be provided due to test bench capacity. Therefore, as a first worst case test all measurements are done with series calibration without any adaptation.

However, the PCSP-equipped engine requires several seconds of cranking, the engine finally revs-up and the start-up gradient is almost comparable to the series calibrated spark plug (Fig. 23). This very positive result encourages to use additional effort for a parameter optimization also at −30 °C as one of the next steps.

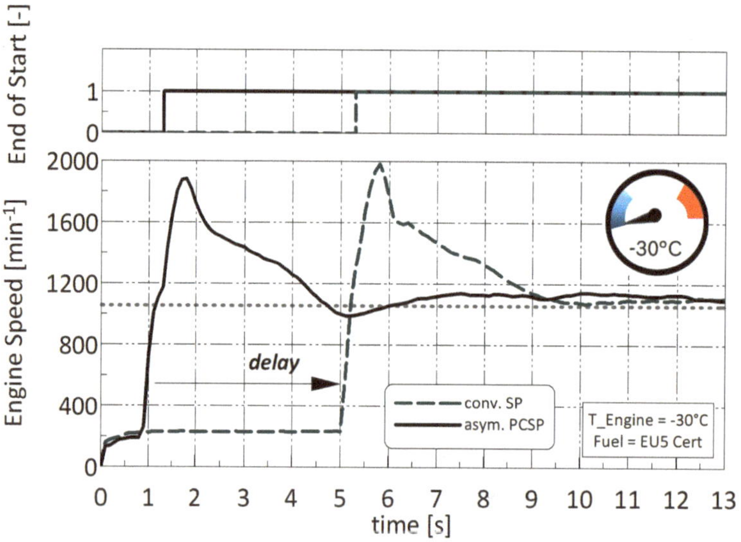

Fig. 23. Engine start in vehicle at T_{amb} = −30 °C using series calibration

7 Dimensioning PCSP Service Lifetime

Compared to other pre-chamber approaches the main advantage of the Bosch PCSP is that it can be completely removed from the cylinder head. Hence, component handling of a PCSP is identical to a conventional spark plug.

Therefore, the pre-chamber spark plug can be designed with a specific service lifetime interval, whereas other pre-chamber concepts must be designed for the entire lifetime of the engine. This allows for a cost-effective design. Extensive endurance tests have been conducted to model the wear behavior of the electrodes and the cap with holes. This enables a targeted design of the electrodes to keep the use of precious metals as low as possible and thus to keep the cost of the component as low as possible.

7.1 Overview Endurance Test Cycle

The test cycle used for the durability test is derived from a real vehicle driving profile. This profile is shown in Fig. 24. The maximum engine speed in this cycle is $n = 6100\ \text{min}^{-1}$ and the maximum load is IMEP = 29 bar. The vehicle test was again transferred to the engine dynamometer for efficiency reasons. The durability test was performed for several samples with different configurations of the electrode design. For each variant, the test was run up to 400 h to be able to quantify the electrode wear over a sufficient engine running time. After every 100 h, computed tomography images of the component were taken to evaluate the electrode wear and to collect model training data. In addition, the peak ignition voltage was measured continuously throughout the test and analyzed at recurring operating points. This allows continuous monitoring of electrode wear in real time.

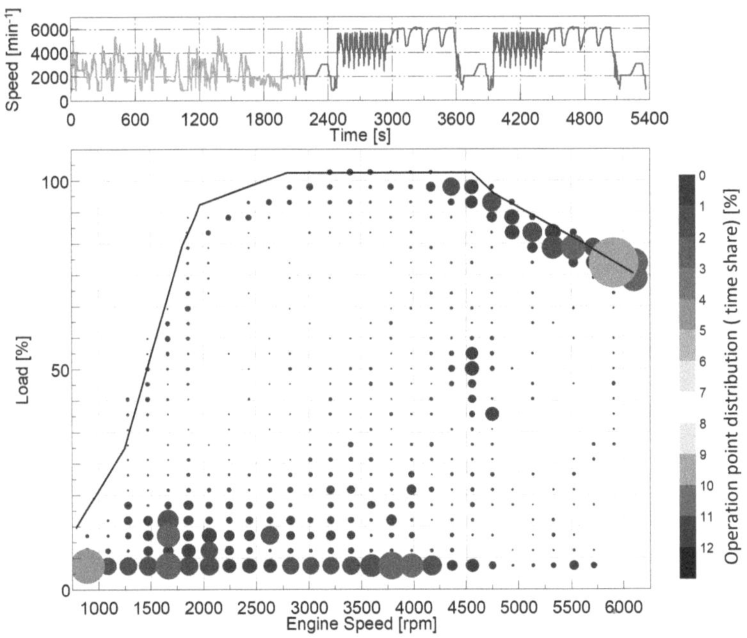

Fig. 24. Endurance test cycle used for service life evaluation

7.2 PCSP Endurance Cycle Test Results

Figure 25 illustrates the results of an endurance test on two distinct PCSP samples employing different electrodes. The graph depicts the ignition voltage over time, alongside images of the electrode before and after the test.

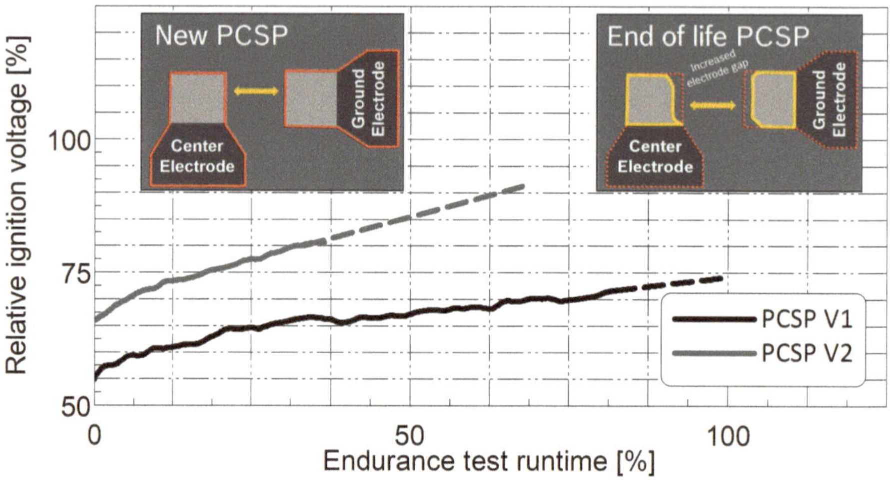

Fig. 25. Ignition voltage during endurance cycle test and schematic CT-images of two PCSP samples

The variations among the test samples were due to differences in the initial electrode gap and the setup of the precious metal on the center and ground electrode pins. The disparity in electrode gap can be detected by the difference in ignition voltage level, while the variation in electrode design can be identified by the gradient of the ignition voltage over run-time.

Based on the results of an elaborate and costly multi-sample test, an electrode wear model for pre-chamber spark plugs has been established, which will allow the electrode concept for future customer designs to be designed without the need for initial endurance testing.

8 Industrialization of Pre-chamber Spark Plugs

The fundamental challenge in the industrialization of a complex product such as the PCSP is to find the optimum balance between function, manufacturability, and costs.

In order to counter this area of conflict, a maximum of "standardized components" from the existing spark plug production are used within the product development. The base spark plug is therefore almost identical to a modern J-gap spark plug from the Bosch portfolio.

The first step in realizing the new functionalities is to define suitable materials (see Fig. 26). The selected materials form an optimum of functional fulfillment (e.g. heat conduction, oxidation resistance, etc.) and manufacturability (formability, machinability, weldability, etc.). Possible manufacturing processes were then evaluated in order to demonstrate the functionality. The final design is thus an iterative derivation from the interaction of the functional requirements in engine operation and the process development with the requirements to manufacture these features cost-effectively and robustly. In the following, the process is explained handling the example of the PCSP cap (1) and the new concept of the lateral ground electrode (2).

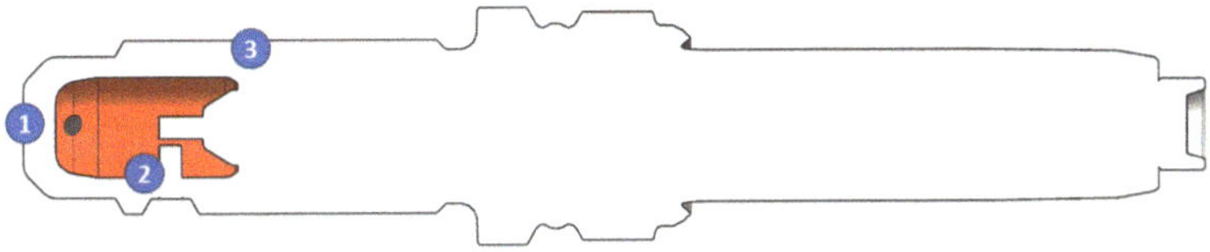

Fig. 26. Schematic diagram of Bosch pre-chamber spark plug with components: cap (1), ground electrode (2) and housing (3)

In the case of the PCSP cap (1), the main challenge is the choice of material. The basic requirement for the production process here is to realize a customer-specific, complex hole pattern with variable orientation in the engine—usually using a machining process. At the same time, the material must be able to withstand high engine loads (temperature, combustion gas, etc.) and still be reliably joined to the joining partner of the base spark plug. Within these limits, only a limited combination of materials is possible, which generally only allows a compromise for the processes. In order to implement this compromise on an industrial scale, a high degree of expertise is required in the production of the components. Using the example of the PCSP cap, both the drilling process and the joining process must therefore be fully exploited within their limits in order to manufacture the product in the required quality.

The example of the lateral arrangement of the ground electrode (2) is a good way of illustrating the complex influence of two competing design and functional requirements on the selection of a suitable process chain. The electrode gap of the PCSP must be set in a novel arrangement with high precision and flexibility. At the same time, the electrode gap may only change after production due to regular wear. So the electrode must be fixed robustly. A classic functional separation of these requirements was carried out along the process chain and in the design. The requirements were separated into the steps of positioning and fixing the electrode. In order to limit the cost increase that is automatically associated with such a separation of processes (higher complexity, more machine use, etc.), the complexity in the design and the individual process had to be reduced to a minimum. Here, too, an attempt was made to encounter this challenge with the highest possible number of standard processes or standard components. In addition, the development of robust processes (e.g. with regard to the adjustability of the electrode gap) is absolutely essential, as parts not in specification at the end of the value stream are always associated with correspondingly high costs.

Conclusion: The accompanying process development for the production of a pre-chamber spark plug was completed as part of a platform development project in 2023. To this end, capable manufacturing concepts for the new interfaces cap, ground electrode and connection technology were developed and validated on a small series scale. A transferability for a series application with precise specifications for the production technology is available, so that the production of pre-chamber spark plugs on a large volume scale is prepared.

9 Summary and Outlook

Bosch conducted a platform development project of a pre-chamber spark plug suitable for high volume production. The project was successfully completed by the end of 2023, enabling series production to commence from mid of 2026 in collaboration with our future project partners.

The paper demonstrates the development process, beginning with the design of the pre-chamber spark plug, and leading to a possible implementation in mass production. A variant frame was established, comprised of symmetrical and asymmetrical PCSPs, featuring a variable pre-chamber volume.

The benefits of using pre-chamber spark plug technology include reduced engine knock, improved combustion stability, increased $\lambda=1$ performance, and high external EGR tolerance. The effects are explained and have been confirmed by engine test results.

Challenging aspects of engine operation with PCSP like catalyst heating, low-load operation and engine cold start have also been detected and possible solutions are discussed and verified by engine measurements.

Bosch pre-chamber spark plug technology is an attractive measure for increased efficiency, e.g. in dedicated hybrid engines. It can be implemented as single (PCSP only) or dual-spark (PCSP + conventional SP) concept. For a successful application, the implementation of a pre-chamber spark plug requires a complete system approach that includes accompanying measures, e.g. a model based knock detection. Bosch PCSP technology will be available on a mass production scale from mid-2026.

References

1. M. Blankmeister, M. Alp, E. Shimizu: Passive Pre-Chamber Spark Plug for Future Gasoline Combustion Systems with Direct Injection. In 4th International Conference Ignition Systems for Gasoline Engines, Berlin (2018)
2. M. Krüger, J. Schwarzer, E. Schünemann, J: Tophoven, J. Gömmel, V. Neubert: Dedicated Hybrid Powertrains for Passenger Car Applications: Evaluation of Gasoline Engine Concepts and Powertrain Topologies. In: 44th International Vienna Motor Symposium, pp. 2023–13, Vienna (2023).

X-in-the-Loop and Reinforcement Learning for Emission Control Development

Mario Picerno, Sung-Yong Lee(✉), und Jakob Andert

Teaching and Research Area Mechatronics in Mobile Propulsion, RWTH, Aachen University, Aachen, Germany
{picerno_ma,lee_sun,andert}@mmp.rwth-aachen.de

1 Introduction

The advent of stringent vehicle emission regulations, aimed at reducing pollutants emissions and protecting the ecosystem, necessitates investments in innovative powertrain technologies. These technologies not only cater to consumer demands but also aid in cost reduction. However, the consequent complexity of modern vehicle controllers often leads to less-than-optimal solutions, and challenges skilled experts, who struggle to fully understand the interactions between the myriad of components of these systems. X-in-the-Loop (XiL) methods with progressively reduced degrees of virtualization are ideal for supporting vehicle development. In particular, the utilization of HiL simulations is common for the virtual calibration of Engine Control Units (ECUs), by providing an environment with reproducibility, automation and safety [1]. XiL can therefore function as a resilient and economical tool for the development of RL-based control functions, facilitating the transition to practical implementations. In response, the automation of control development, aided by Machine Learning (ML) techniques, emerges as a promising solution [2]. Among them, Reinforcement Learning (RL) being such a data-driven approach, devises strategies for complex control problems with minimal input from system experts [3].

The use of RL in automotive applications has been gaining momentum by providing innovative solutions to complex problems [4–8]. In real-world RL applications, several obstacles are recognized, driven by factors such as latency, transparency, safety or data availability [9]. Although it is already a well-established practice in robotics to conduct sim-to-real experiments [10], the implementation of RL for powertrain control from simulations in the real world faces the limitations of modelling complex and non-stationary processes. This presents considerable challenges to a seamless transition to real systems. Here, Transfer Learning (TL) comes as an effective methodology, enabling knowledge to be transferred to a target domain in order to improve performance without requiring extensive additional training. This technique is

A. Heintzel (Hrsg.): Int.Engine.Congr. 2024, Proceedings, S. 219–230, 2025.
https://doi.org/10.1007/978-3-658-47966-4_16

implemented in RL with a simulation model representing the source domain and a real-world application as the target domain. These RL-based methods facilitate easier transitions through various stages of virtualization and adaptations to different system variants, significantly reducing the effort required in these processes.

Previous work focused on the development of RL-based powertrain control procedures in a fully virtualized environment [11], and validated the performances on a HiL simulator with real ECU and actuators [12]. Then, the methodology to transfer the training to the HiL was meticulously described [13]. This contribution validates the transfer procedure and the capabilities of additional RL agents between XiL stages through TL.

The framework is applied to the problem of Exhaust Gas Re-circulation (EGR) control, where the RL agent controls the two EGR actuators (high- and low-pressure) while prioritizing vehicle efficiency, nitrogen oxides (NO_x), and soot emissions reduction.

The structure of the paper is organized as follows: It begins with an introduction that provides foundational insights into RL. Subsequently, the method of applying the Proximal Policy Optimization (PPO) algorithm to EGR control is discussed. The concludes by comparing the proposed policy and traditional model-based ECU controllers. This comparison serves to evaluate the robustness, accuracy, and overall potential of the framework.

2 Fundamentals of Reinforcement Learning

Reinforcement Learning (RL) revolves around the concept of training an agent by enabling it to interact with an environment, which is characterized as a time-discrete Markov Decision Process (MDP). This MDP consists of a state space (S), action space (A), transition probability function (P), and reward function (R). In each time step, an agent observes a state $s_t \in S$ and selects an action $a_t \in A$ according to its policy π_θ: $S \times A \rightarrow [0, 1]$ with adjustable parameters θ (see Fig. 1). This results in the environment going into a different state $s_{t+1} \in S$ with a probability of $P(s_{t+1}|s_t, a_t)$. The reward $r_t = R(s_t, a_t, s_{t+1})$ corresponds to the quality of the agent's decision. A sequence of interactions from a starting state $s_0 \in S$ to some final state $s_T \in S$ is called a trajectory (τ). The objective is to iteratively update the policy parameters θ such that the return $R(\tau) = \sum_{t=0}^{T-1} \gamma^t r_t$ is maximized, with γ being the discount rate, for discounting future rewards. When the policy is implemented using an Artificial Neural Network (ANN), θ consists of the edge weights and biases of that network [3].

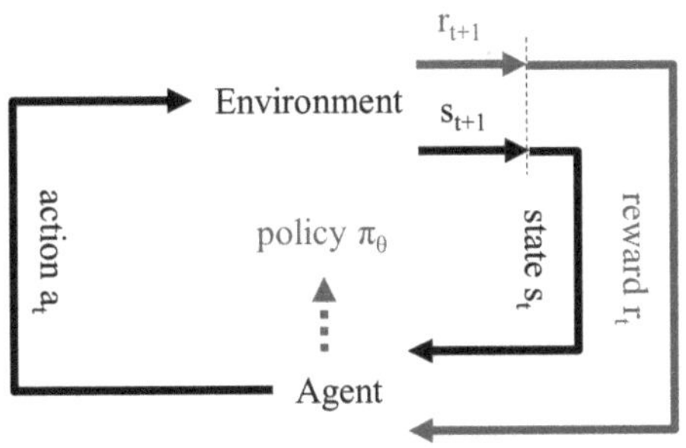

Fig. 1. The agent-environment interaction in an MDP.

The possession or development of an environment model by the agent is a critical determinant in the decision-making process of a RL algorithm. For the acquisition of optimal behaviors in uncertain environments, two fundamental theories can be applied: model-based and model-free algorithms. Unlike model-based RL, which uses predictions of the environment's behavior to determine control actions, model-free RL methods just rely on sampling experiences without an explicit model. As a result, they do not necessitate a pre-defined prediction model. In this study, the model-free PPO algorithm is utilized [14]. It has a small number of hyperparameters and is characterized by its robustness and stability throughout the training process. As an on-policy approach, it necessitates the production of training samples in accordance with the most recent policy of the agent. This causes a relatively lengthy training period and inefficient sampling, but it is offset by its stability and ease of tuning.

Transfer Learning (TL) in ML is focused on applying knowledge acquired from one ML problem to a different, yet related problem. Successful implementation of TL generally requires less data and time than training a model from scratch, often resulting in enhanced model performance [15]. In RL, TL has the capability to address some of the challenges related to training in real-world environments. A significant application is in simulations, where a large volume of data can be safely gathered and used to train RL agents for subsequent deployment in the real-world scenarios. TL can be classified into different categories based on the amount of training samples needed in the target domain. These categories include zero-shot and few-shot transfer, which require no or very few target domain samples, and sample-efficient transfer, which requires a significant number of target domain samples [16].

3 Implementation of RL in X-in-the-Loop

The following : provides an analysis of the X-in-the-Loop (XiL) configuration, with a particular emphasis on the XiL-based simulation framework. Additionally, it delineates the protocols for integrating the RL agent and conducting its training.

3.1 Set-Up of Reinforcement Learning-in-the-Loop (RiL)

Both RL and XiL simulations require interactions with models or physical components that are both accurate and computationally efficient. The integration of real hardware facilitates the adaptation of RL agents to real-world scenarios by decreasing the discrepancy between simulation and real-world conditions. Hence, the XiL-based incremental advancement through virtualization stages could coincide with the transmission of the agent's policy between various environments.

The application of RL to a specific control task still necessitates a certain degree of domain knowledge. The developer is required to define the MDP through the design of a reward function, actions, and a meaningful set of observations. The integration of RL in an XiL framework requires the design of an input/output (IO) interface through which state space, rewards and actions can be transmitted to both the virtual controller in MiL and the ECU in the HiL test bench (see Fig. 2). The state space, as defined by specialists in the field, must include all pertinent information that is essential for addressing the particular use case. Therefore, the IO interface is intentionally structured to mirror the behavior of a control function that the agent is tasked with emulating in the target domain, more precisely within the ECU. This allows for the seamless transfer of the policy that has been developed in the MiL environment to the HiL for the purposes of validation, additional training, and potentially to the vehicle. Nonetheless, it is important to acknowledge that the quality and duration of training are directly influenced by the manner in which an agent's IO interface is defined.

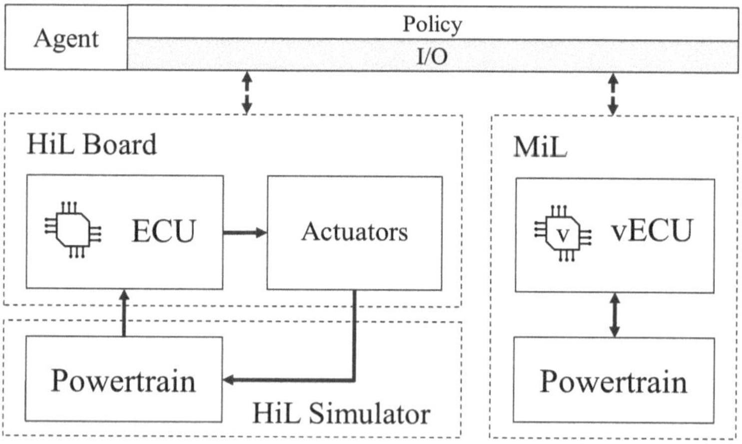

Fig. 2. Global simulation frameworks for air path control in MiL and HiL environments.

The powertrain models, which are utilized in both MiL and HiL simulations, include driver, vehicle, transmission, aftertreatment system, and engine models. A physics-based Mean Value Engine Model (MVEM) simulates the reference engine. The physics-based technique integrates calibrated functions with physical equations, hence minimizing calibration tasks to only the maps. The same approach is followed for the semi-physical modeling technique, employed to predict NO_x and soot emissions within the cylinders [17].

The metrics considered for the EGR control problem are engine speed, pressure, accelerator pedal position, and the temporal deviations over two consequent time steps of these variables, to discern the dynamics of the process. Furthermore, factors including gear selection, coolant temperature, and vehicle velocity are considered. Min-max normalization is performed on all chosen raw signals to the interval [–1, –1] in accordance with their reference operational ranges. In order to reduce simulation instability throughout the procedure, the agent controls the rate of valve opening and closing as opposed to directly controlling the percentage of opening. The hyperbolic tangent function adjusts the specified action to achieve the required velocity of the EGR valve, with a range of ±50% per second. The developed agent policies are represented by a fully-connected feedforward ANN comprising 3 hidden layers, with 16 nodes in each layer. The policy generates the action and its corresponding Standard Deviation (SD), for exploration and is also a measure for the training progress. Table 1 provides a summary of the IO interface with the agent.

Table 1. IO interface specification

Type	Signals
States	Engine speed
	Engine speed variation
	Boost pressure
	Boost pressure variation
	Desired boost pressure
	Desired boost pressure variation
	Boost pressure control error
	Accelerator pedal position
	Accelerator pedal position variation
	Vehicle speed
	High pressure EGR valve position
	Low pressure EGR valve position
	Gear position
	Engine coolant temperature
Reward	Boost pressure control error
	NO_x emissions
	Soot emissions
	Failure flag
Actions	High-pressure EGR valve velocity
	Low-pressure EGR valve velocity

The goal of the agent, as expressed in the reward r, is to optimize engine performance while minimizing emissions, specifically NO_x and soot emissions. The engine performance is ensured by penalizing the absolute difference (ΔB) that exists between the desired and measured boost pressure. The reward function is therefore described by

$$r = -f_{\mathrm{NOx}}\, m_{\mathrm{NOx}} - f_{\mathrm{soot}}\, m_{\mathrm{soot}} - f_{boost}|\Delta B| - F_{\mathrm{fail}} \quad (1)$$

where f_{NOx}, f_{soot}, and f_{boost} are the weighting factors of NO_x emissions, soot emissions and boost deviation, respectively. As the agent functions without any limitations, the reward function includes an indicator to consider model and software failures in the ECU (F_{fail}). Table 2 summarizes the list of the weighting factors for the reward calculation. The aforementioned weighting factors were acquired on the MiL platform through iterative training and fine-tuning.

Table 2. Weights of the reward function

Symbol	Weight description	Value
f_{NOx}	NO_x mass flow	0.155 s/g
f_{soot}	Soot mass flow	0.9 s/g
f_{boost}	Boost pressure control error	0.0016 1/kPa
F_{fail}	Failures penalization	20

3.2 Learning and Experiencing Cycle Interface: LExCI

The Learning and Experiencing Cycle Interface (LExCI) is a RL framework specifically made for training with embedded systems (see Fig. 3). With its Master-Minion architecture, it divides the training setup into the learning domain and the data generation domain which communicate with one another through TCP/IP and a custom protocol. The Master interfaces RLlib while the Minion is connected to the embedded device, the simulation models, and the physical components of the environment [18].

After the settings and hyperparameters of the training (e.g. the architecture of the ANN) have been defined, LExCI starts working in cycles: The Master fetches the current version of the policy from RLlib and sends it as well as other training-related parameters to the data generation domain. There, the Minion uses the updated information to override the default control function in the virtual controller in MiL, or ECU on the HiL. The Minion additionally sets various other, miscellaneous options through the API of the software interfacing the models. Within the HiL test bench, this software is the automation software capable of reading data from the ANN embedded in the plant models, and then writing data through the calibration software in the ECU. The Minion then proceeds to generate data and, once enough experiences have been saved, sends them back to the Master. The Master, in turn, passes them on to RLlib for updating the policy and the training cycle starts again.

When transferring to the HiL, the MiL-Minion is replaced with one that is connected to the HiL. Despite the fact that the plant model is identical in both virtualization levels, the agent performs the same task in environments that display variances, e.g. as a result of having physical actuators or differences in software between the real ECU and the virtual controller.

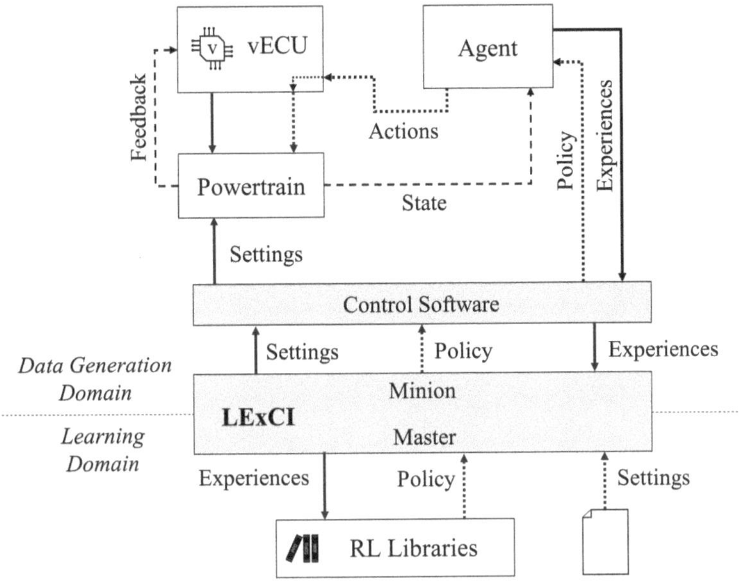

Fig. 3. LExCI framework for air path control in MiL and HiL platforms.

4 Use Case in Diesel Emission Control

This section introduces the MiL training results, the transfer of policies to the HiL platform, and the subsequent results obtained by comparing the agent to the ECU.

4.1 Model-in-the-Loop (MiL)

The MiL platform is a risk-free domain. It may develop mature policies within a few hours, depending on the chosen training segments and computing resources. The MiL environment functions as a platform for algorithm testing and hyperparameter adjustment, which would require a significant amount of time on the HiL where training cannot be accelerated. The training was conducted using the PPO algorithm with randomly selected segments of 300 seconds from a pool of available driving cycles such as WLTC, simulated on a quad-core Intel Xeon W2123 processing unit.

As a result of the specified training samples produced in each cycle, the training progress of PPO agents was rapid. Convergence time is defined as the quantity of training cycles required to bring the validation reward within 5% of its maximum value during training. Additional evidence of effective training can be observed in the transient performance of the agent, as measured by boost pressure deviation and pollutant emissions.

The scatter diagram in Fig. 4 illustrates the MiL validation performances of the complete training in relation to the reference virtual controller. One point corresponds to one out of every five agents among the 220 policies that were evaluated in total. The illustration of the cumulative engine-out NO_x and particulate emissions for each agent is provided. The average absolute boost error is proportional to the size of each circle.

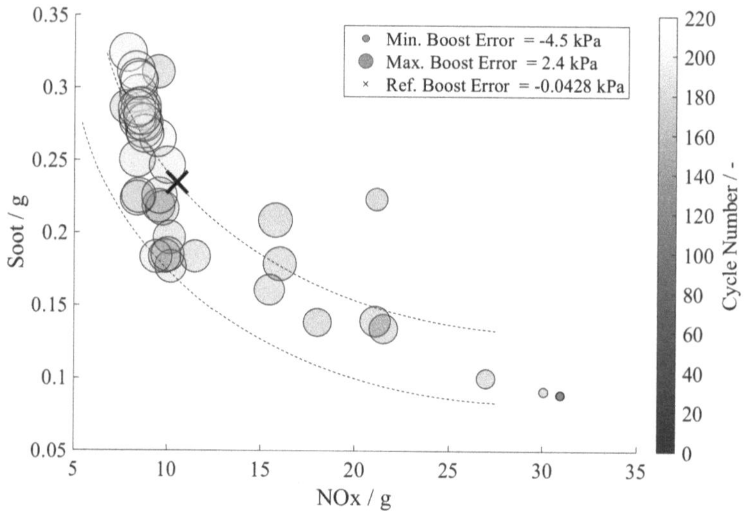

Fig. 4. Validation results for engine-out NO_x and soot emissions, and boost error for each agent.

Throughout the training process, the boost error attains its highest value of 2.4 kPa, whereas its lowest value is equivalent to the reference value of -4.5 kPa. During validation cycles, an increased boost pressure error is evident in the area of 8–9 g of NO_x. High boost error results in low NO_x emissions due to the oxygen concentrations produced by the low in-cylinder mass; however, this is at the expense of soot emissions and boost accuracy. As the training progresses, the boost pressure error is systematically reduced until the developed policies converge toward consistently higher NO_x values and lower soot, ultimately resulting in comparable soot and NO_x emissions to the reference. Typically, any cycle that occurs inside the two dashed lines potentially corresponds to a more effective emissions trade-off. The PPO algorithm exhibited significant stability, a critical factor in HiL setups where training iterations are time-intensive.

4.2 Transfer Learning to Hardware-in-the-Loop (HiL)

From the many available agents resulting in lower emissions than the reference virtual controller, one agent was selected for the HiL transfer. Nevertheless, the policy's validation on the HiL platform, without continued training, resulted in less effective emissions balancing compared to the MiL platform. This implies that the emission weights of the reward variables need to be calibrated before training the agent in the HiL environment. Employing a well-trained agent is advantageous. By utilizing the well-established policy of the agent, the reward elements can be optimized for the HiL. Thus, the emissions trade-off has been re-located to a position lower than the reference ECU.

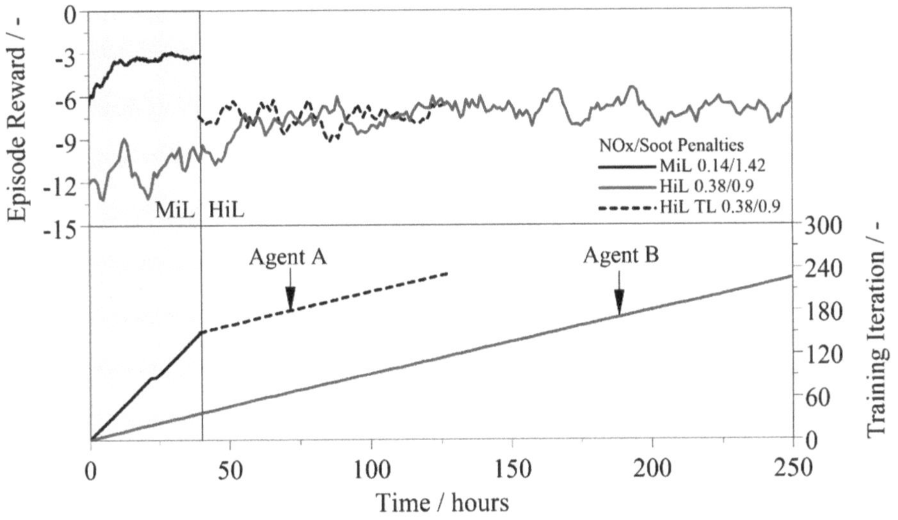

Fig. 5. Episode reward and training iteration number for the MiL, TL, and full HiL trainings based on the WLTC.

Figure 5 illustrates the reward obtained during an episode and the training iterations in function of the actual time for two scenarios: a complete HiL training and a training that first began in MiL and was then transferred to the HiL platform. The observed patterns in the mean episode rewards suggest that the agents taught on the HiL platform provide lower values. This is a direct result of the reduced weighting factors assigned to the rewards in the HiL training. Two agents are identified as the first agents achieving overall lower emissions compared to the ECU and with a deviation between target and actual vehicle speed being below 0.5%. Upon analyzing the training duration, the curve depicting the HiL-only training method reveals that the first optimal agent (Agent B) is achieved after 185 hours. On the other hand, the first optimal transferred agent can be observed after 70 hours (Agent A with approximately 30 hours of training in HiL), emphasizing the time saved in HiL operation. The strategy employed by Agent A is subsequently evaluated against the control strategy of the actual ECU on the HiL system to determine its effectiveness in governing physical actuators.

Figure 6 shows the normalized EGR position of the valves and the relevant emissions in function of the time for 200 seconds of a WLTC validation. The examination of the valves' positions reveals that, during the time interval of 1175 to 1300 seconds, the agent opts for an alternative variation of the high-pressure valve (HP-EGR) in comparison to the reference ECU. However, it follows a similar approach for the low-pressure valve (LP-EGR). In contrast, the actuation after 1300 seonds is comparable between the two valves, and both deviate from the reference controller. The different control of the valves, as opposed to the reference ECU calibration, did not result in any divergence from the required speed, especially during vehicle launch. The lower EGR rate in this segment resulted in reduced soot emissions overall, but it also led to increased nitrogen oxide (NO_x) emissions, as can be noted. Nevertheless, the total emissions at the end of the cycle exhibited a decrease of 3.2% in soot emissions and 2.2% in NO_x.

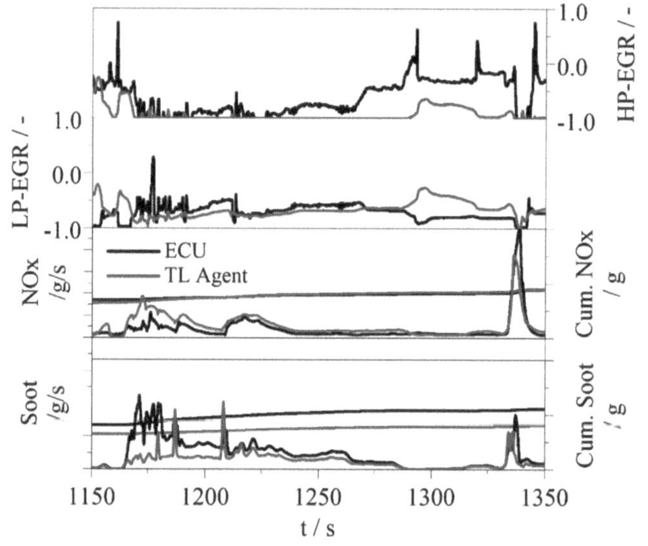

Fig. 6. Normalized EGR position of the valves and relevant emissions for 200 seconds of a WLTC validation for the selected agent and ECU.

5 Conclusion and Outlook

This study discussed the implementation of Reinforcement Learning (RL) into the XiL-based approach for developing powertrain controllers. Our technique utilizes a strong pipeline that involves training and validating RL agents in many domains. This is allowed by the LExCI framework, an open-source solution integrating various target systems, including state-of-the-art HiL systems, for RL-based control tasks.

By employing several layers of virtualization, we are able to transfer multi-action agents from simulated environments to real-world HiL situations. Specifically, the agents that are obtained via pre-training in MiL and fine-tuning on HiL with TL, surpass the performance of a series ECU. We have examined the benefits of TL on training time for EGR control. A HiL-trained agent was compared to a transferred one from MiL to HiL. While the first agent also outperformed the reference ECU in terms of reducing pollutant emissions, the results clearly show that the second agent, which incorporates MiL pre-training with TL, significantly speeds up the training process, while delivering a reduction in emissions of over 2%. The HiL system effectively served as a real-world environment for the agent, enabling it to manipulate real hardware and explore control strategies without he risk of causing engine damage or compromising driver safety in prototype vehicles.

Following studies will primarily concentrate on verifying the effectiveness of the suggested methodology through tests on a real vehicle. This presents additional complexities, such as the need for highly efficient training procedures, as the decreased capacity to automate training methods intensifies the challenge of producing training data, while also amplifying safety concerns. Furthermore, the utilization of the methodology in different applications, such as the management of hybrid or electric vehicles, can unlock the full potential of RL when combined with XiL simulations. This can lead to faster development and evaluation of intelligent embedded systems in larger and real-world settings. The policy acquired by RL agents can then function as a data-driven optimization resource, utilizing the knowledge to enhance existing software calibration. This is expected to generate cost-effective, and environmentally-friendly solutions in several sectors.

References

1. Lee, S-Y, Andert, J, Neumann, D et al.: Hardware-in-the-Loop-Based Virtual Calibration Approach to Meet Real Driving Emissions Requirements. In: SAE International400 Commonwealth Drive, Warrendale, PA, United States 2018 :SAE Technical Paper Series
2. Garg, P, Silvas, E, Willems, F: Potential of Machine Learning Methods for Robust Performance and Efficient Engine Control Development. IFAC-PapersOnLine 54, 189–195 2021
3. Sutton, RS, Barto, AG: Reinforcement learning: An introduction, 2nd ed 2018)
4. Shih, P, Kaul, BC, Jagannathan, S et al.: Reinforcement-learning-based output-feedback control of nonstrict nonlinear discrete-time systems with application to engine emission control. IEEE Trans Syst Man Cybern B Cybern 39, 1162–1179 2009
5. Shin Young Bae, Dong Hwi Jeong, Yeonsoo Kim et al.: Urea Injection Control Based on Deep-Q Networks for SCR Aftertreatment Systems
6. Text="CE:[CROSSREF]Score: 0.56crsid: xF6k0dtype: bibarticleAuthors: Bo Hu, Jie Yang, Jiaxi Li, Shuang Li, Haitao BaiYear: 2019JournalTitle: ProcessesJournalAbbTitle: ProcessesArticleTitle: Intelligent Control Strategy for Transient Response of a Variable Geometry Turbocharger System Based on Deep Reinforcement LearningVolumeID: 7IssueID: 9FirstPage: 601BibArticleDOI: 10.3390/pr7090601" Hu, Yang, Li et al.: Intelligent Control Strategy for Transient Response of a Variable Geometry Turbocharger System Based on Deep Reinforcement Learning. Processes 7, 601 2019

7. Ganesh, AH, Xu, B: A review of reinforcement learning based energy management systems for electrified powertrains: Progress, challenge, and potential solution. Renewable and Sustainable Energy Reviews 154, 111833 2022
8. Gaiselmann, G, Altenburg, S, Studer, S et al.: Deep reinforcement learning for gearshift controllers in automatic transmissions. Array 15, 100235 2022
9. Dulac-Arnold, G, Levine, N, Mankowitz, DJ et al.: Challenges of real-world reinforcement learning: definitions, benchmarks and analysis. Mach Learn 110, 2419–2468 2021
10. Zhao, W, Queralta, JP, Westerlund, T: Sim-to-Real Transfer in Deep Reinforcement Learning for Robotics: a Survey
11. Picerno, M, Koch, L, Badalian, K et al.: Turbocharger Control for Emission Reduction Based on Deep Reinforcement Learning. IFAC-PapersOnLine 56, 8266–8271 2023
12. Koch, L, Picerno, M, Badalian, K et al.: Automated function development for emission control with deep reinforcement learning. Engineering Applications of Artificial Intelligence 117, 105477 2023
13. Picerno, M, Koch, L, Badalian, K et al.: Transfer of Reinforcement Learning-Based Controllers from Model- to Hardware-in-the-Loop (25/10/2023)
14. Schulman, J, Wolski, F, Dhariwal, P et al.: Proximal Policy Optimization Algorithms (20/07/2017)
15. Zhuang, F, Qi, Z, Duan, K et al.: A Comprehensive Survey on Transfer Learning. Proc IEEE 109, 43–76 2021
16. Zhu, Z, Lin, K, Jain, AK et al.: Transfer Learning in Deep Reinforcement Learning: A Survey. IEEE Trans Pattern Anal Mach Intell 45, 13344–13362 2023
17. Lee, S-Y, Andert, J, Pischinger, S et al.: Scalable Mean Value Modeling for Real-Time Engine Simulations with Improved Consistency and Adaptability. In: SAE International400 Commonwealth Drive, Warrendale, PA, United States 2019 :SAE Technical Paper Series
18. Badalian, K, Koch, L, Brinkmann, T et al.: LExCI: A Framework for Reinforcement Learning with Embedded Systems (05/12/2023)

EPA'27, Tier 5, Euro 7—Divergent Ambitions Future Challenges and Technology Approaches for EU and US for Commercial Vehicles

Anton Arnberger(✉), Hannes Noll, Robert Gunja, und Hannes Wancura

AVL List GmbH, Graz, Austria
{anton.arnberger,hannes.noll,robert.gunja,
hannes.wancura}@avl.com

1 Introduction EU Pollutant Emission Legislation

The commercial engine emission legislation stages Euro I to Euro V were characterized by steadily aggravated nominal limits. The ESC (European steady state test cycle) ensured a proper emission compliance in steady state conditions for HD long haul truck applications. Euro IV brought the implementation of the hot started cycle ETC (European transient cycle), which brought the need to homologation under strong transient conditions with the successful goal to limit soot and NO_x emissions under real driving conditions. However, cold start conditions and low load operation was of minor or no relevance. As a result, commercial vehicles in inner city operation (such as city buses, garbage trucks and medium duty trucks in pick up and deliver cycles) have been a burden for the city air pollution (Fig. 1).

A. Heintzel (Hrsg.): Int.Engine.Congr. 2024, Proceedings, S. 231–248, 2025.
https://doi.org/10.1007/978-3-658-47966-4_17

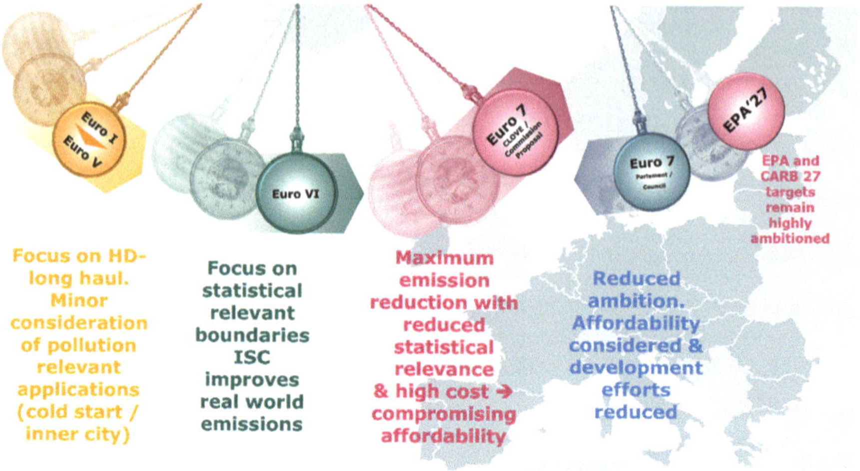

Fig. 1. Characteristics of EU commercial vehicle pollutant legislations and proposals [Fraidl 2023]

In this respect, Euro VI with the focus on statistically relevant operation conditions introducing a representative cold/hot cycle and ISC (in service conformity), was a huge success for the air quality. Ten years after the introduction of Euro VI, still the fleet emissions of the EU-27 countries are ramping down due to the drop out of old commercial vehicles. The constant updates of Euro VI legislation (A~E) brought more and more the focus to the ISC and low load operation, which was of course an additional contributor to the process. Fig. 2 shows the NO_x emission reduction of the EU-27 HD fleet under the circumstance having no further emission update after Euro VI E.

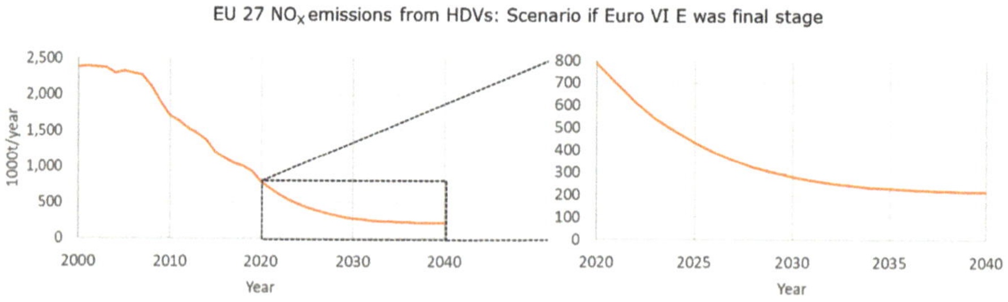

Fig. 2. NO_x emission from HDVs calculated by NEMO Fleet Model [Hausberger 2023]

The Euro 7 CLOVE—and the initial European Commission proposal considered maximum emission reduction with limited statistical relevance. The high costs for hardware and validation would have compromised the product costs for OEMs who at the same time are supposed to work on establishing ZEV vehicles into the fleet such as BEV, H_2-ICE and FCEV.

Parliament and Council proposals brought both, significant relaxed pollutant emission levels and less stringent extended boundary conditions. The initial plan to replace ISC by RDE (real driving emissions) was discarded. Even though these new Euro 7 boundaries bring less costs and validation efforts with retarded introduction by most probably 2029 to the EU OEMs, EPA and CARB still follow their plan introducing ultra-low NO_x limits to the US market by 2027. On top EPAs requirement on OBD monitoring stays more challenging than in EU. This divergent ambitions of those two important HD markets makes a common engine or a common Exhaust Aftertreatment System (EAS) architecture difficult or almost impossible.

2 Future Pollutant Emission Reduction (Euro 7 / EPA, CARB 2027)

2.1 Euro 7 Proposal 11/2022

In November 2022 the European Commission published the long-expected Euro 7 proposal for light-duty and heavy-duty vehicles. For heavy-duty vehicles the proposal not only tightened the limits for already regulated pollutant species, but also introduced limits for new pollutant species such as Nitrous Dioxide (N_2O) and Formaldehyde (HCHO). Regarding the Particulate Number (PN) limit the cut-off diameter was shifted from 23 to 10 nm alongside the reduction of the limit value itself. In total 3 sets of limits were available focusing on the reduction of cold start emissions (100th percentile limit) and covering short trips (budget limit). The third set of limits (90th percentile/hot conditions) targeted the reduction of exhaust emission when the system has reached its typical operation temperature range. Additionally, the NO_x emissions during engine idling were limited to 5 g/hr. An overview of the proposed limits can be seen in the table below (Table 1).

Table 1. Euro 7 emission limits according to EU Commission proposal in November 2022

Pollutant emissions	Cold emissions 100th percentile	Hot emissions 90th percentile	Emission budget for all trips less than 3xWHTC long	Optional idle emission limits
	per kWh	per kWh	per kWh	per hour
NO_x in mg	350	90	150	5000
PM in mg	12	8	10	
PN_{10} in #	$5x10^{11}$	$2x10^{11}$	$3x10^{11}$	
CO in mg	3500	200	2700	
NMOG in mg	200	50	75	
NH_3 in mg	65	65	70	
CH_4 in mg	500	350	500	
N_2O in mg	160	100	140	
HCHO in mg	30	30		

The limit values listed in Table 1 were applicable to the cold (100th percentile) and the hot phase (90th percentile) of the WHTC cycle (world harmonized transient cycle).

The evaluation of the system out emissions follows the moving average window method which considers the work of the WHTC cycle for the respective engine as a reference. Each second of the trip a work-window opens and closes after the vehicle has conducted the work of 1 WHTC cycle. The cumulated emission species mass (number in case of PN10) is then divided by the reference work. The 90th percentile limit allows to remove windows with the highest emission values. The budget limit is designed to make sure that even during short trips the tailpipe emissions are controlled in contrary to the preceding EURO VI-E legislation only evaluating trips with a duration longer than 4 times the work/CO_2 emissions of the WHTC cycle.

Besides the reduction of the limits for the individual emission species the boundary conditions for the ISC testing procedure are adapted in a way that basically all driving conditions are valid trips. This means that the focus of the development activities will have to be put on the vehicle testing environment including possible worst case driving scenarios.

Furthermore, the proposal includes the introduction of On-Board-Monitoring (OBM). The OBM system must be capable to detect emission exceedances and report them to the OBD port and over the air.

In the published document for the EU Commission the introduction date for new types and all types is set to 07/2027.

2.2 Euro 7 Draft Agreement

After different Euro 7 scenarios, based on alternative proposals provided by the European Commission and the European Parliament, were discussed throughout 2023, in December 2023 a draft agreement was reached. The concluded Euro 7 agreement suggests less ambitious limits, or no limit values compared to the initial proposal provided by the European Commission (Tables 1 and 2). In Table 2 a comparison between Euro VI E and Euro 7 regarding the emission limits in the WHTC c/h and the RDE limits is shown.

Table 2. Limit values in the WHTC cold/hot and RDE limits for EURO VI E and Euro 7 (draft agreement)

Pollutant emissions	WHTC		RDE	
	Euro 7 Draft Agreement WHTC c/h combined (1/7 vs. 6/7) 20.12.2023	EU VI Limits WHTC c/h combined (1/7 vs. 6/7)	Euro 7 Draft Agreement 20.12.2023 ISC c/h combined (1/7 vs. 6/7)	EU VI Limits ISC combined (1/7 vs. 6/7)
-	per kWh	per kWh	per kWh	per kWh
NO_x in mg	200	460	260	690
PM in mg	8	10	-	-
PN_{10} in #	$6x10^{11}$	$6x10^{11}$ (PN 23)	$9*10^{11}$	$9.78*10^{11}$ (PN 23)
CO in mg	1500	4000	1950	6000
NMOG in mg	80	160 (NMHC)	105	240 (NMHC)
NH_3 in mg	60	10 (ppm)	85	-
CH_4 in mg	500	500	650	750
N_2O in mg	200	N/A	260	-

The RDE testing boundary conditions in the Euro 7 regulation are largely derived from Euro VI E.

2.3 EPA/CARB 2027 Legislation

In July 2023 the California Air Resources Board (CARB) announced a Clean Truck Partnership with the Truck and Engine Manufacturers Association (EMA) and the leading US truck manufacturers. This partnership includes the adoption of the regulation regarding the NO_x emissions proposed by the EPA in late 2022. Alongside the reduction of the limits for all regulated pollutant species, the regulation introduces the Low Load Cylce (LLC) to the certification procedure. The intention of this is ensure that exhaust emission control systems maintain activity especially in the low load range. The limit values for the pollutant species (Table 3) are slightly higher than initially proposed by the CARB in their HD Omnibus regulation proposal of 2020.

Table 3. Emission limits according to the EPA/Carb legislation

Duty Cycle	NO_x [a] [mg/hp-hr]	HC [mg/hp-hr]	PM [mg/hp-hr]	CO [g/hp-hr]
SET & FTP	35	60	5	6.0
LLC	50	140	5	6.0

[a] An interim NO_x compliance allowance of 15 mg/hp-hr applies for any in-use testing of Medium HDE and Heavy HDE

In addition to the limits for the testbed protocols an NO_x idle limit of 10 g/hr is included. Similar to the European vehicle testing approach (PEMS/ISC) it is planned to introduce NO_x Standards for Off-Cycle emissions in the United States. The evaluation method in the US case is to use time-based windows with a duration of 300 s. The brake specific emission value for NO_x emissions is then calculated by the sum-over-sum method. Depending on the average CO_2 emission rate in the individual window, it is assigned to 2 different bins (idle and low/medium and high) with different limit values (Table 4).

Table 4. Off-Cycle NO_x standards according to EPA/Carb legislation

Off-Cycle Bin	Normalized average window CO_2 emissions rate	NO_x limit for Medium HDE and Heavy HDE with In-Use Compliance Allowance
Bin 1	≤ 6 %	10 g/hr
Bin 2	> 6 %	73 mg/hp-hr

The limit for the low/medium/high bin is valid at 25 °C ambient temperature and above. Below 25 °C correction factors apply down to a temperature of 5 °C. Below that ambient temperature the limit values do not apply.

2.4 Tier 5 Update

As the legislation process for the on-road sector is close to finalization CARB has shifted its focus to the off-road segment and plans the introduction of the next emission legislation step called Tier 5. In the latest Tier 5 related workshop, held in October 2023, an update regarding the proposed emission standards, introduction dates and in-use emission evaluation were published. Table 5 shows a consolidated picture of Tier 5 as currently planned. For the power categories between 56 to 130 kW a 2-phase introduction and 130 to 560 kW the Tier 5 interim standard will be introduced in 2029. The main difference between the today active Tier 4 final standard is the reduction of the PM limit from 0.02 g/kWh to 0.005 g/kWh in the NRTC c/h and the RMC (Non Road Transient Cycle, Ramped Mode Cycle). This 75% reduction of the PM limit mandates the integration of a DPF (Diesel Particulate Filter) into the aftertreatment system The NO_x emission limits will be reduced to 0.22 g/kWh in the interim stage and to 0.04 g/kWh in the final stage. The Tier 5f standard will be introduced in 2034+ for the smaller power category and in 2033+ for the power category between 130 and 560 kW. Alongside the reduction of the emission standards in the NRTC and RMC a Low Load Cycle (LLC) will be introduced in the T5f stage.

Table 5. Proposed Tier 5 emission standards as published in October 2023

Duty Cycle	Power Category	Emission Standard	Implementation Period	NO_x Limit [g/kWh]	PM Limit [g/kWh]	NMHC Limit [g/kWh]	CO Limit [g/kWh]
NRTC c/h & RMC	56≤kW≤130 (75≤hp≤175)	Tier 5 interim	2031-2033	0.22	0.005	0.19	5.0
		Tier 5 final	2034+	0.04		0.080	
	130≤kW≤560 (175≤hp≤750)	Tier 5 interim	2029-2032	0.22	0.005	0.19	3.5
		Tier 5 final	2033+	0.04		0.080	
LLC	56≤kW≤130 (75≤hp≤175)	Tier 5 final	2034+	0.06	0.005	0.19	
	130≤kW≤560 (175≤hp≤750)		2033+				

To make sure that NO_x and PM emissions of off-road vehicles are in a similar range as the laboratory standards while they are operating in the field, CARB plans to monitor the in-use performance of the emission control system. CARB proposes to evaluate the NO_x emissions based on NO_x sensor data using the 3-bin evaluation method. The time-window based evaluation is like what is planned for the on-road sector 2.3 but the windows are then sorted in 3 different bins (idle bin, low bin, high bin). The NO_x limit for the individual bin is based on the NRTC/RMC limit (high bin), the LLC limit (low bin) and the idle limit (idle bin). Considering a conformity factor of 1.5 + NO_x sensor margin. The acceptance criteria proposed by CARB is that 75% of an engine family should fulfill the in-use standards. For the DPF the in-use compliance is based on the PM sensor fault code ratio. The acceptable threshold according to CARB is 5%. If NO_x or PM requirements are not fulfilled, an engine family will be flagged for PEMS testing. Details about the PEMS testing procedure are currently not available.

2.5 Comparison

The different available emission legislation scenarios for the off-road and on-road sector show that the limits for the laboratory test protocols as well as the in-use emission limits are more stringent in the US. The expected NO_x standards, especially in the FTP c/h, but also and the NRTC c/h not only mandate a strong focus on cold-start emission behavior, but also an optimization of NO_x conversion in the warm phase of the testbed cycle to very low levels below 20 mg/kWh. By the introduction of the low load cycle the legal authorities intend to make sure, that the NO_x conversion efficiency remains high even under low load operating conditions.

Figure 3 compares the previously discussed emission legislations. The diagram shows a possible combination of cold and hot start emission to reach the respective combined NO_x emission limit in the cold and hot test cycle and the in-use emission limit. For better comparability all values are displayed in g/kWh and the in-use emission limit for the T5 legislation is shown as an average value of the low- and the high bin. The EPA emission limit already includes the compliance allowance considered for medium and heavy heavy-duty engines.

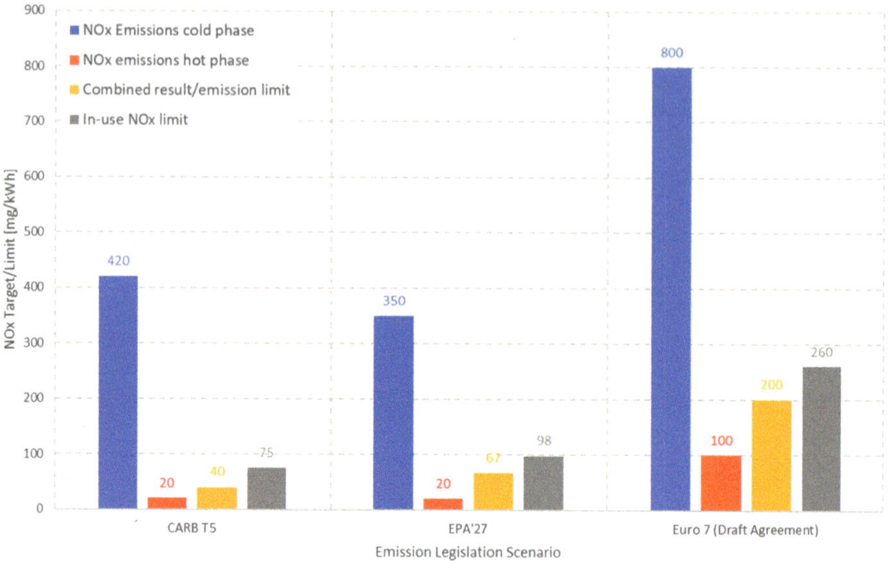

Fig. 3. Comparison of NO_x limits for different emission legislation scenarios

3 Diesel Engine and Aftertreatment System Technology

3.1 Diesel Engine Development / Raw Emission Behavior

On February 14th, 2023, EU published an updated roadmap for CO_2 reduction for commercial vehicles. The HD fleet emissions shall by reduced by 45% in comparison to 2019 reference. This targets the increase of zero emission vehicles on the hand, on the other hand also the further improvement of the Diesel engine efficiency as the diesel remains the most important propulsion system in the next decade.

Similar situation on the north American market, where GHG2 standards will force the HD diesel engine to 432 gCO_2/bhp-hr in SET. This target makes a sweet spot efficiency of roughly 49.5% necessary.

In previous publications, different thermodynamic concepts were presented [Wieser 2020] [Horvath 2020] which promise to achieve the BTE 50% target while maintaining raw NO_x emission < 10 g/kWh. A prerequisite of such engine concepts is the highest charging efficiency combined with high EGR rates for in-cylinder NO_x reduction. Due to the limitation of conventional high pressure EGR systems the proposed solutions either demand the introduction of a low pressure EGR system or the implementation of a serial power turbine to generate the required EGR rate. Modern high-efficiency exhaust aftertreatment systems comprising of a two stage SCR catalyst indicate highest NO_x conversion rates. These encouraging test results show that even engine concepts with a raw NO_x emission > 10 g/kWh have the potential to meet future emission legislation requirements. [Dreisbach 2021]

Figure 4 describes the potential of a high raw NO_x concept by optimizing closed and open cycle for non-EGR operation, by showing different NO_x-BSFC trade-offs measured on a multi cylinder engine (MCE). Red represents a Diesel engine with HP EGR and VGT turbocharger. Increasing raw NO_x emissions from 10 to 14 g/kWh on fixed H/W gives a moderate BSFC improvement in a magnitude of 2%. The orange, dotted line considers the same turbocharger H/W but optimized combustion system in view of best NO_x-BSFC trade off without EGR, means the combustion was optimized for non EGR operation. As running w/o EGR does require less boost pressure and allows positive gas exchange a dedicated layout brings further benefits—this is shown in the green line. At the end the BSFC improvement is 5% by optimizing for an operation on 14 g/kWh (for this particular example).

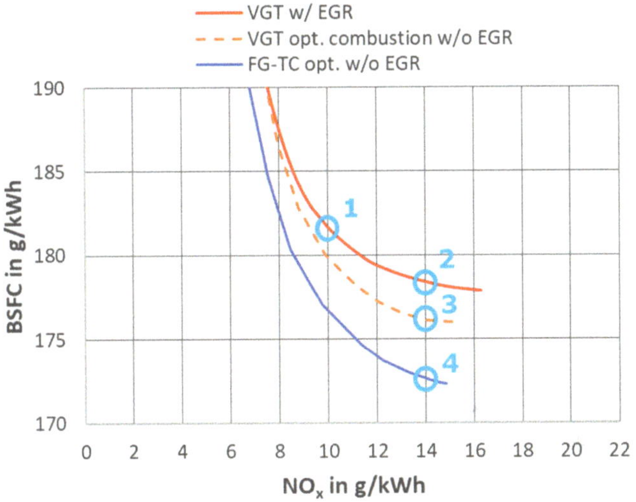

Fig. 4. Measurement campaign: different NO_x-BSFC Trade offs

The second example, comparing EGR with non-EGR concepts, is a simulation-based DOE optimization. Target was to achieve 50% BTE in the sweet spot of both concepts (Fig. 5).

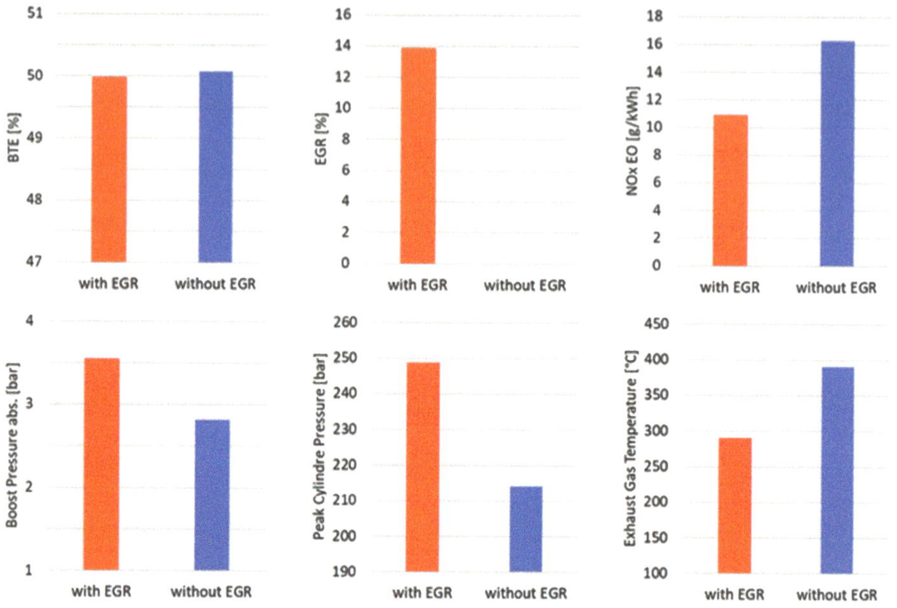

Fig. 5. DOE Results comparing EGR and non-EGR concepts [Dreisbach 2021]

The benefit of Concept "with EGR" is lower raw NO_x emission and consequently a lower development effort for the EAS. High boost pressure levels are applied which lead to a high cylinder mass and consequently to high PCP demand. Due to this high PCP requirement, special attention must be paid to achieving the lowest possible engine friction. The non EGR concept has lower PCP demand due to lower boost pressure and hence better open cycle efficiency.

3.2 Diesel Aftertreatment System Technology

Base on the divergent ambitions of the different emission legislations described in chapter 2.5, it will hardly be possible to use common aftertreatment system architectures in EU and US, at least not when taking affordability into account.

The on-road test cycles have a lower average load than the NRTC which makes it more challenging to rapidly heat-up the aftertreatment system. Consequently, the EPA 27 scenario, the most critical in terms of cold start emission behavior, causing an impact on engine technology and aftertreatment architecture. Possible technology configurations are shown in Table 6. For combustion concepts with cooled EGR the EPA 27 case mandates a dual stage SCR architecture, the Tier 5 and the Euro 7 case seems feasible with a single stage SCR architecture. Without cooled EGR, EPA 27 is highly challenging and requires additional external heat sources (electric heater, burner etc.). The dual stage SCR architecture will be required for Euro 7 and Tier 5 as well as the potential to decrease the NO_x raw emission level from the engine side is limited in the absence of cooled EGR.

Table 6. Aftertreatment architecture for different emission legislation scenarios

Emission Legislation	Aftertreatment Architecture with cooled EGR	Aftertreatment Architecture without cooled EGR
EPA 2027	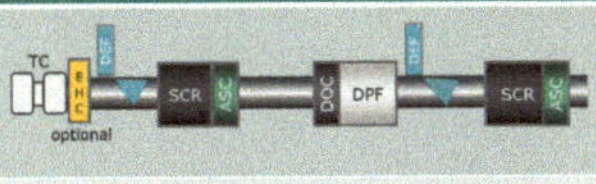	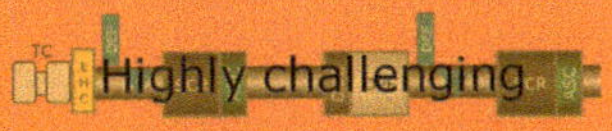 Highly challenging
Euro 7 Draft Agreement		
CARB Tier 5	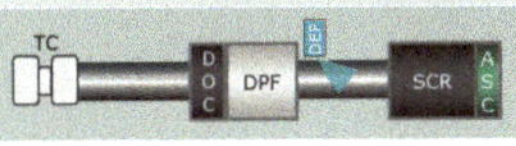	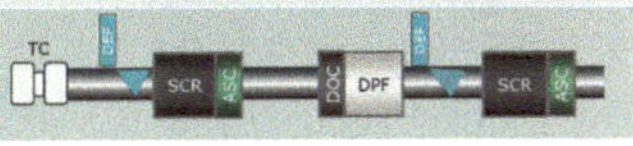

3.3 Hydrogen Engine

Based on the ambitious targets for CO_2 emission reduction in the heavy-duty segment, various manufacturers are looking into the development of H_2-ICE incl. the aftertreatment system to comply with future EU and NA emission standards. Fast way to the market will be the lean burn spark ignited engine, fueled with neat hydrogen. Raw NO_x emissions of those concepts usually vary, based on power density, specific torque, charging technology and transient calibration. Possible aftertreatment concepts for both, Euro 7 and EPA'27 are shown in Table 7. All concepts consider an oxidation catalyst upstream and a Cu-Zeolite SCR as it is beneficial due to its NH_3 storage capability compared to the Ti/V-SCR technology. To limit the N_2O formation on the SCR component a Ti/V SCR coating can be a viable option, however latest findings show that N_20 limits can be met with Cu-Zeolite technology only as well. As the particulate filter will be mandatory for PN10 compliance for Euro 7, it is part of the system and located downstream SCR. Optionally the PF and the Zeolite can be combined in an SCR-on-filter component. For EPA'27 the particulate filter can technically be avoided as the PM raw emissions of lean burn SI hydrogen engines are sufficiently low, however a filter might be used for public acceptance.

Table 7. Aftertreatment architecture for the commercial hydrogen engine

Emission Legislation	Lean Burn SI
EPA 2027	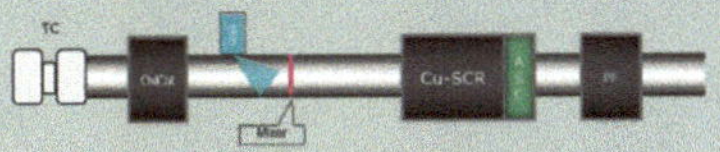 PF optional
Euro 7 Draft Agreement	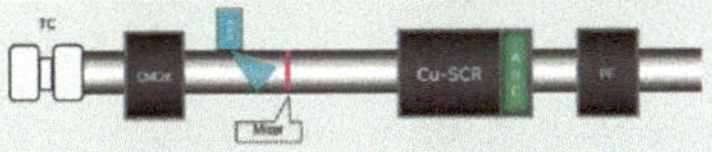 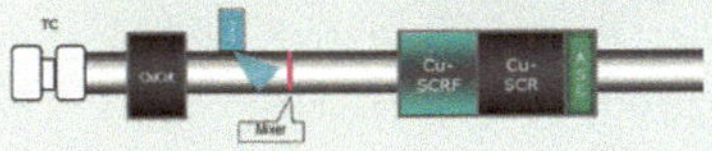

4 OBD and its Impact on the Technology Choice

Additionally to the target fulfilment of pollutant emissions by lowest possible GHG emissions, the chosen technology must be OBD compliant as well. A high NO_x engine out emission strategy is the most promising approach for significant minimizing fuel consumption. Therefore, NO_x reduction systems must ensure high conversion efficiencies over the entire useful life. Conversion performance deviations caused by aging and poisoning effects can lead to increased NO_x TP emissions without having clear characteristics for fault detection. From OBD point of view the distinguishing between the OBD SCR part (BPU—best performing unacceptable) and the worst performing acceptable (WPA) component is going to be the most challenging OBD task.

OBD threshold limits (OTL) can be defined in two different ways. The European regulation defines absolute OBD threshold limits for selected pollutant criteria [ENVI 2023]. In USA on-road regulation (2024 onwards) the OBD threshold limits are absolute limits for NO_x and PM and partially relative factors for NMHC and CO, which varies from case to case [CARB 2022]. The USA CARB non-highway regulation is in a planning phase and the rule makers are targeting to integrate the on-road OBD requirements including in-use monitoring activity ratio requirements to the upcoming Tier 5 legal text. The draft documents are showing high probability for including the on-board monitoring diagnostics (OBMD) requirements as well [CARB 2023]. Despite the different OBD legal requirements, the SCR monitoring challenges for future Euro and US OBD regulations will cause significant development efforts (Table 8).

Table 8. OBD threshold limits and potential OBM threshold limits for different emission legislations [ENVI 2023, CARB 2022, CARB 2023]

Pollutant Emissions	Euro 7			US 2027 and subsequent MY			Tier5 2033 and subsequent MY (e.g. $130 \leq kW \leq 560$)[1)]		
	OBD Thresholds WHTC c/h AVL Assumption	Ratio OBD to Emission Threshold	OBM Thresholds ISC Emissions	HDOBD Thresholds FPT/RMC	Ratio OBD to Emission Threshold	OBM Thresholds ISC Emissions	On OBD Thresholds NRTC Draft 31.10.23	Ratio OBD to Emission Threshold Draft 31.10.23	OBMD Thresholds ISC Emissions Draft
	per kWh	-	-	per kWh	-	-	per kWh	-	-
NO_x in mg	1.200	6x	2,5x	299	6x	-	400[2)] or 3x /2x[3)]	1,8x [4)] 10x	600 3B-MAW bins B and C
NH_3 [-]	-	-	2,5x	-	-	-	-	-	-
PM in mg	25	3,125x	Status: state of health	22,4	6x	-	40[2)] or 3x /2x[3)]	8x[2)] or 3x /2x[3)]	40
NMHC [-]	-	-	-	2x std.	2x	-	3x /2x[3)]	3x /2x	-
CO [-]	-	-	-	2x std.	2x	-	3x /2x[3)]	3x /2x	-

The qualification of OBD limit parts needs to be demonstrated in the legal cycles. In this cycle even the OBD limit parts achieve still considerable NO_x conversion efficiency levels. The performance difference between the before mentioned BPU and the

WPA SCR systems are so minimal, that it cannot be done robustly with state-of-the-art OBD monitoring solutions. For an assessment of the OBD concept feasibility for high efficiency aftertreatment solutions, more detailed investigation of SCR conversion efficiency monitoring approaches is required.

AVL conducted a feasibility study in a Software In The Loop (SiL) environment, using a representative EAS demonstrator with technology that meets US MY27 emission limits (2-stage SCR catalyst system incl. electrically heated catalyst). To reduce the task complexity the following simplifications and boundaries were set:

- One aftertreatment model representing identical hardware specifications
- Simplified 2-stage dosing strategy: dosing amount on Close Coupled SCR (ccSCR) is ramped down to zero, when Under-Floor SCR (ufSCR) is above 250 °C
- No urea dosing adaptation functions
- An engine model was used, with two different NO_x EO trace sets for FTP hot (7 and 15 g/kWh total cycle emissions)
- Simplified CSERS in combination with EGR control which lead to predominantly higher NO_x emissions during FTP cold.

The in-vehicle deterioration of the SCR catalyst can be a result of chemical and thermal processes and mechanical impacts. The predominant damage mechanisms on SCR catalysts are the thermal damage of the SCR via long term high temperatures occurring during the diesel particulate filter (DPF) regeneration events and the chemical harm via poisoning by inorganic compound and oil intrusion into SCR [Sciencedirect 2020]. Deterioration by mechanical impacts is negligible, therefore they are not considered in the virtual damage models. Also, the chemical damage mechanisms aren't considered in the testing environment because the variety of inorganic compounds in fuels and their impact on SCR catalysts cannot be comprehensively modeled and the parametrization data for aging model creation prerequisites extensive hardware durability testing campaign upfront.

In the present study the deterioration of conversion drop is simulated by manipulating performance factor which reflects the thermal damage of SCR catalyst. The severity of the catalyst damages is not linearly reflecting the conversion efficiency drop, but relative performance difference correlates with reference data. The target SCR catalyst damage is defined by approaching the OBD threshold in the combined Federal Test Procedure (FTP) cold/hot cycle. Like expected for the system with moderate NO_x EO level more severe damage stage (here defined as damage state 4) has been determined. For the high NO_x EO level system less severe damage (defined as damage stage 2) leads to cycle NO_x TP emissions close to the OBD threshold (Fig. 6).

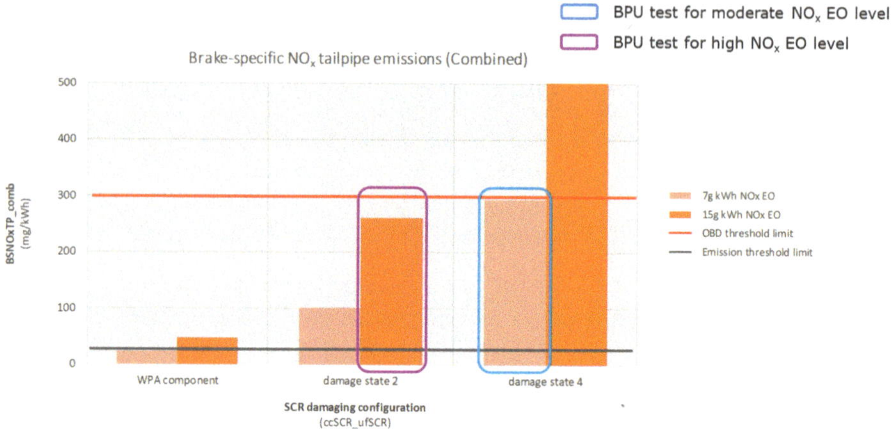

Fig. 6. Average TP NO_x results in legal cycle with deteriorated SCR catalysts

Based on the NO_x TP emission cycle results (see Table 9) no state-of-the-art OBD monitoring approach appears to be feasible to differentiate the WPA SCR and the BPU SCR catalysts as the SCR conversion efficiency separation is very low. Best case separation that was demonstrated achieved ~ 12% points.

Table 9. Average SCR conversion efficiencies with 7 g/kWh and 15 g/kWh NO_x EO system

Parameter	Unit	ccSCR			ufSCR			SCR system
7 g/kWh NO_x EO	-	FTP cold	FTP hot	combined	FTP cold	FTP hot	combined	combined
Avg. SCR efficiency WPA	%	74,3	71,9	98,0	92,2	99,6	99,9	99,7
Avg. SCR efficiency BPU	%	62,1	62,1	93,0	81,5	90,3	96,3	95,9
SCR efficiency difference	%	**12,2**	**9,7**	**5**	**10,7**	**9,3**	**3,6**	**3,7**

Parameter	Unit	ccSCR			ufSCR			SCR system
15 g/kWh NO_x EO	-	FTP cold	FTP hot	combined	FTP cold	FTP hot	combined	combined
Avg. SCR efficiency WPA	%	68,1	62,5	98,0	93,8	99,7	99,9	99,7
Avg. SCR efficiency BPU	%	65,1	60,7	96,0	88,6	96,1	98,5	98,2
SCR efficiency difference	%	**3,0**	**1,8**	**2,0**	**5,2**	**3,6**	**1,4**	**1,5**

In order to also prove the likelihood of the algorithms to finish within the legal cycle, investigation of the enablement criteria was performed predominantly with the NO_x and NH_3 traces from the engine model. The engine with conventional SCR strategy of 7 g/kWh NO_x EO seems feasible in the SiL environment.

As of the lower separation, of the non-EGR engine with 15 g/kWh NO_x EO detection will be much more challenging.

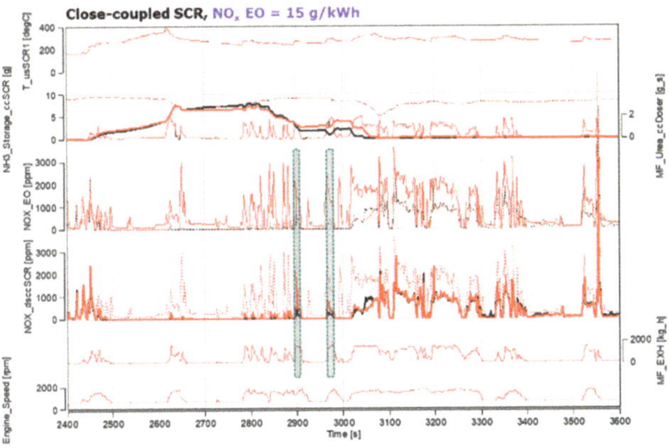

Potential diagnostic areas (NOx eff.)

Parameter	Unit	WPA SCR	BPU SCR
NO_x_EO	ppm	1403,3	
NO_x dsccSCR	ppm	192,3	421,2
Local SCR efficiency	-	**0,863**	**0,700**

Fig. 7. Traces of ccSCR in hot FTP for a high NO_x concept (focus on NO_x signals)

The NO_x signal traces before and after the ccSCR are showing just two, potentially promising, detections areas with a local SCR conversion efficiency difference of more of 16%, which were useful for NO_x efficiency monitoring method. (See Fig. 7)

However, via AVL NH_3 load-based monitoring method [De Monte 2018], the NH_3 and N_2O slip through the ccSCR and its ASC can be detected in multiple detection areas. Mathematical integration of the NO_x signals seems to be also a further prospective monitoring characteristic for determination of the WPA and BPU SCRs. The integration function starts at minimum ccSCR upstream temperature of 250 degrees Celsius. Until the proposed calculation stop (marked in the Fig. 8 with an ellipse) the BPU ccSCR releases more than 10 g unconverted NO_x emissions compared to the WPA ccSCR catalyst.

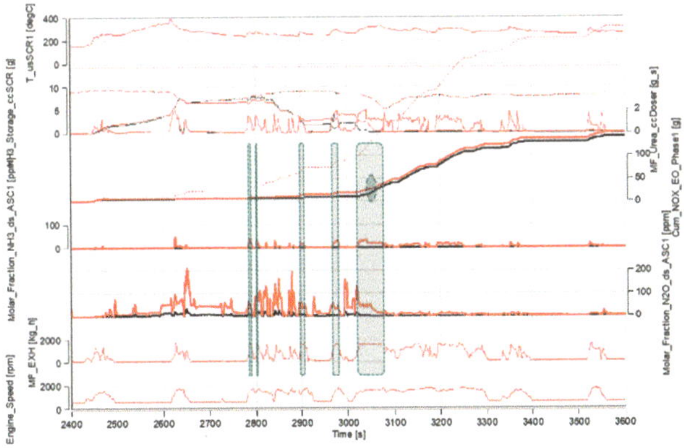

Potential diagnostic areas (NH_3 slip)

Potential diagnostic areas (stop of integration window)

Parameter	Unit	WPA SCR	BPU SCR
NH_3_ds_ASC1	ppm	0,0	18,3
N_2O_ds_ASC1	ppm	2,6	35,1
NO_x window	**g**	**14,6**	**24,8**

Fig. 8. Traces of ccSCR in hot FTP for a high NO_x concept (focus on NH_3 load)

In contrast to the ccSCR, for the ufSCR the method using NO_x efficiency monitoring shows more promising results than the NH_3 load-based monitoring method. (compare Fig. 9 and 10)

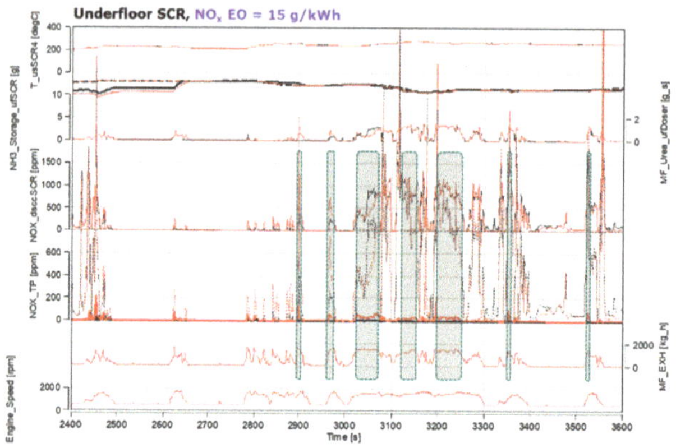

Potential diagnostic areas (NO_x eff.)

Parameter	Unit	WPA SCR	BPU SCR
NO_x_EO	ppm	1004,6	1017,4
NO_x dsccSCR	ppm	1,3	39,7
Local SCR efficiency	-	**0,999**	**0,961**

Fig. 9. Traces of ufSCR in hot FTP for a high NO_x concept (focus on NO_x signals)

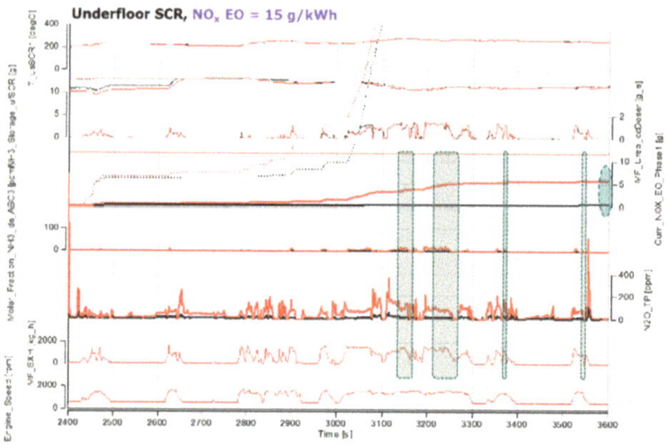

Potential diagnostic areas (NH_3 slip)

Potential diagnostic areas (stop of integration window)

Parameter	Unit	WPA SCR	BPU SCR
NH_3 dsASC3	ppm	0,0	16,6
N_2O TP	ppm	14,6	97,0
NOX window	**g**	**0,5**	**5,8**

Fig. 10. Traces of ufSCR in hot FTP for a high NO_x concept (focus on NH_3 load)

All in all, two different monitoring approaches for closed-coupled and underfloor SCR systems seem necessary to sufficiently robust detect OBD limit parts, at least for high NO_x EO approaches.

5 Summary

Based on the divergent ambitions of the different emission legislations, EU and North American manufacturers will not make use of common aftertreatment system architectures, at least not when taking affordability into account. EPA'27 mandates a dual stage SCR architecture, whereas Euro 7 seems feasible with single stage SCR, at least if an engine concept with cooled EGR applies.

Europe's fleet emission reduction goal forces OEMs to further push engine efficiencies to highest levels. Non-EGR concepts with high raw NO_x emissions deliver high efficiency at low boost- and moderate peak cylinder pressure. Consequently, also the EU market will see dual stage SCR systems.

North Americas GHG2 limits are also targeting a sweet spot efficiency of roughly 49.5%. However, the implementation of a non-EGR concept with high raw NO_x emissions is significantly more complex than under EU boundaries. Firstly, an additional external heat source (e.g.: electric heater, burner etc.) is required to fulfill the more stringent emission limits. Secondly, more stringent OBD threshold limits compared to EU apply. Having clear separation for fault detection and finding sufficient OBD monitoring frequency in the field will be the highest challenge of such a concept in NA.

Focusing on a global future engine, for sure the concept using EGR will be less risky, nevertheless publications indicate that there is the intention to establish high raw NO_x engines in the US as well. [Int 2023] For such concepts, AVLs approach is to mitigate the risk of unclear SCR catalyst detectability by using combined OBD monitoring methods based on NO_x and NH_3-signal traces analysis. The combination improves both, WPA and BPU separation and monitoring frequency.

References

§ 1971.1. On-Board Diagnostic System Requirements—2010 and Subsequent Model-Year Heavy-Duty Engines https://govt.westlaw.com/calregs/Document/I3DF2824074D611E-D90EEB837E3125243?viewType=FullText&originationContext=documenttoc&transitionType=CategoryPageItem&contextData=(sc.Default) 22.11.2022

De Monte M., Noll H., Mannsberger S., "SCR control strategies with multiple reduction devices for lowest NO_x emissions" in SAE Heavy-Duty Diesel Emissions Control Symposium, Gothenburg, 2018.

Dreisbach at al., 50% Efficiency on a Combustion Engine—Status and Measurement Results, 8. Internationaler Motorenkongress 2020, Baden-Baden, 2020

Fraidl at al; Global View on Future PassCar Powertrains: Is Europe Coming Back to an Efficient Fact Based Approach ?; 19th Symposium „Sustainable Mobility, Transport and Power Generation"; Graz, September 28–29, 2023

Hausberger at al; Fleetmodel emission calculation using TUG-NEMO emission fleet model, Technical University Graz

Horvath A., Seitz H.F., Gelter J., Raser B.: Der zukünftige Motor für schwere Nutzfahrzeuge—Grundmotorkonzept zur maximalen CO_2-Reduktion, MTZ 10|2020

https://www.internationaltrucks.com/engines/s13integrated; 23.12.2023

NO_x reduction behaviour in copper zeolite catalysts for ammonia SCR systems: A review https://www.sciencedirect.com/science/article/abs/pii/S1385894719326658 09.01.2023

Proposal for a REGULATION OF THE EUROPEAN PARLIAMENT AND OF THE COUNCIL (2022/0365 (COD))

Tier 5 Rulemaking Workgroup: Off-Road On-Board Diagnostics (OR-OBD) Proposal https://ww2.arb.ca.gov/sites/default/files/2023-10/T5%20OR-OBD%20Workshop%2010_23_23%20clean.pdf 31.10.2023

Wieser M., Hofer F., Theissl H., et. al.: 50% Brake Thermal Efficiency—the realization of a vision. 7. Internationaler Motorenkongress 2020, Baden-Baden, 2020

Efficient Emission Reduction by the Thermal Management System CatVap® with Conventional and Synthetic Fuels

Artur Müller[1(✉)], Fabian Feldhaus[1], Robin Thannimmotil[1], Georg Hüthwohl[1], Robert Szolak[2], Paul Beutel[2], und Florian Rümmele[2]

[1] Albonair GmbH, Dortmund, Germany
{artur.mueller,fabian.feldhaus,robin.thannimmotil, georg.huethwohl}@albonair.com
[2] Fraunhofer Institut für Solare Energiesystem ISE, Freiburg, Germany
{robert.szolak,paul.beutel, florian.ruemmele}@ise.fraunhofer.de

1 Introduction

The different proposals for the planned Euro 7 emissions regulation suggest lower limits for nitrogen oxide (NO_x) emissions.

A certain temperature level of the exhaust aftertreatment system (EATS), in particular the SCR catalytic converter, is necessary for nitrogen oxide reduction. The SCR reduction reactions can only take place above this so-called light-off temperature, which is dependent on the catalyst substrate. In addition, the decomposition reactions of the reducing agent (urea) are temperature-dependent; they are proceeding fairly slow at lower temperatures. For a further nitrogen oxide emission reduction, the focus is on operating points with low exhaust gas temperatures, such as engine cold start or partial load operation. This will ensure a proper operation of the SCR system over the entire engine operating range. The use of additional heating systems is a cost effective and suitable solution.

A known disadvantage of current EURO VI exhaust gas aftertreatment systems (see Fig. 1) is the temperature loss of the SCR catalyst during low load operation, where the engine exhaust gas temperatures are quite low. As a result, nitrogen oxides cannot be completely reduced under the various conditions of a real drive operation (RDE).

A. Heintzel (Hrsg.): Int.Engine.Congr. 2024, Proceedings, S. 249–263, 2025.
https://doi.org/10.1007/978-3-658-47966-4_18

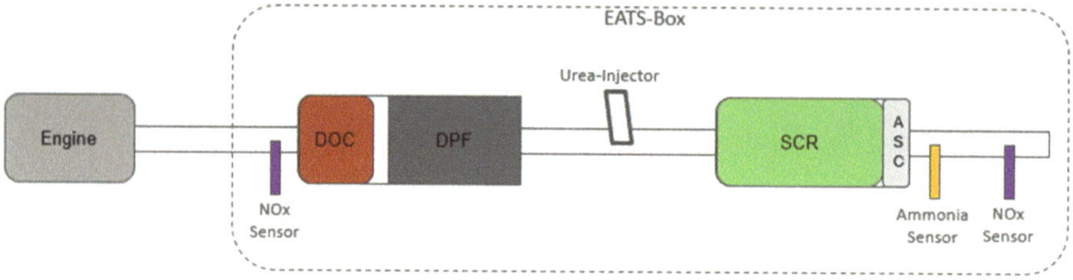

Fig. 1. Typical EURO VI EATS layout

The CatVap® system, developed by Fraunhofer ISE and Albonair, is an adequate heat source to counteract the cooling effects and ensure the effectiveness of the exhaust gas aftertreatment system.

On the one hand, this system enables the exhaust gas aftertreatment system to heat up quickly and effectively after an engine cold start and, on the other hand, ensures that a certain temperature range is maintained in which an efficient and effective nitrogen oxide reduction is possible.

As shown in Fig. 2, the CatVap® system can be installed directly upstream of the EATS and thus ensures a rapid heating of the downstream components, independently of the engine operation. The required target temperature for efficient nitrogen oxide conversion is therefore not only reached earlier, but also kept within the required temperature range.

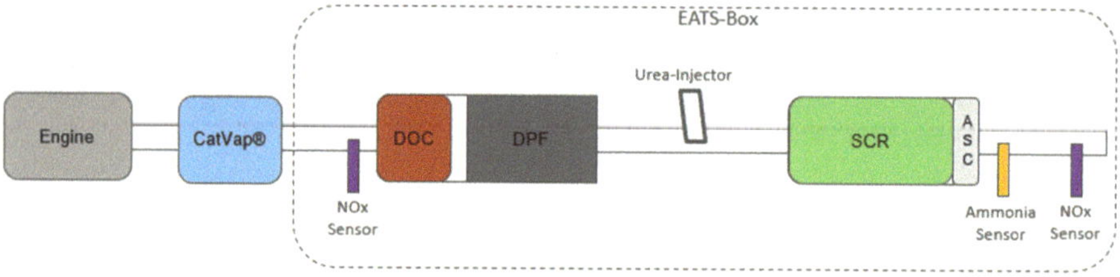

Fig. 2. Possible EATS configuration with CatVap® and an Euro VI BOX for future Euro 7 emission requirements

Furthermore, a configuration with a Close Coupled SCR system is also conceivable (Fig. 3). The advantage of this variant is the immediate heating of the SCR system close to the engine before other EATS components, such as DOC or DPF, are supplied with heat. In this case, the nitrogen oxide reduction can be shifted towards colder exhaust gas temperatures, which ultimately increases the NO_x conversion rate due to the overall system.

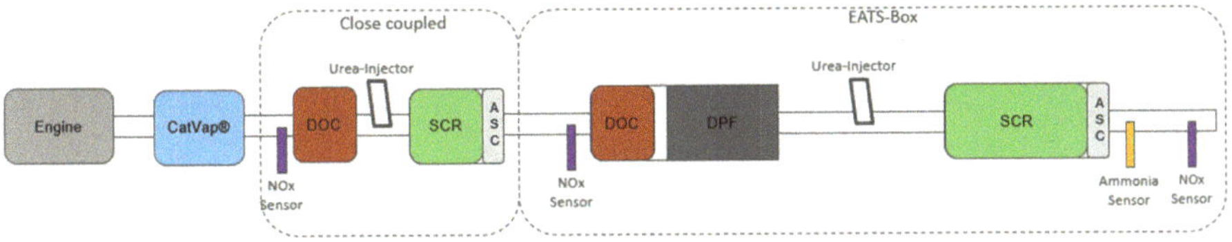

Fig. 3. Possible EATS configuration with CatVap® and a *cc*-SCR solution for future Euro 7 emission requirements

2 Albonair CatVap® System

The CatVap® system is an innovative heat supply system which was developed by Fraunhofer Institute for Solar Energy Technology and Albonair GmbH in collaboration with, Emitec Technologies GmbH and Thomas Magnete GmbH.

The system does not emit harmful secondary emissions (e.g. soot particles or NO_x) and is able to provide additional heat on purpose efficiently. It is based well known and proven series components that are available on the market.

CatVap® is an abbreviation for "CATalytic eVAPoration process".

The provided heat is catalytically generated by the fuel and the engine exhaust gas, without any need for an additional air supply. Started by electrically heated catalysts, the system then operates without any need for additional electrical power demands, whereby thermal outputs of up to 20 kW can be supplied (up to 100 kW with DOC support). Conventional fuels such as diesel or renewable, synthetically produced hydrocarbons can be used.

The CatVap® (Fig. 4) contains an inlet flange through which a portion of the engine exhaust flow is guided into the intake chamber. From there, the exhaust gas flow is routed through a perforated plate ("exhaust gas plate"). Due to the inserted holes of this plate the exhaust gas is accelerated and passed through the electrically heated catalyst to the fuel plate, where it absorbs evaporated or rather catalytically processed fuel. CFD simulations were used to determine the number, size and orientation of the holes to ensure a defined distribution.

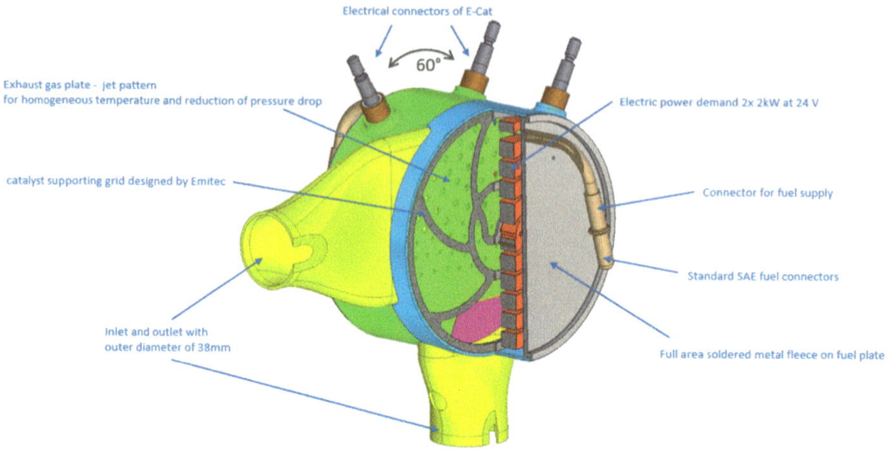

Fig. 4. CatVap® design

The electrically heated catalytic converter was provided by Emitec Technologies GmbH. It is stabilized by a support structure and has a precious metal coating, which is necessary for the catalytic conversion of the fuel.

After the metered diesel fuel has been (catalytically) processed, the gas mixture, which is now heated, enters the outlet area.

The process described is shown in Fig. 5:

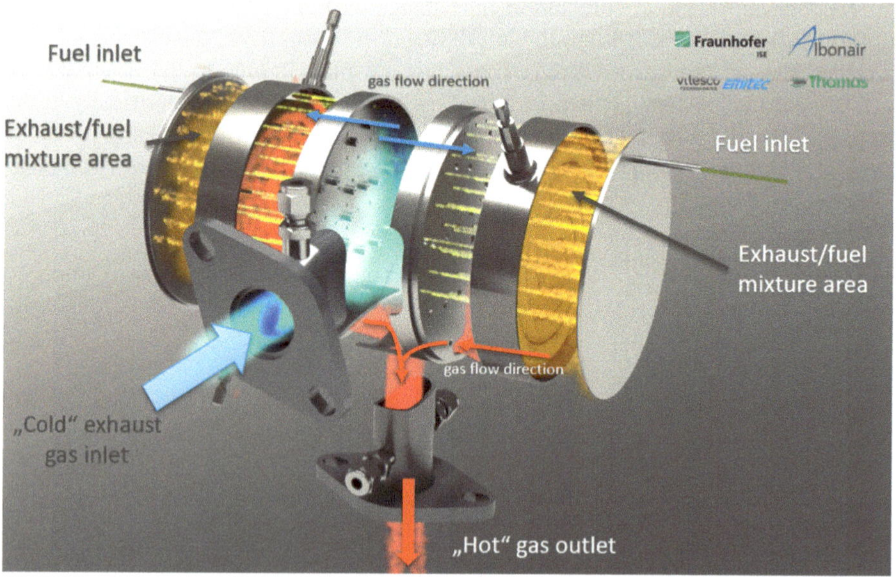

Fig. 5. Functional sequence inside of CatVap®

To achieve the fuel's evaporation temperature inside the CatVap® during an engine cold start, the system is heated for a certain period of time by means of the electrically heated catalyst. Currently, this can take up to 150 s and depends on the given boundary conditions [3].

3 Operating Modes and Technical Data of CatVap®

The CatVap® can be run in different operating modes:

Heating Mode
In "Heating Mode", the maximum heating output is released at the maximum exhaust gas mass flow. Here, a mass flow of up to 150 kg/h can be passed through the CatVap®, whereby a thermal output of up to 20 kW can be generated. Depending on the inlet temperature and residual oxygen content, outlet temperatures of up to 600 °C can be achieved.

DOC-Mode (Diesel Oxidation Mode)
During the "DOC mode", more fuel is metered into the CatVap® than can be catalytically converted, while the exhaust gas mass flow is very low (3...20 kg/h).

The portion of fuel that is not catalytically processed, is being reformed to hydrogen, carbon monoxide and also cracked into short-chained alkenes. As a result of the created hydrogen and short-chained alkenes, the light off temperature in the DOC is significant lower, starting with temperatures above 160 °C. A heat output of up to 40 kW can be released. The temperature level downstream of the DOC is limited to about 400 °C.

Heat Retention Mode
Running the "Heat Retention Mode", the control unit calculates the necessary heat demand that must be supplied to the EATS to achieve a desired target temperature. In combination with a calibrated and stored λ-map for the air/fuel ratio, the exhaust gas mass flow through CatVap® is regulated according to the required heat output.

DPF Regeneration Mode
The "DPF Regeneration Mode" is similar to the "DOC Mode": more fuel is dosed into the CatVap® than can be catalytically converted. This emulates the function of a conventional HC Doser or similar operating strategies like late fuel injection.

Up to 100 kW of thermal output can be provided in this mode, in order to perform a sufficiently hot DPF regeneration.

DPF regeneration is usually initiated by the engine control unit (ECU), whereby a necessary temperature level can be specified, too. This temperature level is controlled solely by the CatVap® for the entire duration of regeneration. After the DPF regeneration is completed, the system returns to the previous operating mode.

No further thermal management measures (e.g. late post-injection) are required when using regeneration mode.

The following technical operating data can be summarized:

- max. heat output in Heating Mode: 20 kW
- max. heat output in DOC Mode: 40 kW
- max. heat output in DPF Regen Mode: 100 kW
- min. heat output in Heat Retention Mode: 2 kW (deactivated, when the engine exhaust temperature reaches a certain value)
- max. electrical power: 4 kW (only during heat up phase at start-up)
- max. current during electrical heat up: 180 A at 24 V
- max. duration of electrical heat up: 150 s
- max. exhaust mass flow through CatVap®: 150 kg/h
- pressure loss over the entire EATS for a 3″ pipe: 40 mbar at 1.000 kg/h and 450 °C

4 Experimental Setup and Procedure

After promising test results performed in 2022 on the engine test bench, on-road tests were initiated, using a medium-duty commercial vehicle with a gross weight of 7,5 t to demonstrate further potential of CatVap® system.

The test setup on the vehicle is based on a typical EURO VI arrangement of exhaust gas aftertreatment components (see also Fig. 2 above):

A diesel oxidation catalyst (DOC) is followed by a diesel particulate filter (DPF) and further downstream the SCR system, consisting of a DEF/AdBlue® injector, SCR catalyst (copper monolith) and an ammonia slip catalyst (ASC). Integrated in a common housing, these elements form the so-called EATS box.

The CatVap® system was installed upstream of this box.

The CatVap® system used in this setup (Fig. 6, upper picture) consists of an exhaust gas flap (1), which is commonly used for medium duty (MD) applications. This flap ensures the required split of the exhaust gas mass flow by varying the attack angle. This allows the exhaust gas back pressure to be regulated, whereby a higher pressure leads to a larger exhaust gas mass flow through the bypass (3) and the CatVap® (4). The diameter of the exhaust gas flap is the same as that of the exhaust pipe.

The bypass flap (2) is used only in two stages. It is fully opened in normal operation and is only particularly closed if very low exhaust gas mass flows (3…20 kg/h) have to be realized via the bypass (3). This is only necessary in DOC or DPF Regeneration Mode, by closing the bypass flap to 85%, what results in increasing pressure losses. This procedure enables very precise regulations of the exhaust gas mass flow. In future designs, the bypass flap is not required anymore.

To minimize heat losses, the CatVap®, as well as the exhaust pipe behind the exhaust flap and the bypass pipe behind the bypass flap were thermally insulated. The downpipe from the engine was not insulated.

The lower picture in Fig. 6 shows the future CatVap® design which will have an integrated bypass system, only one active exhaust flap and minimize heat losses. It is much more compact than the current prototype.

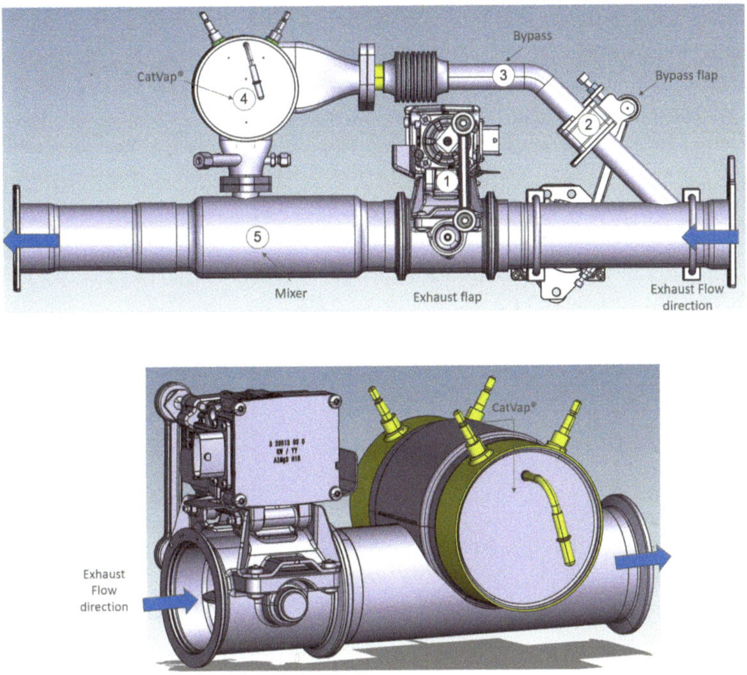

Fig. 6. 3D Modell of current CatVap® prototype and future CatVap®

The vehicle is equipped with a 5,2 l four-cylinder medium-duty diesel engine (Euro 6d), with a max. engine power of 140 kW and a torque of 550 Nm. The engine's own exhaust flap was deactivated during these test series. Other thermal management measures were deactivated except for the purpose of DPF regeneration. The tests were performed with standard diesel fuel.

Figure 7 shows the test setup equipped to the vehicle.

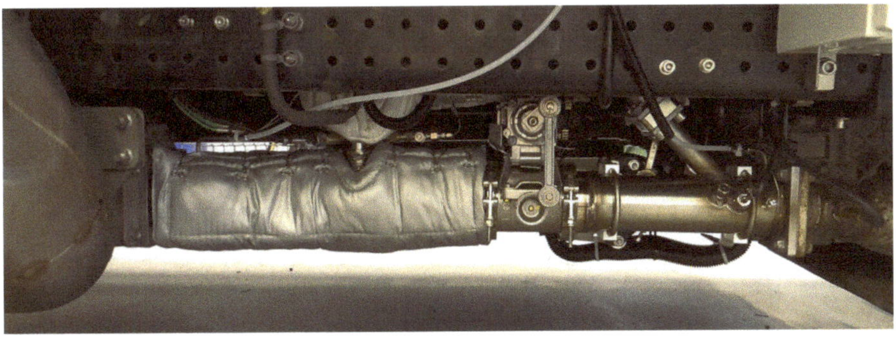

Fig. 7. CatVap prototype *on* test vehicle

The road trials were carried out in accordance with the EURO VI RDE test specifications for vehicle class N2.

The route driven led through the middle of the Ruhr-area with a share of 45% urban, 25% rural and 30% highway roads and is shown in Fig. 8.

Fig. 8. RDE Route through the „Ruhr-*area*"

The RDE testing series were proceeded during November 2023 with ambient temperatures of about 0 °C. Before each test cycle, the vehicle was parked outside overnight to ensure complete cooling and cold-start conditions.

The exhaust emissions were measured at the tailpipe by a PEMS system (mobile FTIR), which was provided by IAG nG.

For comparison purposes, additional RDE trials were carried out with a deactivated CatVap® system to get data for reference (baseline measurements).

5 Test Results and Evaluation

In the following diagrams, the baseline measurements (CatVap® deactivated) are represented by black lines, while the measurements with activated CatVap® are illustrated by red lines. Dashed lines reflect the measured temperatures before the DOC (TbDOC) respectively before the SCR catalyst (TbSCR); solid lines represent measured temperatures after the DOC (TaDOC) and accordingly after the SCR (TaSCR) (Fig. 9).

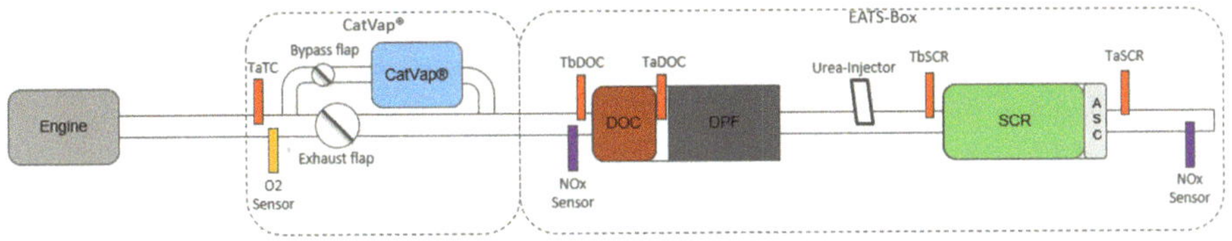

Fig. 9. Exhaust system w/ sensors

The calculation of the emissions was done with respect to the proposed Euro 7 emission regulation [1, 2] with a power threshold of 6% for valid Moving Average Windows (MAW).

During the first 600 s of the performed RDE test cycles, the temperature after the turbocharger (TaTC) is most of the time below 200 °C. The system supported by CatVap® manages to reach the necessary light-off temperature of the SCR catalyst after 270 s, while the base system was unable to raise the SCR to the necessary temperature, even after 600 s; see Fig. 10.

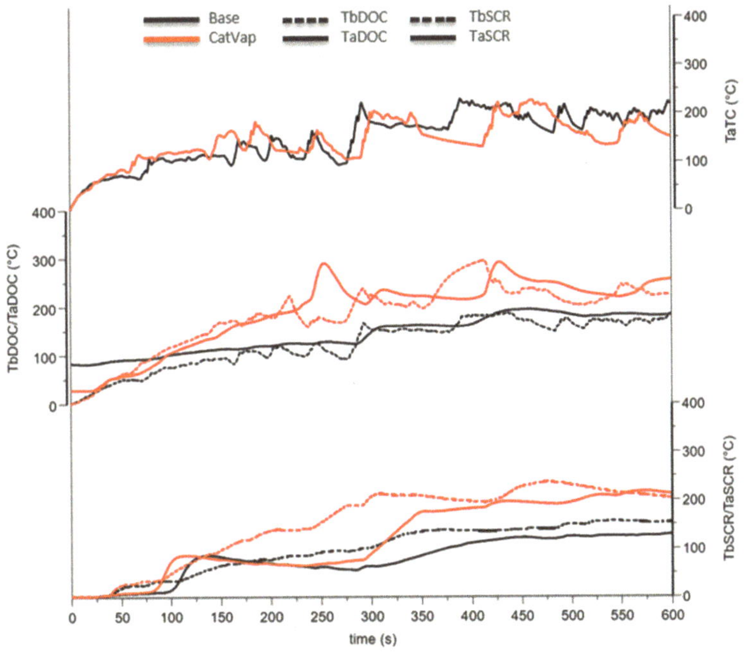

Fig. 10. RDE performance—First 600s of test cycle

With CatVap® activated, the temperatures after DOC (TaDOC) and before SCR (TbSCR) are found continuously above the critical SCR light off temperature of 190 °C, during the entire RDE cycle, as shown in Fig. 11.

In comparison, the results of the basic system (no use of CatVap®) indicate that is not able to deliver sufficient heat to the exhaust aftertreatment system. Therefore, the SCR cools down in this case. Due to this, the efficiency of the SCR decreasing and a larger slip of NO_x (see Fig. 11, too) is caused.

In certain cases, the SCR temperature dripped so much, that the AdBlue injection was no longer enabled.

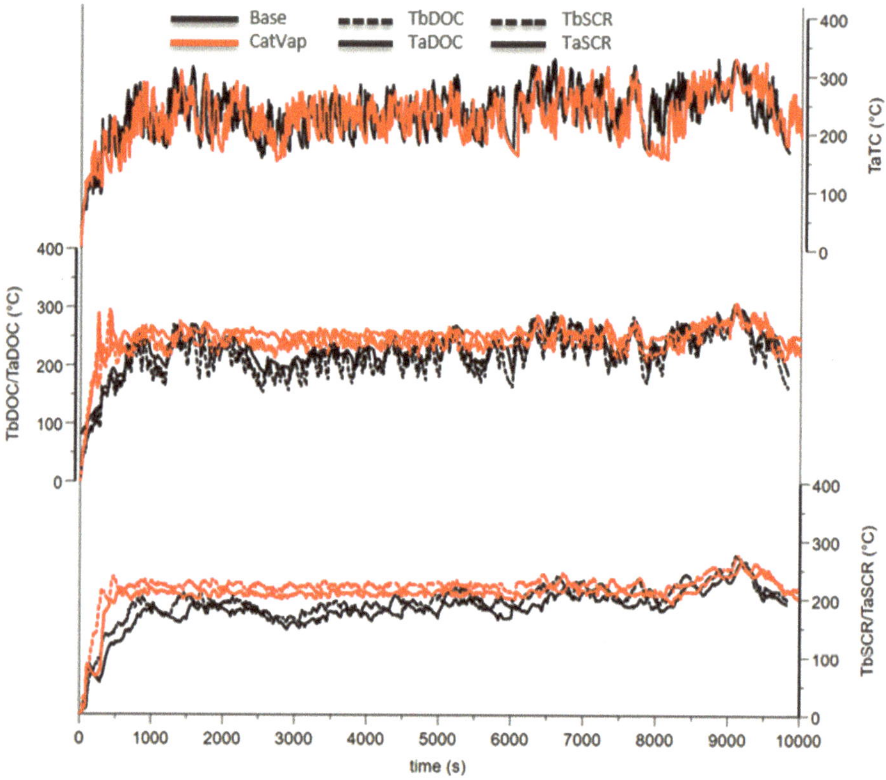

Fig. 11. RDE performance—complete test cycle

As mentioned before, Fig. 12 shows the NO_x emissions of the entire RDE test cycle. Nitrogen oxides were more effectively reduced by using the CatVap®, as the SCR catalyst was heated up faster than the system without the CatVap®. Ensuring a continuously higher temperatures in the SCR results in a high conversion rate during the complete test cycle.

In Fig. 11 it is clearly obvious, that the basis application is able to reduce nitrogen oxides much later and that there are considerable nitrogen oxide breakthroughs due to insufficient heat in the SCR.

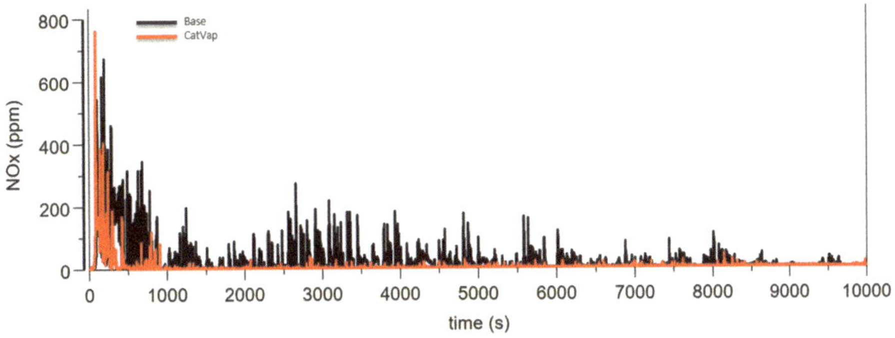

Fig. 12. RDE performance—NO_x emission

The described behavior and capabilities of the supplied heat by the CatVap® results in a higher cumulative mass of nitrogen oxides regarding the test cycle (Fig. 13): the system with activated CatVap® emits 72,83% less nitrogen oxides in relation to the same amount of work (4× WHTC work).

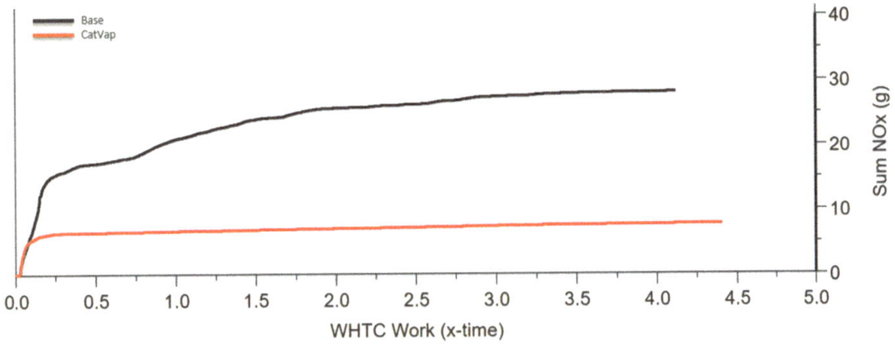

Fig. 13. RDE performance—cumulated NO_x emission

During the starting process, the CatVap® is heated electrically until the chemical catalytic process is stabilized. This took 135 s in the performed RDE test cycle and is shown in the below diagram (detailed view of the first 600 s. Value 1 means enabled, value 2 means disabled) of Fig. 14. After that period, no electrical power is consumed by the catalytic converter (see Fig. 4 and Fig. 5), except in cases where the CatVap® outlet temperature dropped below 300 °C. Cases like that are for example longer non-use periods, after which the system has been restarted. In such cases, the system consumes electrical power only for a few seconds. This procedure ensures that the catalytical process is restarted immediately and the exhaust system can be supplied with sufficient heat.

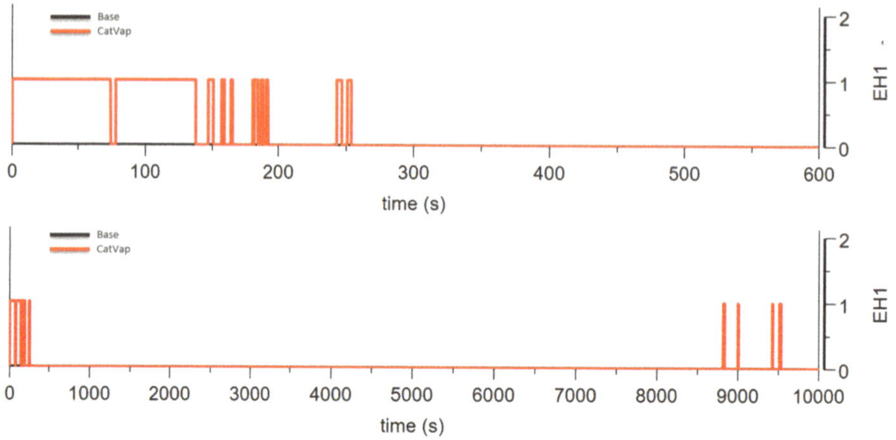

Fig. 14. RDE performance—CatVap® electrical heater activation time

Figure 15 shows the generated heat (in kW) during the entire RDE test cycle. Up to 25 kW of heat is generated during the start phase, to achieve a fast and sufficient temperature level for effective NO_x reduction.

Once the target temperature has been reached, the supplied heat is reduced to the necessary minimum. If the exhaust gas temperature is high enough, the CatVap® is deactivated and reactivated when temperatures are dropping. This "power on demand" application is intended to prevent superfluous provision of heat and reduce fuel consumption to a minimum.

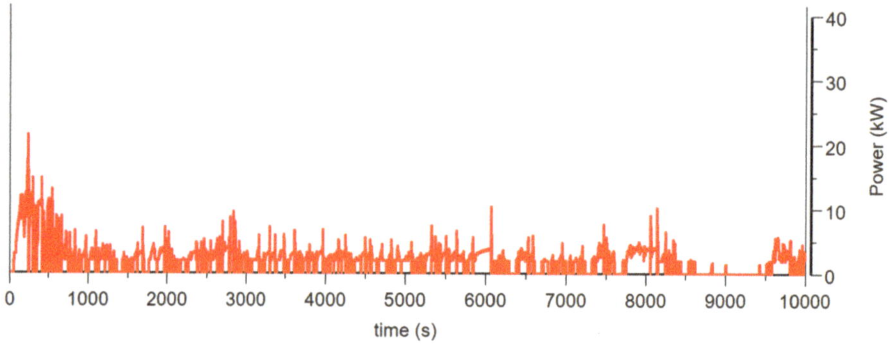

Fig. 15. RDE performance—CatVap chemical thermal power

The result of the emissions analysis is shown in Fig. 16. Without CatVap®, the NO_x emissions are far above the current proposed Euro 7 limit of 260 mg/kWh. With the support of CatVap®, NO_x emissions are 24% of the limited value. All other emissions, such as CO, CH_4, N_2O and NH_3, meet the development target value of a maximum of 80% of the respective proposed Euro 7 limit value. Therefore, they do not pose a problem.

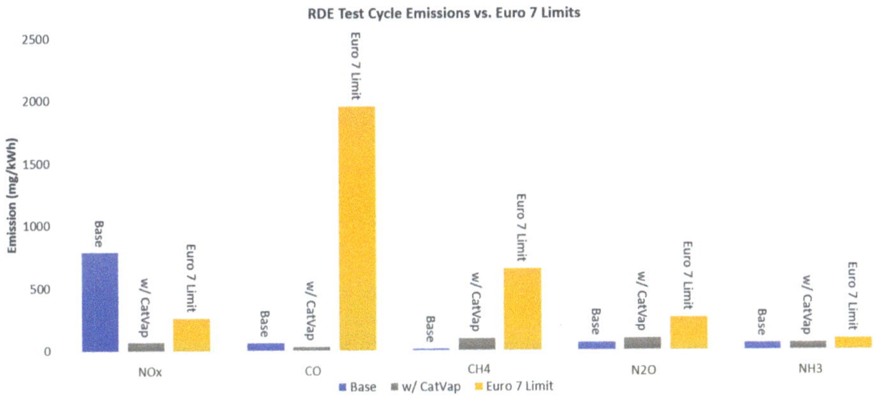

Fig. 16. RDE performance—Overview of emissions

The fuel consumption determined via the cumulative CO_2 emissions (Fig. 17) is 1,61% slightly higher with CatVap at 4-times WHTC work than with the basic measurement. This small increase can be compensated by adjusting the engine control accordingly.

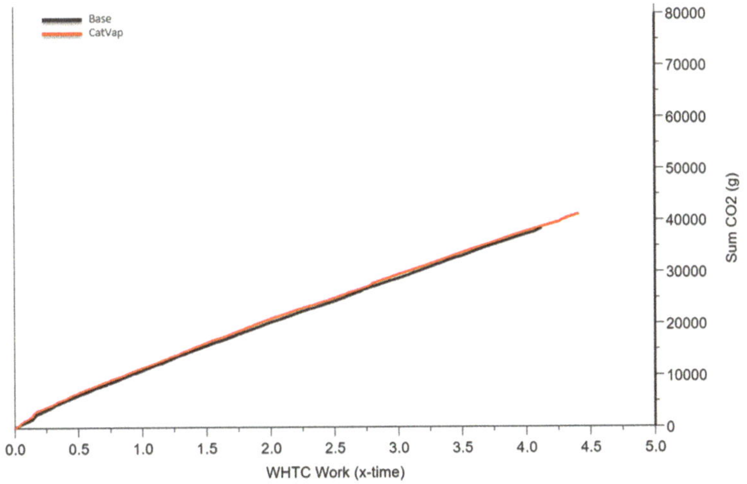

Fig. 17. RDE performance—cumulated CO2 emission

6 CatVap® Tests with "Hydrotreated Vegetable Oil" (HVO)

CatVap® is designed as a system that can be operated with several fuels, which each variant delivering the same heat output.

For the purpose of verification, hydrotreated vegetable oil (HVO) was chosen as the test fuel in order to prove that CatVap® can also be capable with alternative fuels to generate the same heat outputs, resulting in comparable temperature curves.

The comparison was carried out during a cold WHTC cycle on an engine test bench.

The following diagram (Fig. 18) shows two measurements with standard Euro 6/ VI diesel fuel (black and green) and one measurement with HVO fuel (red). All measurements were taken with an active CatVap®. Dashed lines reflect the measured temperatures before the DOC (TbDOC) and before the SCR catalyst (TbSCR); solid lines represent temperatures after the DOC (TaDOC) and after the SCR (TaSCR).

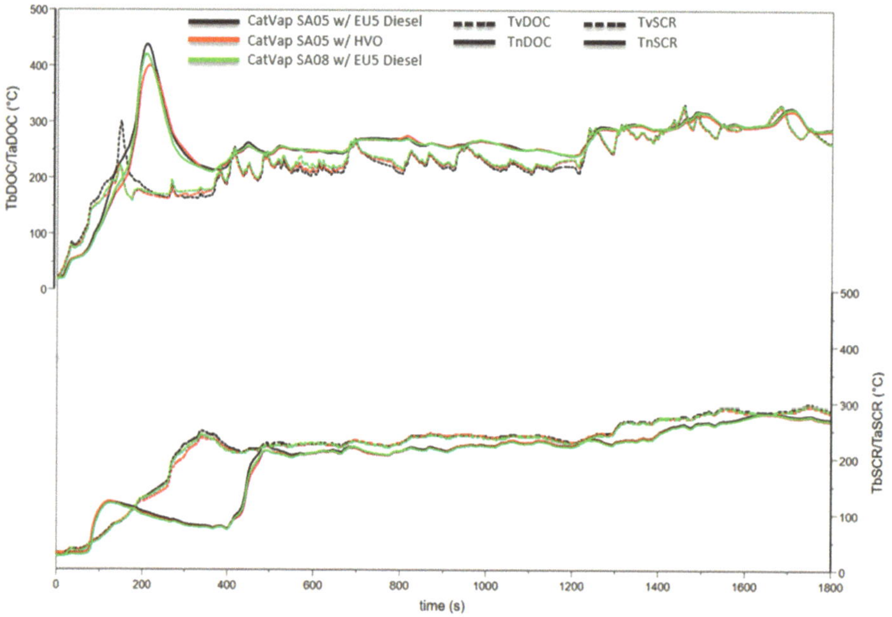

Fig. 18. WHTC Cold Perfomance

As a results out from these tests, it can be found, that there are no significant differences between the individual test runs and different fuels. After approx. 280 s the light-off temperature of the SCR catalyst was reached in all tests, so that the AdBlue® dosing release status was achieved.

These results demonstrate that there are no differences between standard diesel fuel and HVO in the performance of the CatVap® system under test bench conditions.

7 Summary

The RDE test cycles, carried out on a commercial vehicle equipped with a medium-duty diesel engine and a Euro VI exhaust system with a copper zeolite SCR catalyst, confirmed the positive results of previous experiences from the engine test bench. The promising potential of CatVap® is demonstrated successfully.

Under low ambient temperature conditions, during the entire test drives, the heat-up time of the EATS components was significantly reduced.

The amount of emitted NO_x emissions, measured at tailpipe end, was 72,83% lower with CatVap® than in the baseline measurement. In relation to the proposed Euro 7 nitrogen oxide emission RDE limit (Fig. 15) of 260 mg/kWh, CatVap® combined with a standard Euro VI exhaust system configuration was able to undercut the limit values by 76%. The other measured emissions were more than 35% below the proposed limits. The current European Union directives were used to evaluate the emissions, with a performance limit of 6% for valid MAW [2].

The compatibility of CatVap® with HVO fuel was proven. No significant differences were measured during the cold WHTC cycle on the engine test bench in terms of temperature behavior of the exhaust system.

The used prototype showed consistent performance and no signs of failure throughout all the tests. Neither the electrically heated catalytic converter nor the coating used on it failed. The test results could be repeated several times. A total mileage of approx. 10.000 km was completed during the test drives in 2023.

References

[1] 2022/0365(COD) - ANNEXE I to the Proposal for a Regulation of the European Parliament and the Council on type-approval of motor vehicles and engines and of systems, components and separate technical units intended for such vehicles, with respect to their emissions and battery durability (Euro 7) and repealing Regulations (EC) No 715/2007 and (EC) No 595/20091 from 21 December 2023.

[2] UN Regulation No 49—Uniform provisions concerning the measures to be taken against the emission of gaseous and particulate pollutants from compression-ignition engines and positive ignition engines for use in vehicles [2023/64] from 16 December 2023

[3] Szolak, R. et al. (2022). CatVap®—the Efficient Heating Technology for Future Exhaust Aftertreatment Systems. In: Liebl, J. (eds) Heavy-Duty-, On- und Off-Highway-Motoren 2021. Proceedings. Springer Vieweg, Wiesbaden. https://doi.org/10.1007/978-3-658-38105-9_12

Hydrogen Internal Combustion Engine Dyno Test Results with a Driven-Turbo

Thomas Waldron[1(✉)], Jared Brin[1,2], and Hans-Felix Seitz[1,2]

[1] SuperTurbo Technologies, Loveland, USA
{twaldron,jbrin}@superturbo.net, felic.seitz@avl.com
[2] AVL List GmbH, Graz, Austria

1 Introduction

The SuperTurbo is a mechanically driven turbocharger that has been developed for commercial diesel engines to improve efficiency, emissions and performance. When presented with the unique challenges for hydrogen as an engine fuel, it became apparent that the SuperTurbo could specifically address several of the challenges seen on these engines. *Rapid response and precise delivery of airflow can allow a H2 ICE to maintain a desired air fuel ratio (AFR) throughout dynamic operating conditions.*

This paper will primarily focus on a the 13L engine operating with port fuel injection (PFI) and zero exhaust gas recirculation (EGR). Understanding the performance and limitations with this architecture is a key starting point, as it represents the most expedient and cost-effective approach to early implementation of H2 ICE. This paper will cover the WHTC cycle testing and then more specific studies related to steady-state maps and load step transients. The baseline evaluated for the study is a VGT, however analysis of other boosting architectures will also be reviewed.

High excess Air Ratio (or Lambda) is critically necessary to prevent an exponential rise in engine NO_x emissions for hydrogen engines [5]. Fig. 1 below shows this relation as tested on the AVL engine. Holding an optimal Lambda prevents NO_x creation and also can prevent unwanted hydrogen slip conditions often seen with excessively high Lambda. The SuperTurbo, by design, is an on-demand air system that is well suited to consistently control ideal Lamba conditions. This paper will show the challenges faced by traditional turbochargers when exhaust enthalpy is not adequate to maintain high compressor power, especially in transient cycles.

A. Heintzel (Hrsg.): Int.Engine.Congr. 2024, Proceedings, S. 264–282, 2025.
https://doi.org/10.1007/978-3-658-47966-4_19

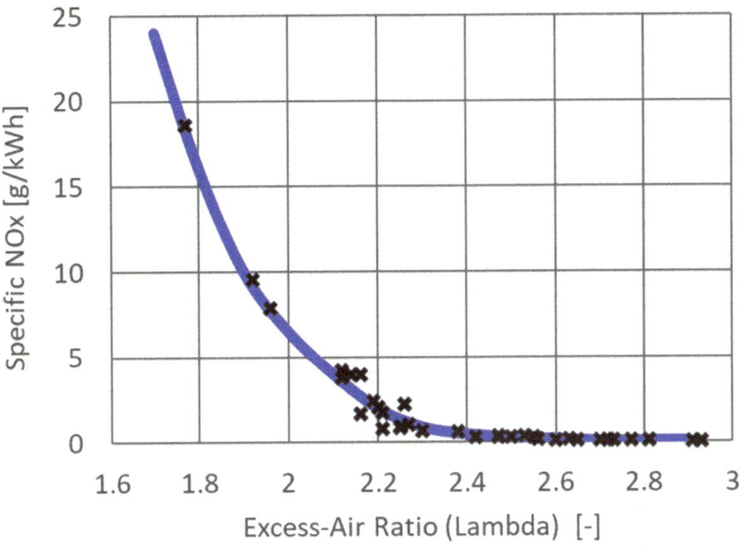

Fig. 1. NO_x vs Lambda from AVL 13L H2 ICE testing

In addition to the problems created by the necessity to maintain lean-burn operation, there are several other areas that must be addressed for H_2 ICE. Reaching high BMEP levels is important for truck and equipment engines that require high relative power versus displacement. Maximizing BTE and BSFC to lower operating costs and extend vehicle range is also important.

2 Background

2.1 Engine

AVL has successfully demonstrated the possibility to convert a 13L CNG heavy duty engine into a hydrogen engine.[4] The demonstrator has proved the feasibility to reach a BMEP of 21bar+ with BTE of 42% and single stage turbocharger. To achieve more flexibility on engine tests in view of boost pressure demands a turbocharger with variable turbine geometry was used. Following baseline VGT testing, the SuperTurbo was then adapted to the engine. Fig. 2 below shows the SuperTurbo installed on the mockup engine for test fitting and also installed on the new test engine in the dyno cell. The SuperTurbo is often installed with a mechanical connection at the PTO, which provides more compact packaging and increased mechanical efficiency, however for the purposes of this test program, a simplified and more expedient connection at the front of the engine was utilized. Interestingly the high airflow demand for H2 requires a larger ratio from the crankshaft to the SuperTurbo, so the normal diesel input ratio of 3:1 was increased to 4:1 (Table 1).

Fig. 2. AVL 13L Engine + SuperTurbo (Mockup and Test)

Table 1. 13L Engine specification

Test Engine		AVL HD IL6 Hydrogen Engine
Properties		
Bore	mm	130,0
Stroke	mm	161,0
Volume/cylinder	ltr	2,14
Swept volume	ltr	12,82
Combustion System		Swirl guided, Central spark plug
Valves and Valvetrain		4V / OHC
Compression Ratio	-	10 : 1
Hydrogen FIE		MPI with ~ 15 bar gas pressure
Charging System		Single stage, intercooled
Air trottle		Intake throttle for part load
Compressor surge protection		Compressor blowoff vave
Performance		
Max. torque	Nm	2030
Max torque speedrange	rpm	900 – 1100
Rated power	KW	276
Rated speed	rpm	1700 – 1800
Turbocharger		**SUPERTURBO**
Compressor		Single Stage, ext. blow-off bypass valve
Turbine		Fixed geometry, free floating
Turbine Stator Variant		10% and 20%
Mechanical drive		belt
Pulley Ratio		4.0:1 and 4.2:1
Mechanical connection		SuperTurbo CVT

In order to show the benefits of the SuperTurbo technology on a hydrogen engine, investigations by means of 1D simulation were performed. A GT-Power (GTP) engine model with H2 combustion (MPI gas injection) was created and calibrated to AVL's hydrogen engine test results. This simulation model was the baseline for the investigations with the SuperTurbo charging technology. The model can calculate a complete engine map (from full load to low engine load of 4bar BMEP) for the investigation of different charging concepts. The hydrogen combustion is modelled based on test data and AVL experience. An engine data based NO_x emission prediction is applied. That simulation effort, as presented last year [9] serves as the basis for the dyno test program and allows for a comparison to the engine dyno results.

2.2 SuperTurbo

The SuperTurbo designed for commercial vehicles covers an engine displacement range from 7L to 15L+. It combines and improves upon the capabilities of a supercharger, turbocharger and turbo-compounder. The internal components are common across this range while the turbine and compressor designs are changed based upon the individual engine requirements. With 90% part commonality, the device is designed to economically scale and cover most of the heavy commercial industry. In order to maximize benefits, this large SuperTurbo is specified with a continuous power limit of 30 kW (primarily for compounding) and an instantaneous power limit of 50 kW (primarily for supercharging). The SuperTurbo is designed to mechanically integrate with the engine through connections at either the PTO or FEAD. The mechanical connection includes a controlled clutch and torsional damper. The core components or subassemblies are designed as cartridge-like units that are contained in a common housing. The two core components are the speed reducing fixed ratio planetary which controls the turbo shaft and the CVT which allows for overall ratio adjustment [8]. The primary control centers around delivering precise boost pressure and MAF. While following air commands the SuperTurbo will be supercharging when turbine power is less than compressor power (example: transient) and likewise compounding when turbine power exceeds compressor power (example: highway cruise or higher engine power).

Fig. 3. SuperTurbo cross section

3 WHTC Results

With future regulations focused on cycle-based compliance, the most important simulation result from this study is the comparative World Harmonized Transient Cycle (WHTC). Fig. 4 shows a comparison of the cycle simulation versus the actual cycle dyno test result for both the VGT and the SuperTurbo as they relate to engine-out NO_x creation. Future NO_x regulations (Euro7, China7, CARB, EPA) [1, 3] are focused on ultra-low NO_x emissions and hydrogen engines will need to meet or exceed the most stringent standards.

The peaks of instantaneous NO_x output during WHTC transient sections with the VGT highlight a fundamental challenge. Holding high Lambda during transient sections is difficult given available turbine power in comparison to desired compressor power. This results in large peaks of NO_x creation as shown in Fig. 4. The accumulated NO_x is also shown in total grams through the cycle. Please note that the laboratory NO_x sensor limits at 5000 ppm, so VGT based NO_x peaks exceeding that value are not shown.

The primary control mechanism utilized is a lambda controller. There is a necessary compromise for the exhaust driven turbo that has to balance the transient response necessary to run a qualified cycle, the available airflow and the fuel limiter. Individual transient examples will be evaluated later in the paper to detail this effect.

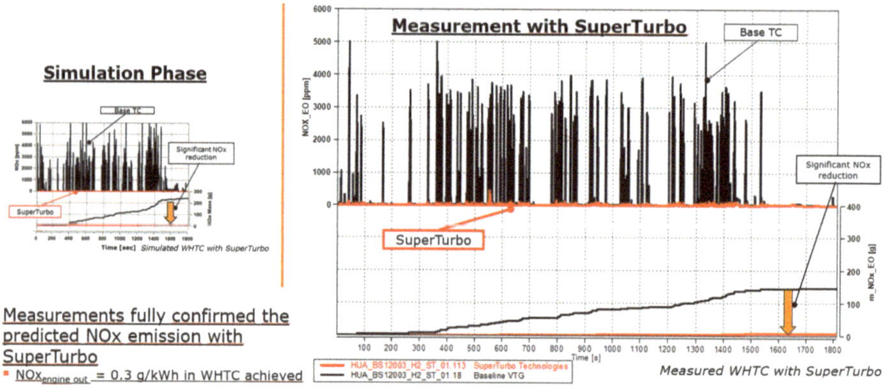

Fig. 4. Comparative WHTC NO_x

Access to supercharging power has demonstrated the ability to lower NO_x and increase transient response + engine work. It could be assumed that pulling supercharging power from the engine would have a large negative net effect on engine efficiency. However, the simulation and engine test have shown that a cycle efficiency penalty is not present.

When supercharging the SuperTurbo is only taking power from the engine to overcome inertia and compensate for inadequate turbine power. The turbine is still actively feeding the transient rise and thus the SuperTurbo is combining exhaust power and crankshaft power to 'efficiently' supercharge.

Table 2 below shows a summary of NO_x emissions and fuel consumption through various cycles for the different boosting systems. It is not surprising that more boosting equates to higher Lambda and thus lower NO_x. Interestingly, more boosting from the SuperTurbo also equates to good fuel consumption. This higher efficiency comes from several sources. The SuperTurbo does turbo-compound through much of the cycle, which mostly negates the cumulative supercharging power draw. The maintenance of lean-burn conditions keeps temperatures lower and reduces heat transfer. The compounding effect can be seen providing better efficiency at a higher loaded cycle like the NRTC.

Table 2. Various cycle NO_x and BSFC

Unit	Test #	BSFC [g/kWh]	EO NOx [g/kWh]	Avg. Lambda [-]	Engine Work [kWh]
WHTC					
VGT	18	85.5	6.05	3.18	26.0
ST	113	87.7	0.30	3.50	25.8
	144	86.8	0.33	3.46	25.9
NRTC					
VGT	20	89.3	3.72	2.55	29.2
ST	103	88.1	0.37	2.98	28.2
	157	87.4	0.45	2.98	28.0
FTP					
ST	105	84.0	0.30	3.13	22.8

Table 2 shows that conceptually when controlled properly the SuperTurbo can result in Euro6 complaint NO_x on the WHTC without aftertreatment. It is still advisable to consider an SCR for different cycles and real-world driving. Euro7 and similar standards will require NO_x aftertreatment, although they could be significantly simplified with these low levels of engine out NO_x.

To further understand what the SuperTurbo is doing through the cycle, Fig. 5 outlines both the power and accumulated work of the device. The simulated 'recovery' cycle with a less aggressive Lambda target has smaller supercharging power peaks and higher levels of turbo compounding. The net effect is >1 kWh recovered back to the engine over the cycle. The simulated 'boosting' cycle with a higher Lambda target has large supercharging peaks into the ~35 kW range and less compounding. Again, the SuperTurbo mechanical specification is 50 kW of instantaneous supercharging power and 30 kW of continuous compounding. The ‘boosting’ control strategy results in a more balanced, yet slightly negative total SuperTurbo cycle work.

The bottom of Fig. 5 pulls data from the WHTC dyno test. The black line indicates SuperTurbo power from the dyno as compared to the simulated blue line. This data is included to show that the SuperTurbo is continuously transitioning between supercharging and compounding during the dyno test cycle and help validate the simulation.

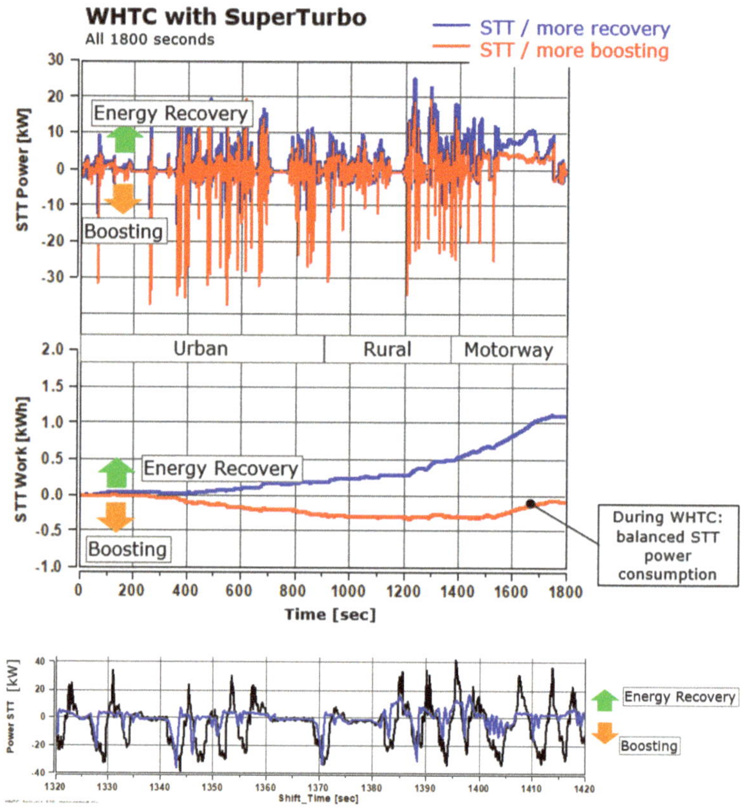

Fig. 5. WHTC SuperTurbo Power (Simulation top and dyno test below)

To better understand the full cycle results, it’s also necessary to analyse specific sections of the cycle in more detail. Fig. 6 shows a dynamic section of the WHTC from the engine test. This section of the WHTC was chosen to highlight intensive gradient load increases. The result shows how the simulation compares to the actual engine test. One noteworthy difference between the initial simulation and the dyno test was the laboratory measured NO_x was lower than predicted relative to laboratory measured lambda values.

The SuperTurbo is able to follow a lambda limiter with relative precision. This higher Lambda limit results in near elimination of the NO_x peaks in this transient section. Additionally, the SuperTurbo is able to better follow the BMEP demand trace of the cycle. This section of the WHTC as shown in Fig. 6 is demonstrative of the full cycle results that were previously shown in Fig. 4.

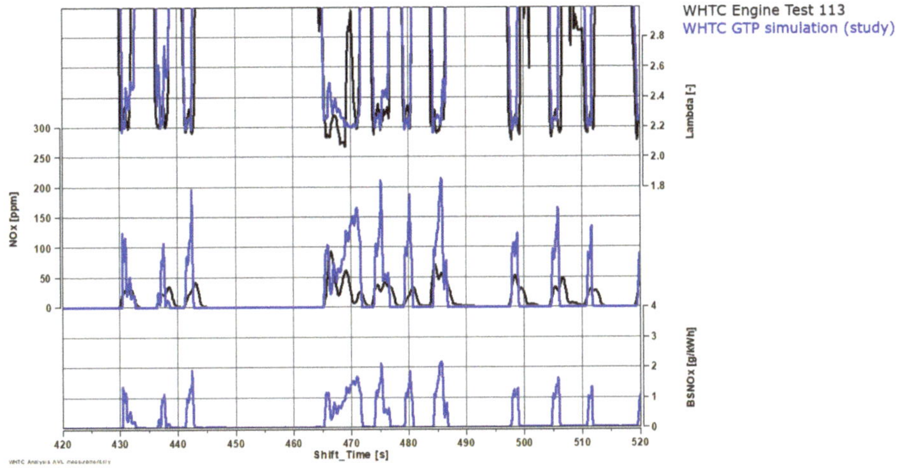

Fig. 6. WHTC Transient Section Lambda Control

The conclusions drawn from the full WHTC H2 ICE simulation are that having a driven-turbo like the SuperTurbo can:

- Achieve ultra-low cycle NO_x
- Improve total engine work and driveability
- Produce good overall efficiency

4 Transient Simulation Analysis

The AVL and SuperTurbo project also included a variety of specific transient response evaluations, some of which will be shared in this paper. Looking at individual transients in more detail can be helpful to understand the full cycle results. The primary constraints and variables to balance against each other in transient scenarios are: Lambda limits, NO_x creation, and torque/power response. Fig. 7 top section shows the simulated VGT sensitivity to Lambda limits and compares them against the SuperTurbo. Fig. 7 bottom section shows the dyno measurements that correlate with this analysis.

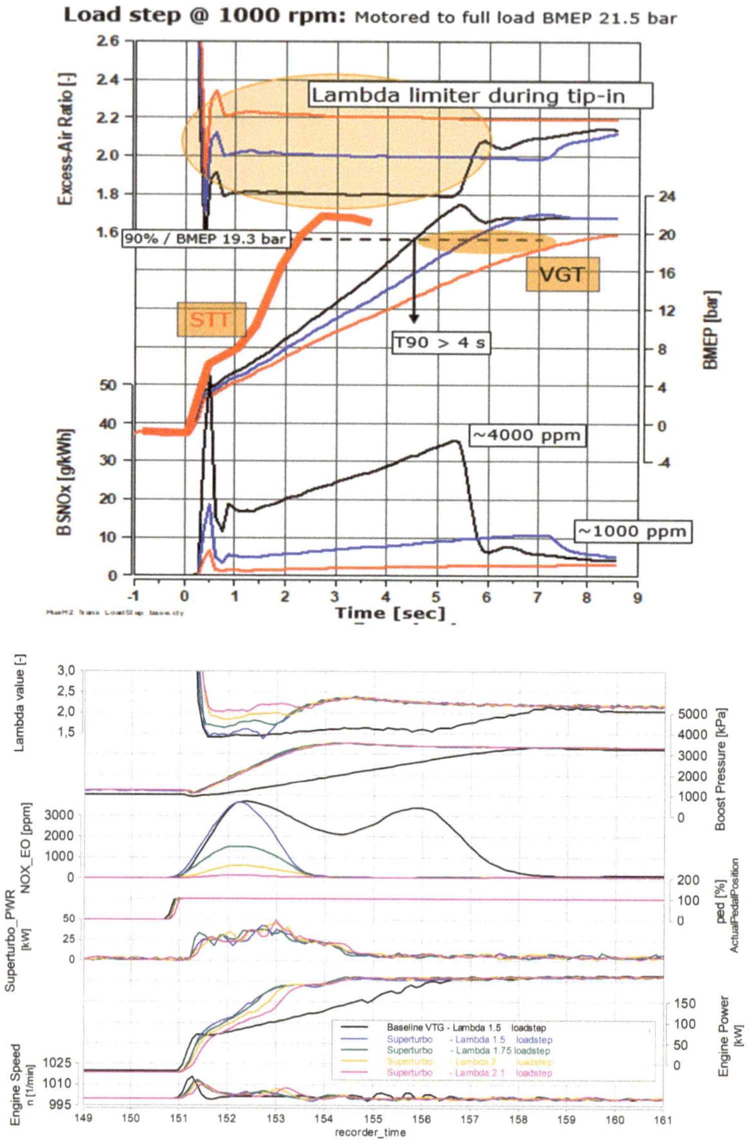

Fig. 7. Variable Lambda transients VGT vs SuperTurbo (Simulation top and dyno test below)

On the engine dyno test the VGT is demonstrating the fundamental problem related to available exhaust/turbine power versus the compressor power needed to maintain high lean burn. The slow rise in boost pressure relates directly to high NO_x output and slow engine response.

Once the SuperTurbo was installed on the engine, it was possible to run transient evaluations with varying lambda limits, all while holding similar fast engine response. This includes lean burn levels to the point where NO_x creation is significantly reduced.

Analysis like these (and others) were completed to show both the flexibility and limits of a controllable driven-turbo. These results show how one can target a transient response time and adjust a NO_x output within a given power limit (and correlating boost response) from an on-demand air system. Noteworthy from these transient tests is the requirement to access significant levels of supercharging power. For ultra-low NO_x the SuperTurbo is utilizing up to it's 50 kW power capacity.

5 Steady State Analysis

Simulation of steady state conditions is fundamental to building the full drivecycle analysis and establishing those models. Properly matching turbine and compressor maps all begin in the steady state environment. The AVL + SuperTurbo simulation project included several iterations of different turbine and compressor options. The turbine design balances compounding power, pumping losses, turbine efficiency and pressure ratio (and ability to drive EGR if desired). The turbine can be designed to target compounding and efficiency at different engine operating points which then have correlating sacrifices at other points. Example: a more restrictive turbine will increase compounding power at lower engine power and speeds, but will have higher pumping losses at rated power. A more free flowing turbine will perform well at rated power, but not maximize exhaust power collection for on-highway cruise conditions. Since the SuperTurbo has reduced concerns over turbine weight and inertia and no concern for overspeed, it allows a different approach to efficient turbine design[7]. Compressor maps for H2 ICE and their need for high Lambda operation are uniquely stressed. Single stage compressors need to maximize pressure ratio and mass flow to facilitate lean-burn along the lug curve, especially for high BMEP applications.

The tuning of the turbine and compressor maps for the AVL 13L engine were focused on the mid rpm ranges with power in the 75–100% range. Fig. 8 (top section) shows a comparison of BTE for the SuperTurbo vs. baseline VGT in the simulated environment. Peak BTE gain aligns with maximum turbo compounding power, however turbine design was specifically modified to add benefits to areas central to on-highway vehicle operation. BTE benefits are the result of both compounded exhaust energy recovery and the ability to tune optimal air flow for combustion efficiency.

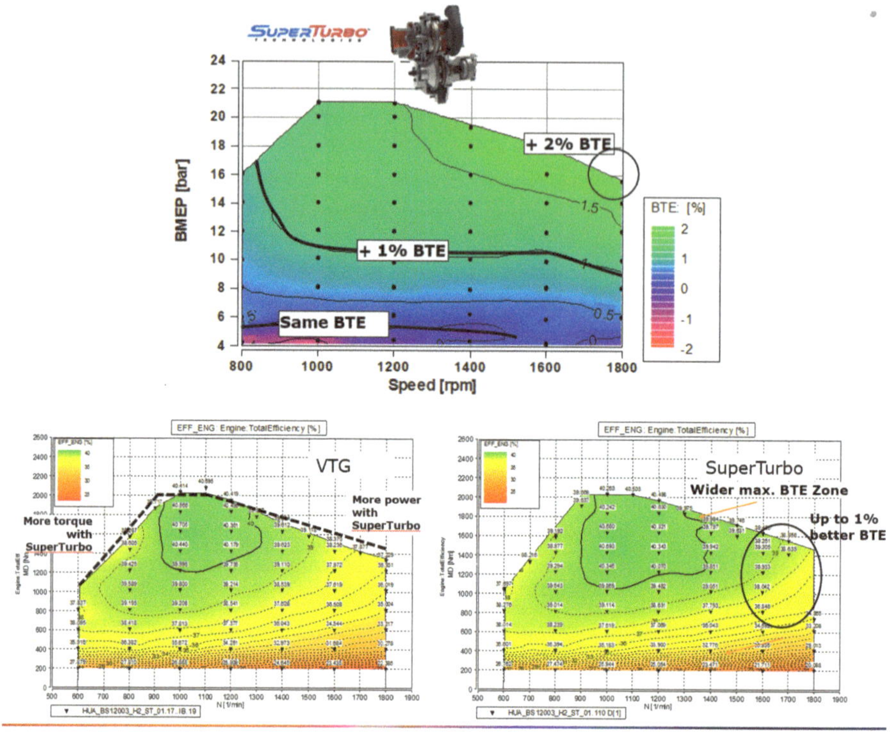

Fig. 8. BTE Comparison of VGT vs SuperTurbo (Simulation top and test below) *non-match color schemes

Figure 8 (bottom section) shows the comparative BTE for the SuperTurbo as compared to the VGT from the engine test. With the ability to increase airflow at low rpm, the SuperTurbo was also capable of increasing the low-end lug curve. Additionally, the compounding power was capable of increasing the rated power condition of the engine. The SuperTurbo was also capable of widening the overall maximum BTE zone.

Understanding the source of steady state efficiency gains also requires clarity on one of the unique functions of the SuperTurbo. The compounding of exhaust power back to the engine is predictable in the steady state environment, and obviously more dynamic in a transient cycle. The simulation and test have shown that there is available exhaust enthalpy to create compounding conditions on the 13L hydrogen engine.

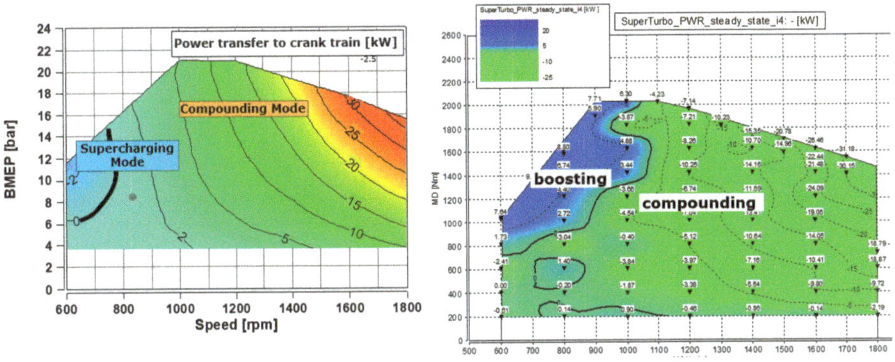

Fig. 9. SuperTurbo Steady State Power (Simulation Left, Dyno Right)

Compounding power will adjust according to several factors, among them: turbine efficiency, turbine pressure ratio and inlet temperature, required compressor power, SuperTurbo speed command, and several others. Fig. 9 (right) shows a steady state analysis of how the SuperTurbo is operating. High levels of compounding, as expected, occur in regions where there is significant exhaust energy. While in-cylinder combustion efficiency is always prioritized, the compounding effect can contribute significantly to overall engine efficiency. There are regions where upwards AFR adjustment can and should be made by accessing supercharging power. In the case of H2 ICE, supercharging should be applied in steady state conditions requiring higher Lambda values not otherwise achievable with standard turbochargers. Low RPM and high load conditions can benefit from efficient supercharging (small adds to turbo power) to reach a more optimal AFR for both combustion and NO_x creation.

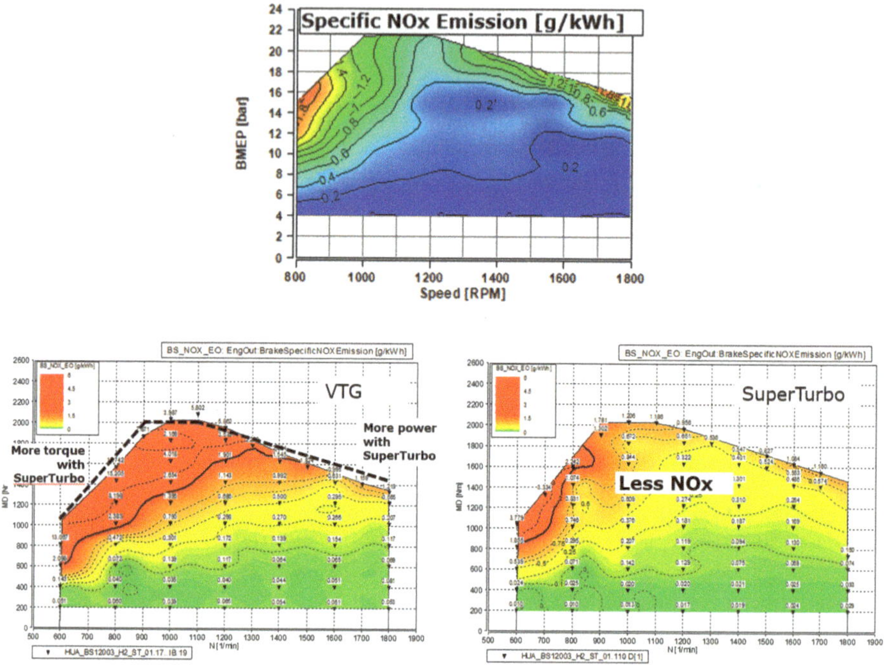

Fig. 10. Steady State NO_x (SuperTurbo simulation top and both systems dyno test below) *non-matching color schemes

Figure 10 shows comparative steady state NO_x maps. As predicted in the simulation, the overall NO_x map can be kept below 1 g/kWh with the SuperTurbo and much of the map is ultra-low. A compressor with adequate mass flow to hold rated power lean conditions is critical, as is the ability to supercharge at high-power low-rpm conditions. All of the simulation results shown up to this point have been based on PFI with no EGR. By design this simulation project was focused on proven and tested architectures for H2 ICE which could also translate to expedient implementation and low total system cost.

6 Additional Analysis

6.1 Improving Efficiency

While this engine test program was focused on validating the NO_x and performance benefits, attention will necessarily shift to improving overall hydrogen engine efficiency. This test program utilized available hardware, including turbines and compressor. Conducting a deliberate design of both wheels to specifically fit the hydrogen engine would improve efficiency. Likewise, the SuperTurbo mechanical integration can progress from expedient to deliberate and reduce associated unnecessary parasitic losses. The proven ability to maintain dynamic lean burn also opens the door to increasing compression ratio and implementing more advanced direct injection. Perhaps the most impactful way of improving efficiency for this engine is in the combustion system. There was as significant drop in rate of heat release when running lean, and this came with a correlating extended burn duration. If this can be adjusted, measurable increases to total efficiency should result. Optimizing the system around the now proven lean burn capability will help push efficiency higher.

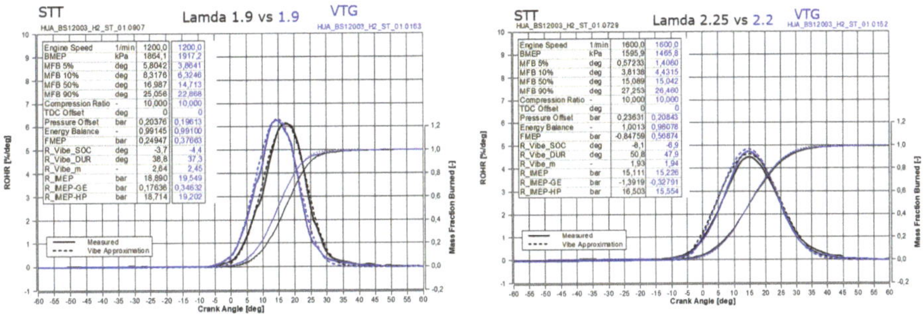

Fig. 11. Lambda dependent ROHR and Burn Duration

6.2 H2 ICE Compressor Design

A single stage compressor has limitations related to the requirements of a lean-burn operating map. At rated engine power, the compressor mass flow will be stressed and efficiency can drop quickly. Likewise, pressure ratios well surpassing 4:1 can be expected. Two stage turbocharging can address this problem, but will come with additional concerns, including: interstage cooling, turbo bypassing, packaging, increased back pressure, control complications, and low aftertreatment temperatures. So, to make a single stage compressor optimal for lean burn hydrogen, widening and extending the compressor map is beneficial, even if that means losing a few points of compressor efficiency. AVL and SuperTurbo are already designing and testing new compressor designs to specifically address these challenges and create a scalable answer. Fig. 12 shows gas stand testing of this new hydrogen compressor.

Fig. 12. Hydrogen specific compressor testing

6.3 Alternative Boosting Analysis

VGT and SuperTurbo are not the only ways to address the challenges for H2 ICE. As such, a more complete analysis of available options is required[10]. All systems should be first evaluated against the fundamental problem of engine transient response versus engine out NO_x. As such, several boosting architectures have been run through the AVL engine model. They include VGT, Two Stage Turbo, 48 V 10 kW electric compressor (paired with a FG Turbo), 48 V 10 kW electric turbo, 350 V 20 kW electric compressor (paired with FG Turbo), and 400 V 20 kW electric turbo. The simulated comparative results from the model can be seen in Fig. 13. The trade-off between transient response and NO_x for each system can be compared and related to an aftertreatment specification necessary to achieve future emission standards. Unsurprisingly, the more power that can be applied to supplement exhaust power, the better the result.

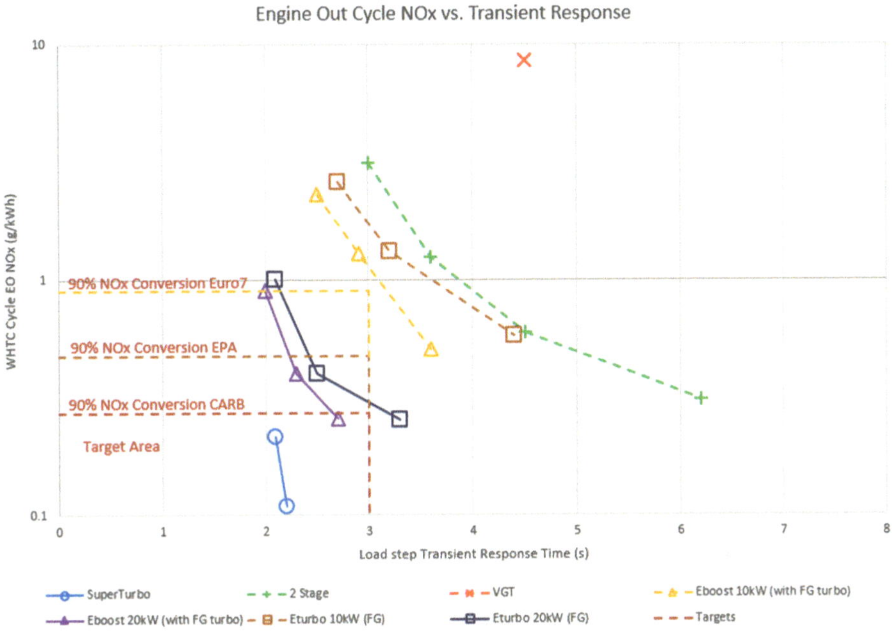

Fig. 13. WHTC NO_x versus 1000 rpm transient simulation

Boosting selection does include more variables than just NO_x and transient trade-offs. Further analysis has been conducted to attempt and capture a more complete picture of all relative aspects related to boosting architectures. Fig. 14 below pictorially captures several of the key considerations. The AVL 13L simulation model was used to calculate and quantify everything from the 12 o'clock to 6 o'clock on the spider diagram. Transient response is a 1000 rpm load step for each system at Lambda 2.2. Cycle NO_x is the full WHTC engine out NO_x with a transient limit of 3 seconds (4.5 for VGT). Dynamic lean limit is the highest lambda achievable with the same 3 second maximum transient. Steady state lean limit is based on the % of the compressor map being used at rated power (the 2 stage has clear advantages here, and the larger FG turbo that accompanies the e-compressor also benefits). Cycle efficiency is overall WHTC ideal simulated BSFC. Steady state efficiency is the BTE during the cruise section of the WHTC. Component cost is manufacturing costs as determined by detailed component level costing program with FEV. System costs (also determined by FEV) include the electrical system costs needed to support electrified boosting and also projected EAS costs that relate to the engine out NO_x profile from the AVL model. EAS temperature is based on data from several programs (VGT has higher temps as it can't hold lean burn while 2 stage will suffer from multiple heat sinks and pressure drops). Electrical system requirements are purely determined via Voltage. Some of displayed data is logarithmic to assist with clarity.

System Costs

Total H_2 ICE system costs can be lower with a driven turbo like the SuperTurbo. The SuperTurbo is more expensive than a VGT alternative, but a fully compliant engine and Exhaust Aftertreatment System (EAS) can be a lower total cost. Fig. 14 below shows the AVL estimate of EAS component reduction for future European emission compliance.

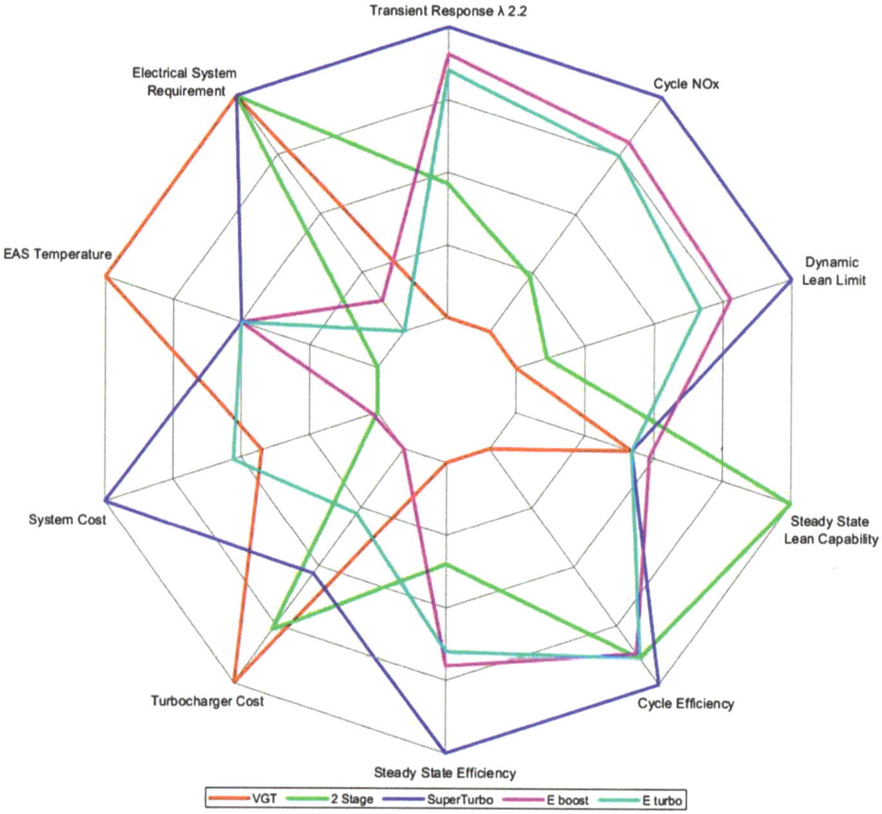

Fig. 14. Boosting selection spider chart

7 Summary and Outlook

This study proposes that H2 ICE system architectures can be produced in the near future at a low net cost and technology risk. This includes PFI, No EGR, and a downsized SCR based EAS. There is also no requirement to change the architecture for mild or high voltage electrification or hybridization of the vehicle.

The results presented here are select portions of current knowledge developed over years of joint simulation and testing efforts by SuperTurbo and AVL on hydrogen combustion. The conclusions of the work is that a mechanically driven turbo with SuperTurbo specifications shows:

- Faster, Diesel-like torque build up can be achieved, even at high Lambda
- NO_x peaks during acceleration can be eliminated
- Significant raw NO_x reduction towards EURO VI emission requirements can be demonstrated
- Simplifications in the EAS system are feasible (single stage SCR, less Urea consumption)
- Good fuel consumption in the WHTC is achieved

Largely based on the results from the program being presented in this paper, the SuperTurbo was selected by an H2 ICE vehicle demonstration consortium to demonstrate the ICE pathway to decarbonization. The Class 8 vehicle is being built now with a mission to promote H2 ICE within the USA market. Among the most critical considerations for the USA market is an H2 ICE vehicle with near zero NO_x emissions while maintaining good engine/vehicle performance, for which the SuperTurbo is uniquely beneficial. Fig. 15 shows the H2 ICE vehicle that is currently being prepared. More details about the full vehicle are and will be available.

Fig. 15. SwRI H2 ICE joint industry project vehicle

Acknowledgements. SuperTurbo Tech would like to thank AVL List and our other OEM partners for their contributions to this work and continued H_2 ICE development.

References

1. Boriboonsomsin, K., Johnson, K., Scora, G., Sandex, D. et al., Collection of Activity Data from On-Road Heavy-Duty Diesel Vehicles. Final Report, California Air Resources Board and the California Environmental Protection Agency, May 2017.
2. Brown, J., and Waldron, T., "Drivecycle Benefits of Controlling Airflow with the Super-TurboTM," SAE Technical Paper 2018–01–0970, 2018, https://doi-org/https://doi.org/10.4271/2018-01-0970.
3. California Air Resources Board (CARB), "California Air Resources Board Staff Current Assessment of the Technical Feasibility of Lower NO_x Standards and Associated Test Procedures for 2022 and Subsequent Model Year Medium-Duty and Heavy-Duty Diesel Engines," White Paper, April 2019. https://ww3.arb.ca.gov/msprog/hdlownox/white_paper_04182019a.pdf
4. Dreisbach, R.; Arnberger, A.; Zukancic, A.; Wieser, M.; Kunder, N.; Plettenberg, M.; Raser, B.; Eichlseder, H. The heavy-duty hydrogen engine and its realization until 2025. In Proceedings of the 42nd International Vienna Motor Symposium, Vienna, Austria, 29–30 April 2021.
5. Ebert, T., Koch, D., Keyou GmbH et al., Effectiveness of H2-specific operating strategy in dynamic engine operation, IVT 18th Symposium 'Sustainable Mobility, Transport and Power Generation', Graz Austria, September 2021
6. Manufacturers of Emission Controls Association (MECA), "Technology Feasibility for Heavy-Duty Diesel Trucks in Achieving 90% Lower NO_x Standards in 2027," White Paper, February 2020. http://www.meca.org/resources/MECA_2027_Low_NOx_White_Paper_FINAL.pdf
7. Sherrill, R., Brown, J., and Waldron, T., "Design and Testing of a Mechanically Driven Turbocharger for Improved Efficiency and Drivability," in 13th International Conference on Turbochargers and Turbocharging, London, 2018.
8. Waldron, T., Brin, J. D'Orsi, N., "Cycle benefits for commercial diesel engines utilizing a SuperTurbo" Aufladetechnische Konferenz, Dresden Germany, September 2020
9. Waldron, T., Brin J., Seitz H, "Benefits of a Driven-Turbo for Hydrogen Internal Combustion Engines" 10th International Engine Congress 2023, Baden Baden Germany, January 2023

Virtual Product Development for the Hydrogen Direct Injection Engine (H_2-DI)

Cyrill Mandanis(✉), Rajagopal Vinod Bethi, Paul Jochmann, Robin Hellmann, und Fabian Köpple

Power Solutions, Robert Bosch GmbH, Schwieberdingen, Germany
{Cyrill.Mandanis,fixed-term.Rajagopal.VinodBethi,Paul.Jochmann, Robin.Hellmann,Fabian.Koepple}@de.bosch.com

1 Introduction

Although H_2 as fuel for internal combustion engines has a more than 200 years long history (cf. [1] and [2]), the perspective to generate H_2 from "green" electricity and such obtain a basically CO_2 and potentially emission free fuel for internal combustion engines has accelerated the research and development activities significantly in the last years. Meanwhile announcements of engine manufacturers to start the series productions of monovalent H_2 IC engines from the year 2024 on, are clearly emphasizing the need and relevance of tools to support the development and optimization process of the engines. Like for gasoline or Diesel engines, the design targets are to achieve high power rates and at the same time a high efficiency, as well as low emissions. In the case of H_2 as a fuel, the near zero emission target seems to be reachable but can also be ensured for all operation conditions by an adequately designed exhaust gas aftertreatment system.

A. Heintzel (Hrsg.): Int.Engine.Congr. 2024, Proceedings, S. 283–299, 2025.
https://doi.org/10.1007/978-3-658-47966-4_20

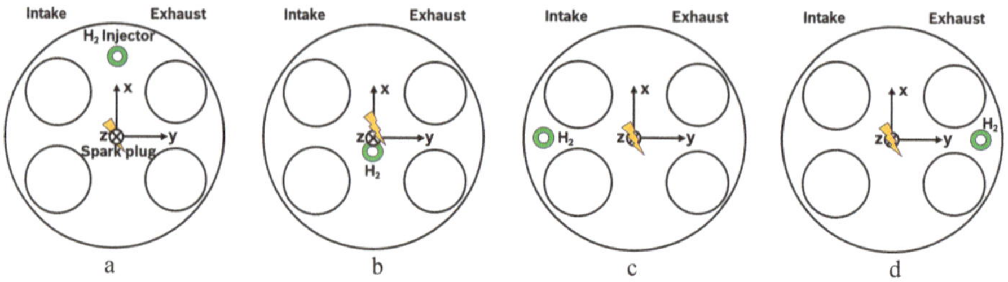

Fig. 1. Possibilities for positioning the H_2 injector in the combustion chamber. The flash symbol is marking the typical central positioning of the spark plug

To reach the mentioned targets regarding engine power, efficiency and emissions, a main scope is to ensure the highest homogenization level of hydrogen and air. Looking at the engine, the main parameters influencing the homogenization are the engine geometry (e.g., bore and stroke), position of injector (e.g., central or lateral, for more details cf. Fig. 1). The specific engine geometry is determining the type and the level of charge motion (Swirl, Tumble X and/or Tumble Y depicted in the schematic of Fig. 2). Especially the design of the intake ports, the combustion chamber roof type (e.g., pent-roof or flat deck) but also the piston bowl shape has a major impact on the charge motion, which is together with the momentum induced by the hydrogen direct injection, one of the main factors for achieving a good homogenization, like shown and mentioned in [3] and [4].

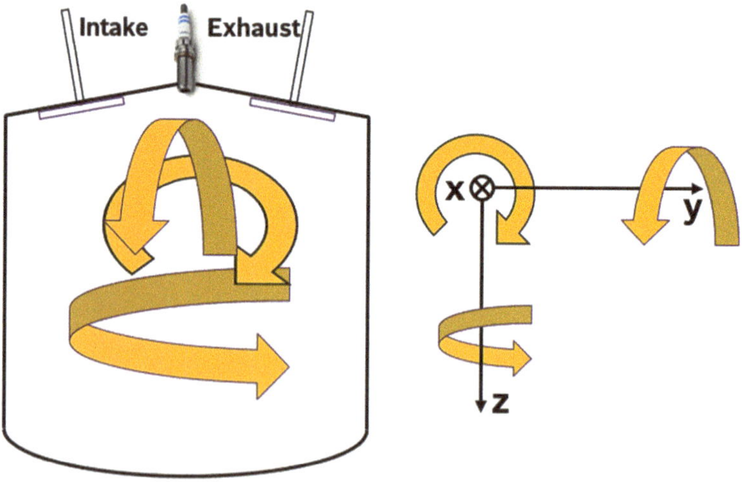

Fig. 2. Exemplary engine geometry and charge motion types: Tumble X, Tumble Y and Swirl

Particular engine parameters are fixed through design aspects like packaging, material properties as well as manufacturing aspects and the design of the fuel injection system must be matched to the engine to reach the best results. Typical design parameters of the H_2 fuel injection system are rail pressure level, nozzle and jet guiding cap design, needle lift, opening and closing dynamics of the injector. While some of these parameters are also fixed in an early stage of the injector design process (e.g., dimensions, maximum rail pressure, needle lift), jet guiding caps can be adapted to the specific engine requirements. From this a typical and recurring task in the matching process between engine and fuel injection system is the optimization of the jet guiding cap layout to the specific engine. Considering the high numbers of parameters as well as the broad range of the parameter values, a high dimensional and large parameter space must be evaluated to find the optimum injector. For scanning and evaluating the huge design parameter space at Power Solutions of Robert Bosch GmbH, a CFD-based and AI supported workflow has been established for enabling a virtual product development for the jet guiding caps. In this paper the cornerstones of the workflow will be introduced together with an exemplary application, which should demonstrate the advantages and benefits of this approach.

2 Methodology

The investigations into jet guiding cap performance involves a robust 3D Computational Fluid Dynamics (CFD) workflow. The objective is to identify and optimize jet guiding cap designs for enhanced mixture formation, promoting clean and efficient combustion. To address computational demands, a methodology prioritizing fast, accurate results was chosen. Instead of costly dynamic engine simulations, a simpler approach was employed as a first step: preliminary investigations in a pressurized chamber using a time-resolved 3D CFD setup with a static mesh. This expedites the identification of promising jet guiding cap designs. The chamber simulation workflow is highly automated and flexible, enabling multi-dimensional Design of Experiments (DoEs). The following section contains details about the automation tool developed for chamber simulations, DoE strategy and dynamic engine simulation setup.

2.1 Automation Workflow for Chamber Simulation

In order to evaluate each jet guiding cap design in a chamber setup, a four-step automation workflow has been developed, culminating in the extraction of various Key Performance Indicators (KPIs), as depicted in Fig. 3.

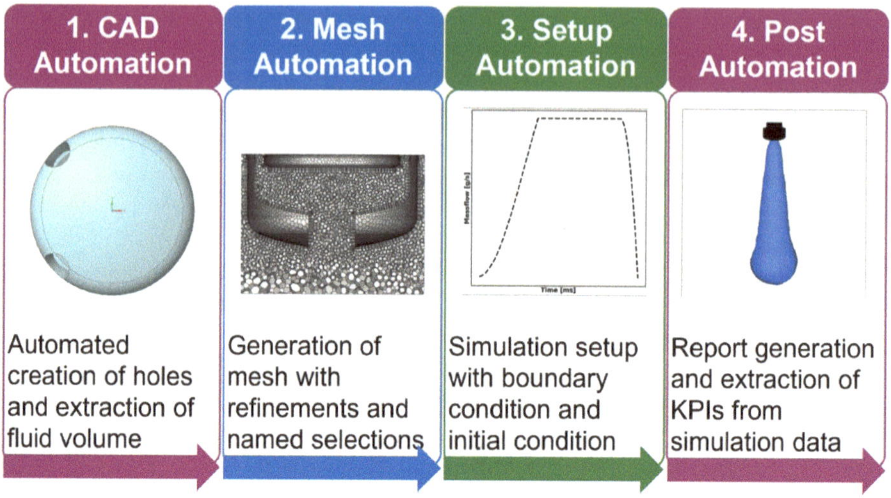

Fig. 3. Step-by-step description of automated chamber workflow

Due to the interdependency of these steps, a common framework was established to streamline the development process. For the framework, the Python programming language was selected for its versatility in automation and data engineering. The workflow is adaptable, handling variations in base geometry and simulation setup. Its flexibility, modularity, and user-friendly GUI allows the systematic and holistic analyses of multiple jet guiding caps.

The first step involves the creation of geometry, guided by a set of input parameters specifying hole positions, diameters, and adjustments to the base geometry, such as the cap's positioning relative to the chamber. Once the fluid domain is extracted, meshing is carried out. The meshing process incorporates automated controls for mesh sizing, initialization domains, and refinement areas. The resulting volume mesh, featuring polyhedral elements serves as an input for the subsequent step: setup automation. The setup contains a cylindrical domain with the extracted jet guiding cap fluid volume. The injector with the needle in the fully open position is considered, initializing the cells above the needle with H_2 at the injection pressure. The rest of the domain is initialized with air at chamber pressure. To prescribe a mass flow boundary condition on the inlet surface, the mass flow curve is dynamically generated based on injection pressure, chamber pressure, temperature, needle lift, and hole opening area. A compressible gas, transient setup with k-zeta-f RANS turbulence modeling is utilized and solved using High Performance Computing (HPC). Finally, the post-processing step produces a report containing contour plots of various field variables such as H_2 mass fraction, velocity and absolute pressure, as well as iso-surface representations of the H_2 jet. The Key Performance Indicators (KPIs), such as "Jet Opening Angles" (τ), "Penetration", "Footprints", "Turbulent Kinetic Energy" (TKE) of Air, and "Air Entrained", are extracted from the simulation data files. Utilizing a combination of data processing techniques and measurement algorithms, these five KPI's are evaluated. The KPI's help characterize the performance of the jet guiding caps and form the basis of the design optimization.

2.2 DoE Definition and Sampling Strategy

The starting point to carry out a DoE study of various jet guiding cap designs is defining the parameter space to sample from. The investigated design space represents a virtual design parameter variation without guarantee of manufacturability.

Table 1 lists the parameters varied, their ranges and constraints. The chosen parameters have been identified as crucial in guiding hydrogen jets. Furthermore, their upper and lower limit have been chosen to cover a large and meaningful design space, enabling a large number of unique designs.

Table 1. Description of the parameters for the DoE

Parameter	Description	Range/Values	Constraint
"Arrange"	To define arrangement of the holes	[1–4]	No constraints
"OArea"	The total opening area of the holes	20 to 28 mm^2	No constraints
"Zeta"	The radially outward inclination angle of the holes	0° to 60°	> "$Zeta_{min}$", where "$Zeta_{min}$" is calculated based on the number of holes and "OArea" to avoid overlapping holes
"Eta"	The tangentially clockwise inclination angle of the holes	−15° to 15°	No constraints

The "Arrange" parameter represents the arrangement of the holes on the cap. The holes can either be equally spaced out in 360° creating symmetric designs or in 180° leading to asymmetric designs. In both cases, there is an additional possibility of having a straight hole without any inclination at the center of the cap. Thus, the "Arrange" parameter can have values of 1, 2, 3 or 4 indicating the positioning and arrangement of the holes. The "OArea" denoting opening area, is the sum of the cross-sectional area of the holes. For the sake of simplicity, it is assumed that all holes for a given design share the same diameter. "Zeta" is the radially outward angle with respect to Z-axis. To avoid overlapping holes and produce feasible jet guiding cap designs, the "Zeta" value needs to be constrained. "Eta" is the tangential angle with respect to X-axis. Both "Zeta" and "Eta" define the inclination of the hole axis as shown in Fig. 4.

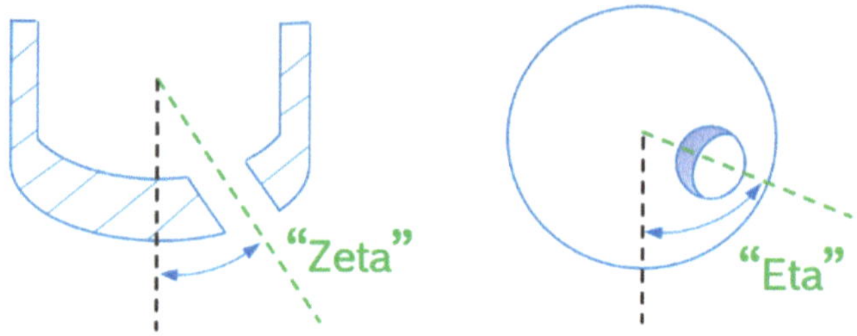

Fig. 4. Depiction of "Zeta" and "Eta" parameters

Finally, due to the complexity and dimensionality of the described parameter space, the sampling method needs to be robust and efficient. These requirements led to the use of the Sobol sampling method as the preferred algorithm to generate 348 jet guiding cap designs [5]. Sobol sampling provides a uniform coverage and evenly spreads out the sampling points resulting in uncorrelated design parameters. Such independent input features enhance the ability to individually assess their impact on KPIs. This approach ensures a clearer understanding of each variable's influence, eliminating confounding effects of interdependence during the KPI evaluation process.

2.3 Engine Simulations

To deepen the understanding of how the chamber KPIs affect the homogenization for a specific engine, a subset of designs from the DoE is selected for in-detail analysis of the mixture formation using dynamic engine simulations. The selection process for these designs is explained in the results part. The engine that is being investigated is a heavy-duty commercial vehicle engine with 6 cylinders and a total displacement of ca. 15 liters, a lens shaped piston and the injector in a lateral position.

The engine simulations are performed for the operating point 1100 rpm and 9.5 bar IMEP. Similar to the chamber simulations, also for the engine investigations, transient, compressible gas simulations with the k-zeta-f RANS turbulence model are conducted. A single cycle is simulated, where a mass flow boundary condition is imposed on the inlet surface of the injector. No combustion model is activated, since the aim of this analysis is to investigate the effect of different jet guiding caps on the homogenization alone.

The performance of a jet guiding cap in the engine in terms of the homogenization is evaluated based on two KPIs, which are both evaluated at ignition timing for the entire mixture in the combustion chamber: the uniformity index for the equivalence ratio and the lambda histogram. The uniformity index is a scalar value, that describes how uniform the mixture in the cylinder is in terms of the equivalence ratio. The uniformity index is calculated with the following formula:

$$UI\ of\ EQR = 1 - \frac{\sum_c \left| eqr_c - \overline{eqr} \right| V_c}{2|\overline{eqr}| \sum_c V_c} \tag{1}$$

where V_c stands for the volume of a cell, eqr_c for the equivalence ratio value in the cell and $\overline{eqr}$ is the volume average of the equivalence ratio. The closer the uniformity index value is to 1, the better is the level of homogenization in the mixture.

The histogram on the other hand is a distribution and adds more context to the uniformity index, because it shows the actual distribution of lambda values for a certain domain and crank angle position. The histogram is obtained by dividing the mixture into different lambda classes ranging from 0 to 8 with a bin size of 0.01.

3 Results

3.1 Chamber DoE Results

The analysis of the DoE results starts with the scatterplots containing all DoE design points shown in Fig. 5. There is one scatterplot for each of the five chamber KPIs. The KPI values are normalized, with the maximum value across all design points corresponding to 1 and the minimum value to 0. The plots employ "Zeta" (0° to 60°) on the x-axis, "Eta" (−15° to 15°) on the y-axis, and marker size proportional to "OArea" (20 mm^2 to 28 mm^2). Fig. 5 therefore gives an initial insight on the influence of the design parameters on the KPIs and can already be used to draw several conclusions. Notably, "Eta" does not appear to significantly impact any of the KPI values. Additionally, design points with higher "Zeta" values exhibit higher Tau values but lower Penetration values, and vice versa. Moreover, deviations in Footprints (i.e., higher values in Footprints KPI) also increase for higher "Zeta" values. Air TKE and Air Entrained do not exhibit clear trends in these plots, and certain points deviate from expected patterns, likely due to unrepresented parameters such as "Arrange" and "Number of Holes".

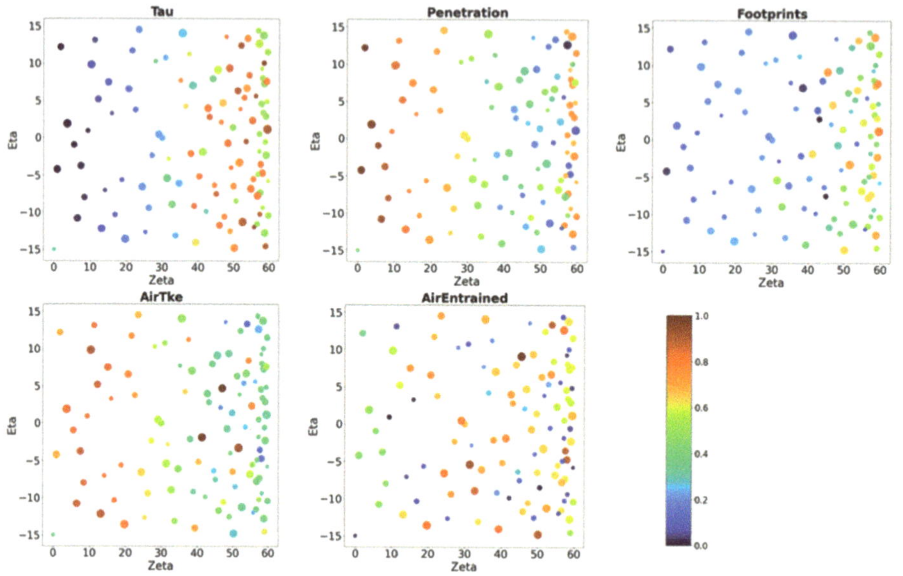

Fig. 5. DoE results: x-axis = "Zeta", y-axis = "Eta", marker size = "OArea" and color = normalized KPI values

3.2 Knowledge Discovery—Chamber DoE

To deepen the understanding of the underlying physical phenomena the in-house Knowledge-Discovery (KD) framework pyMICE2 is applied. The framework was presented by Botticelli [6] and has already been successfully applied to problems related to ICE, e.g. in [7]. A detailed and application-oriented description of the KD framework in the context of high-pressure gasoline injector simulations, a spray chamber and in an engine is given by Hellmann [8]. The methodology presented in [6–8] is applied to the DoE data described in Sect. 3.1.

The coefficient of determination R^2, a global evaluation metric to validate the accuracy of the ML model, is 0.979 for the Tau KPI and 0.963 for the Penetration KPI indicating highly accurate ML models.

Figure 6 depicts the contribution of a single input feature to the Tau KPI (left) and Penetration KPI (right) by means of the relative feature importance. The number of holes "NoH" clearly has the biggest influence on the Tau KPI, followed by the "Zeta" angle. This strongly dominant correlation of "NoH" is also visible in Fig. 5 and can now be quantified using KD. The remaining design parameters show only minor influences. For the Penetration KPI, the "Zeta" angle has the strongest influence at around 42%, followed by "Arrange" and "Noh" at around 28% and 26% respectively. The negligible influence of "Eta" and "OArea" on Tau and Penetration can be confirmed again by the KD framework, cf. visual impression of the scatterplots in Fig. 5.

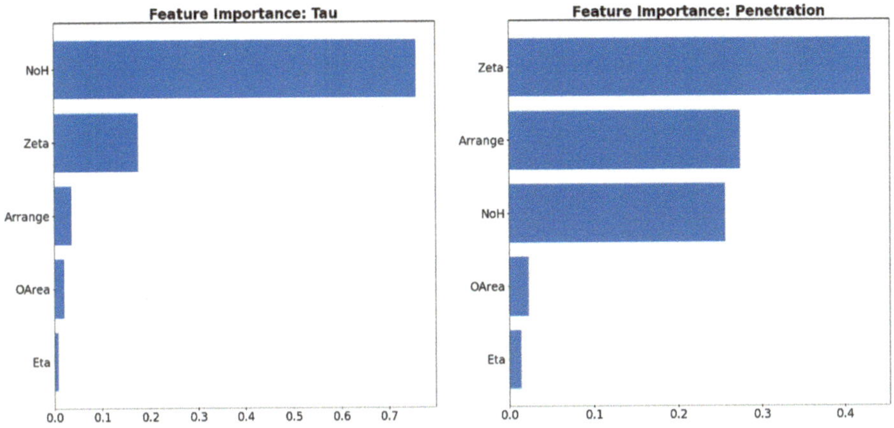

Fig. 6. Feature Importance of the varied 5 design parameters on the KPIs Tau and penetration

Due to the high influence of features "Zeta" and "NoH", their individual interactions with the Tau KPI can be analyzed in more detail using the partial dependencies in Fig. 7. The partial dependency plot on the left for "Zeta" shows how the Tau KPI increases for higher "Zeta" values. For the "NoH" feature, the influence of increasing hole numbers leading to higher Tau angles is more pronounced for lower number of holes, while the slope to the right flattens out, indicating the non-linear behavior so that even more holes do not necessarily increase the Tau KPI.

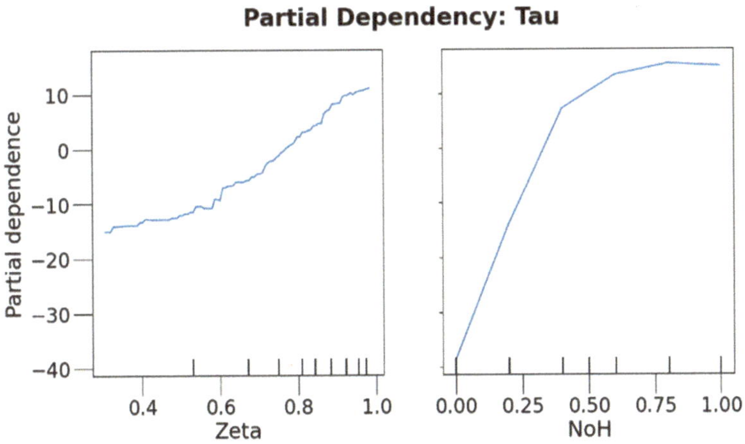

Fig. 7. Partial dependency plot of "Zeta" and "Noh" for the KPI Tau

3.3 Jet Guiding Cap Selection

In order to narrow down from the 348 jet guiding caps of the DoE to a few promising designs for further investigation in dynamic engine simulations, a rather simple approach was used, where for each KPI, except the Footprints, the one design is selected that maximizes this given KPI. This selection process yields the designs shown in Table 2.

Table 2. Overview of jet guiding caps selection

	Cap ID	Description	# of holes	Zeta angle	Hole diameter (mm)
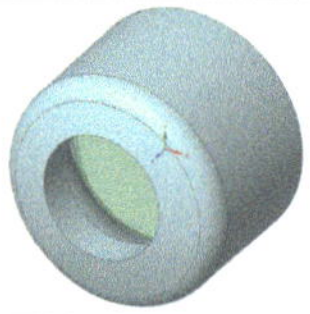	Case0010	Maximizes penetration KPI	1	0°	5.92
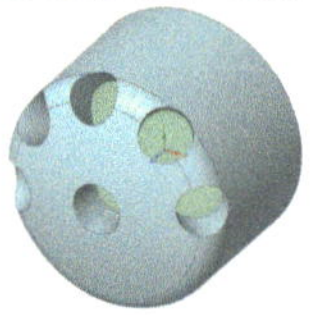	Case0069	Maximizes air entrainment KPI	6	0°/45.47°	2.4
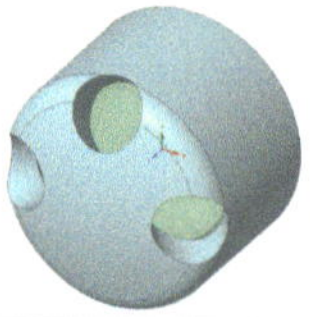	Case0084	Maximizes air TKE KPI	3	41.25°	3.35
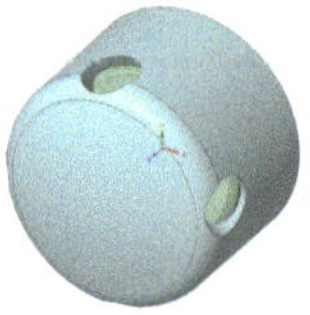	Case0093	Maximizes Tau angle KPI	4	56.25°	2.55

In Fig. 8 the H_2 mass fraction isosurfaces depicting the distribution of the H_2 jet for these four variants are shown. These images give a good impression on the wide range of results that are obtained from this DoE. Fig 8(a) illustrates Case0010 from Table 2, which is a single-hole cap with "Zeta" angle of 0° that maximizes the penetration KPI. Fig. 8(d) illustrates Case0093 from Table 2, a four-hole jet guiding cap with "Zeta" angles of 56.25° that leads to the highest Tau angle value.

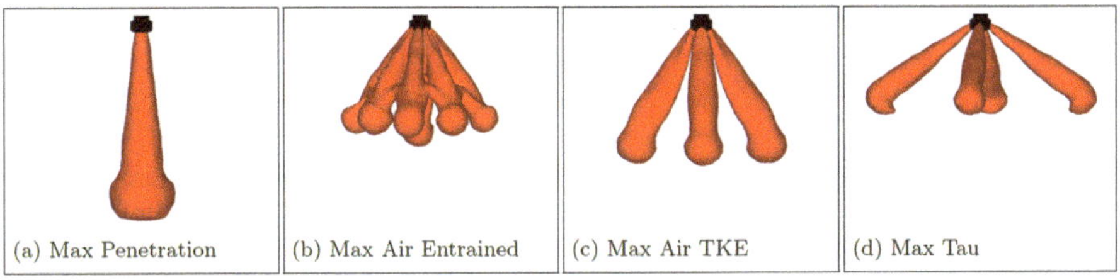

Fig. 8. DoE results: postprocessed isosurface of H_2 Jet for cases with max KPI values

3.4 Engine Simulation Results

This section discusses the performance in the engine of the four maximum KPI jet guiding caps, that were introduced in Sect. 3.3. The start of injection at 542°CA (−178°CA bTDC) and the rail pressure of 25 bar was kept constant in all engine simulations, so that a good comparability between the variants is ensured.

Charge Motion Results The influence of the different jet guiding caps on the charge motion is shown by evaluating the tumble motion around the x-axis and around the y-axis.

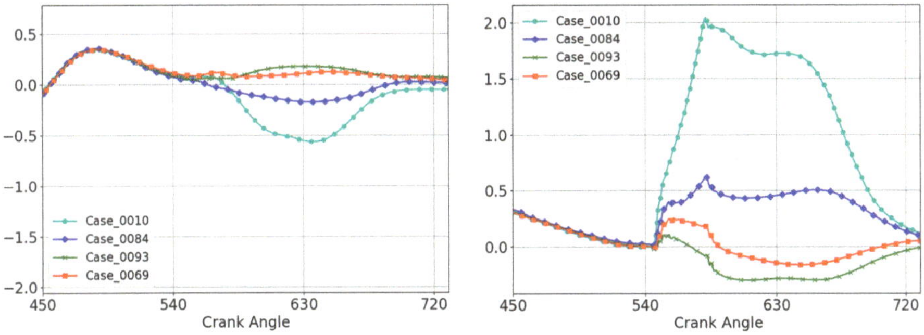

Fig. 9. Tumble ratio [-] around x-axis (left) and y-axis (right)

The results for the tumble around the x-axis are shown in Fig. 9 on the left. The single-hole jet guiding cap Case0010 generates a rather strong negative tumble-x motion, whereas the three investigated multi-hole jet guiding caps are generating rather weak positive or negative tumble-x motion. For the tumble motion around the y-axis shown in Fig. 9 on the right, the influence of the jet guiding cap is even bigger. The lateral injector position in combination with the piston shape of this engine explains why Case0010, that has only a single jet concentrated in vertical direction, generates by far the strongest tumble motion around the y-axis. For the three multi-hole jet guiding caps also this tumble motion is significantly weaker. The result for the mean turbulent kinetic energy (TKE) inside the combustion chamber can be seen in Fig. 10. Since the TKE at the end of compression is mainly a consequence of the break-up of the large tumble vortices into turbulent structures due to the upwards moving piston, the jet guiding cap Case0010 with strong tumble motion in x- and y-direction therefore also leads to the highest TKE values close to top dead center (TDC).

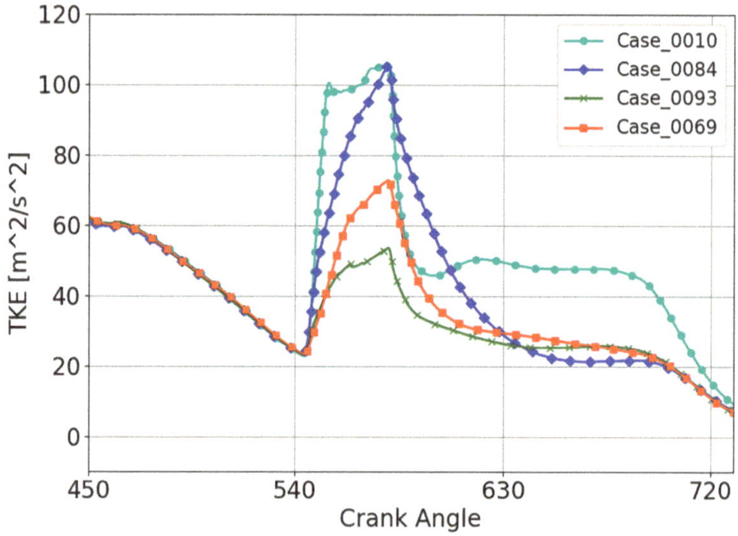

Fig. 10. Mean turbulent kinetic energy in cylinder volume over crank angle position

During the injection, the mean TKE values in the cylinder differ strongly between the four jet guiding caps. The maximum Air TKE cap Case0084 and Case0010 lead to the highest TKE values at the end of injection. During the compression stroke this advantage quickly disappears for Case0084 and around TDC it has a similarly low TKE value as Case0069 and Case0093. Based on this result, it can be concluded that the chamber KPI Air TKE doesn't directly translate to the TKE values in the engine simulation close to TDC but more to the TKE values observed during the injection phase of the engine simulation. Therefore, the Air TKE metric from the chamber simulation for this engine needs to be revised in order to maximize the TKE values at ignition timing.

Homogenization Results For the evaluation of the level of homogenization in the combustion chamber, the uniformity index of the equivalence ratio is discussed first. Fig. 11 shows the evolution of the uniformity index from right before the start of injection at 542°CA up to TDC for the different jet guiding caps. During the injection, from 542°CA to 580°CA, there is a correlation between the number of holes of the jet guiding cap and the uniformity index, where the uniformity index value increases with the number of holes. This can be easily explained, because a multi-hole cap initially distributes the fuel better in the combustion chamber, since there are multiple hydrogen jets, compared to a single-hole jet guiding cap like Case0010, that only generates one hydrogen jet. After the end of injection however the process of homogenization is progressing more slowly for the three multi-hole jet guiding caps (Case0069, Case0093 & Case0084), compared to the single-hole jet guiding cap Case0010. Towards the latest stages of the compression stroke from 670°CA to 720°CA the multi-hole variants again show a steeper increase in the uniformity index curve. At TDC the ranking is as follows: Case0093 reaches the highest uniformity index value with 0.949, Case0010 and Case0069 reach a value of 0.926 and Case0084 clearly has the lowest uniformity with a value of 0.883.

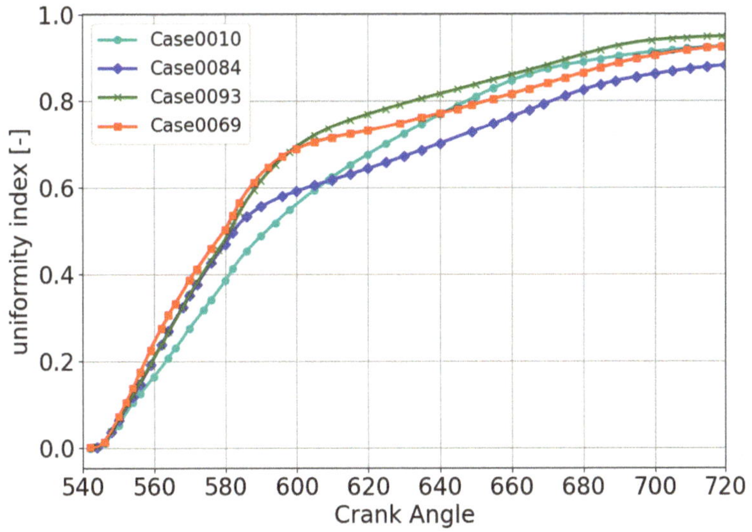

Fig. 11. Uniformity index of equivalence ratio in cylinder volume over crank angle position

In Fig. 12 the lambda histogram in the combustion chamber at TDC is shown for all four jet guiding caps. The histogram confirms the homogenization result that was obtained from the uniformity index, because Case0084 that has the lowest uniformity index value also shows the widest lambda distribution and on the other hand Case0093 that has the highest uniformity index value at TDC also has the narrowest lambda distribution and the highest peak.

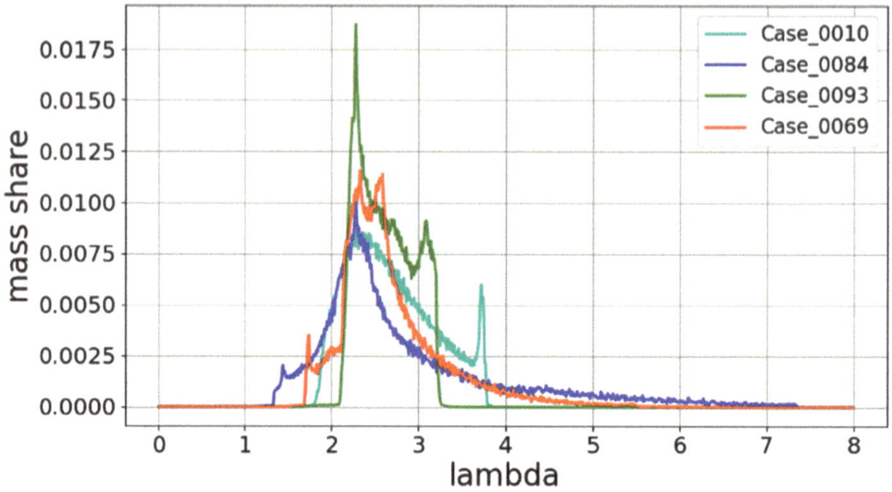

Fig. 12. Lambda histogram for combustion chamber volume at TDC

These results indicate that the Case0093, which is the jet guiding cap that had the highest Tau angle from all variants of the chamber DoE, is significantly benefiting the homogenization. The entire mixture in the combustion chamber at TDC is in a very narrow lambda window between 2 and 3.3 for Case0093.

To better understand the different mechanisms for the homogenization process, the lambda isosurfaces in Table 3 are shown as well as the x- and y-cut planes for the lambda field in Table 4. Case0010 and Case0093 are illustrated, because both show very good homogenization results, but generate these through a different mechanism. From the isosurfaces of Case0010, the tumble motion around the y-axis, that the jet generates can be seen, while for the Case0093 the isosurface images show that three of the jets quickly hit the liner walls and spread the fuel along those walls whereas one jet is hitting the liner on the opposite side of the injector. After the injection at 600°CA the lambda isosurface covers more of the combustion chamber for Case0093 compared to the more concentrated isosurface of Case0010, which explains the higher uniformity index value at 600°CA for Case0093 in Fig. 11. For Case0010, the higher tumble motion compared to all other cap variants, is the main driver of the distribution of hydrogen in the combustion chamber. This can be seen by looking at the lambda fields at 600°CA and 650°CA, where the fuel rich regions are at first mainly visible along the piston and the injector opposite liner and then later on, mainly along the upper walls on the cylinder head and the liner on the injector side.

Table 3. Images of lambda = 1.5 isosurfaces for different crank angle positions. Upper row: Case0010; lower row: Case0093

554°CA	564°CA	600°CA
		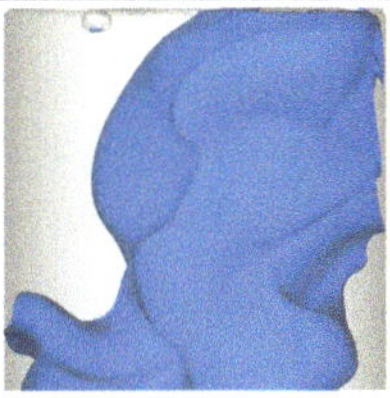
	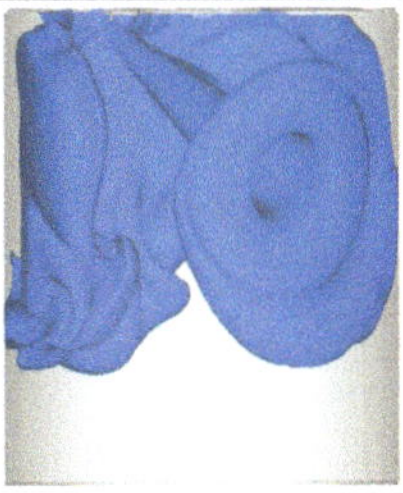	

Table 4. Images of lambda cut planes for different crank angle positions. Upper row: Case0010; lower row: Case0093 (images not in scale)

600°CA	650°CA	720°CA
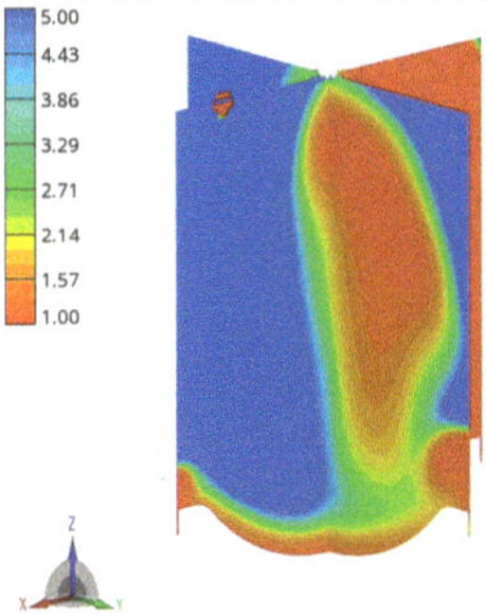	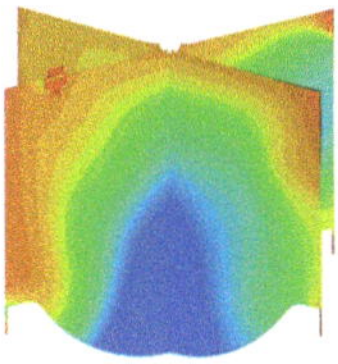	
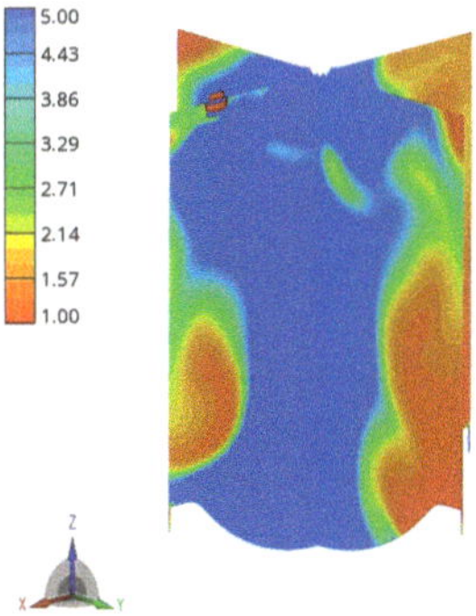	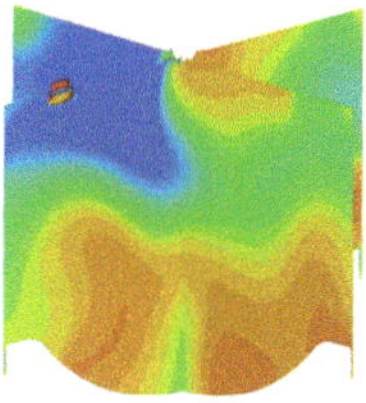	

From this investigation of a small selection of jet guiding caps from the chamber DoE in dynamic engine simulations, some important conclusions can be drawn about the importance of the chamber KPIs for the performance in this engine. The penetration KPI is strongly influencing the tumble around the y-axis, which is beneficial for higher TKE values at ignition and benefits the homogenization. A high Tau angle can lead to even better homogenization, whereas the Air TKE KPI from the chamber does not show a good correlation with the engine TKE values at the time of ignition.

4 Conclusion

In the current contribution a highly automated simulation workflow for pressurized chamber investigations was introduced and applied in combination with the DoE methodology. In a first step 348 different jet guiding caps variants were automatically generated and evaluated according to predefined KPIs. From the evaluation, insights related to the connection between design parameter and KPI were extracted and four designs maximizing a different KPI each were selected. These four jet guiding cap designs were evaluated with dynamic engine CFD for rating the performance regarding homogenization. The comparison is showing different levels of homogenization but also the different underlying mechanisms responsible for this result. One path to high homogenization levels is a high penetration generating high levels of Tumble and the other path is generating high Tau angles leading to a good spatial distribution and low penetration.

Additionally, the chamber DoE results were exploited for the generation of machine learning (ML) models. With this ML models a deeper understanding of the design parameters on the defined KPIs is possible, as well as conducting optimization tasks for the jet guiding cap designs. Finally, it must be pointed out, that the currently available amount of data, as well as probably the number and types of KPI are not sufficient for a holistic inspection of the whole multi-dimensional design space. Therefore, also the currently found variants are not indicating global maxima but revealing only the paths to look for an optimized design. In future the presented approach, supported by improved KPIs, as well as with a higher number of investigated designs will be able to provide even more optimal solutions for jet guiding caps.

References

1. Eichlseder, H., Klell, M.: Wasserstoff in der Fahrzeugtechnik. 2nd edn. Vieweg und Teubner Verlag, Wiesbaden (2010).
2. Verhelst, S., Wallner, T.: Hydrogen-fueled internal combustion engines. Progress in Energy and Combustion Science 35, pp. 490–527 (2009).
3. Chi, Y., Shin, B., Pelzetter, R., Tichy, M., Peppler, M., Hoffmann, S., Morel, V., Schück, C., Jochmann, P., Schünemann, E.: Hydrogen Engine for a Passenger Car Hybrid Powertrain: Attractive Solution for Sustainable Mobility. 44th International Vienna Motor Symposium (2023).

4. Paltrinieri, S., Olcuire, M., Calia, V., Mortellaro, F., Medda, M., Gullino, F., Stapf, K. G., Geiler, J., Jochmann, P., Boee, M., Lippisch, M., Wundling, C.: Experimental and Numerical Investigation of Hydrogen Injection and its Preliminary Impact on High Performance Engines Development, SAE Technical Paper 2023-01-0402 (2023).
5. Burhenne, S., Jacob, D., Henze, G.P.: Sampling based on Sobol sequences for Monte Carlo techniques applied to building simulations. Proceedings of the International Building Performance Simulation Association (2011).
6. Botticelli, M., Hellmann, R., Jochmann, P., Stapf, K.G., Schünemann, E.: AI-Driven Gasoline Direct Injection Development: A Knowledge-Discovery Framework for Comprehensible Evaluations of Complex Physical Phenomena. Internationaler Motorenkongress 2021, pp. 469–484. Springer (2021).
7. Botticelli, M., Hellmann, R., Jochmann, P., Stapf, K.G., Schünemann, E.: Application of Machine Learning to Gasoline Direct Injection Systems: Towards a Data-Driven Development. 2020 19th IEEE International Conference on Machine Learning and Applications (2020).
8. Hellmann, R., Botticelli, M., Jochmann, P., Köpple, F., Schlüter, R.: Machine learning algorithms applied to fuel direct injection development: from gasoline to H_2 engines. 12. Tagung Einspritzung und Kraftstoffe (2023).

Using Drop-In Gasolines with Increased Renewable Potential: A Comparison to Some Typical Fossil Fuels Concerning Emission Testing and Fuel Composition

Hanno Krämer(✉) und Markus Send

Audi AG, Ingolstadt, Germany
{hanno.kraemer,markus.send}@audi.de

1 Introduction

Aiming to the EU greenhouse gas reduction targets in the transportation sector, the introduction of battery electric vehicles powered with CO_2 neutral electricity and vehicles powered with CO_2 neutral hydrogen will be one main pillar. However, due to the high share of CO_2 emissions generated by the existing vehicle fleet, this will by far not be sufficient [1]. Hartung et al. [2] pointed out within a well-to-wheel approach, that the 2030 CO_2 emissions target of the European road traffic sector cannot be met by electrification of the transport sector as the only measure. With the Introduction of fuels with a rising renewable content, the difference between GHG target and the required CO_2 emission limits can be reduced, but a gap will still remain. Looking forward to 2040, the CO_2 target can be met and eventually even overcompensated with a significant increase of renewable fuel production.

But, the existing vehicle fleet can only be addressed via introduction of drop-in renewable fuels that are fully compliant with current fuel standards like the EN228. First proposals of such kind of fuels have been recently published under the label of "blue gasoline" [3, 4]. Some evolutionary steps of these fuels are part of our study. Raising the renewable content within the EN228 should be accompanied by a carefully defined E20 standard that still has to be worked out to be compatible with as many existent vehicles as possible. This measure opens another and comparatively easy pathway to increase the renewable content in the fuel by using Ethanol in concentrations of 15..20..25% that are very well known in countries like the US, Brazil, Paraguay etc. As a next step E20 will be introduced in India [5], while other countries already show increasing interest [6]. Consequently, two intermediate Ethanol fuels are part of our study.

A. Heintzel (Hrsg.): Int.Engine.Congr. 2024, Proceedings, S. 300–311, 2025.
https://doi.org/10.1007/978-3-658-47966-4_21

2 Method

Today's fuels that are sold at the filling stations comply with the current fuel standard EN228, but they differ significantly concerning their thermodynamic behavior described by RON/MON and their burning behavior with respect to engine emissions i.e. NOx, HC, CO_2 and especially PN. Thus, on the way to CO_2 reduced fuels with elevated renewable content particular attention has to be paid to achieve high octane numbers accompanied by low gaseous emissions and sooting tendencies of the fuel candidates during engine operation.

In this context interesting candidates with increased renewable content should always be compared to well-known compositions like certification fuel or first-fill quality fuel. A comprehensive study of this type was given 2023 by Krämer et al. [7] where very promising future fuel candidates like POSYN and MTO were compared to filling-station-like fuels (RON95 E10, RON98) and to certification fuels of the emission standards EU5cert and EU6cert. In this paper, fuels with realistic intermediate renewable content of 20..50% are compared to a set of the above cited well known and well evaluated fuels of the previous study with again experimental focus on gaseous and PN-emissions in the same test cycles.

2.1 Investigated Fuels

In the following, fuel selection and their individual characteristics are described in more detail

MTO is a fuel proposal by Audi based on 80% iso-butene derivatives [8]. MTO can be produced following a second-generation microbiological pathway [9, 10] that starts with C5 or C6 sugars which are in a first reaction step converted to iso-butene. Different reaction paths using the addition of hydrogen or methanol then lead to iso-butane, MTBE (=methanol+ isobutene), iso-octane and iso-dodecane. A complementary pathway via Methanol-to-Olefin (MTO) processes is under investigation, seeking the potential to produce greater quantities of that kind of fuel. Audi's MTO fuel is within the borders of EN228 and, as an alkylate fuel, it is also supposed to improve greatly PN-emissions compared to classical fossil fuels containing roughly 30% aromatics. Additionally, the comparatively low final boiling point of 184 °C should help to diminish wall wetting effects in the engine and their resulting PN-emissions.

EU5cert and **EU6cert** are two quite different conventional certification fuels, one fulfilling the past and the other fulfilling the actual emissions testing fuel specification. From EU5cert to EU6cert changes were made i.a. concerning the ethanol content, the aromatics content, the final boiling point and the octane rating of the fuels. As a result, EU6cert represents a rather modern type of fuel accompanied by a low sooting tendency. In contrast EU5cert represents a fuel composition dominated by heavy aromatics leading to the highest YSI value in this study, accompanied additionally by a high final boiling point resulting in an elevated risk of wall wetting and resulting pool fires. Therefore, EU5cert is frequently used as a benchmark fuel with a high sooting tendency.

RON95E10 and **RON98** are typical German market fuels from the Ingolstadt region with different octane ratings and differences especially concerning the ethanol content. In this study RON95E10 is a "real" E10 containing about 10% Ethanol while the RON98 quality represents a Premium Plus fuel where the higher octane rating is achieved using a slightly higher aromatic content but particularly by using about 14% Ether instead of ethanol. In sum they both are very similar concerning other PN-emissions related fuel properties like aromatics content and final boiling point, thus giving us a good impression if i.e. the difference between zero and 10% ethanol content makes a remarkable difference concerning PN-emissions. Both fuels fulfill EN228 fuel standard.

E25 is a very well known fuel type that is characterized by an ethanol content of roughly 25%. Home of E25 is Brazil where it is used since the 1970ies as standard fuel quality but E25 is also found for example in Paraguay and Thailand. Hence, E25 has a regular place in Audi's engine development procedures and here especially in the application department as well as in endurance runs. Besides the high mechanical load due to the nearly ideal ignition timing of this heigh RON fuel, E25 has a measurable lower volumetric heating value compared to other fuels. This induces prolonged injection times in order to keep the same power level of the engine. This can lead to wall wetting and consequently increased PN-emissions. This effect will be more probable at the high engine loads of the RDE cycles.

E20_worst in this study is a new research fuel at Audi based on our very PN-intensive EU5cert fuel (as described above). E20 is splash-blended to its 20% ethanol content just by adding the missing 15% of ethanol to the EU5cert. The result is a fuel that combines the PN-pushing attributes of EU5cert with the PN-pushing properties of 20% ethanol like prolonged injection time and eventually worse mixture formation due to the increased cooling effect of ethanol evaporation. These effects compete with ethanol's characteristic to be itself a very clean burning component (YSI is only 10.3). So, the main question is which effects regarding the resulting PN-emissions are in sum predominant and if E20_worst is at the end a worse or a cleaner performing fuel compared to our EU5cert.

R33 in this study refers to "blue gasoline" and is a fuel with increased renewable content developed by a consortium of industrial partners from oil and automotive industry [3, 4]. R33 means roughly 33% renewable content with consequential CO_2 savings of more than 20%. R33 combines 2nd generation ethanol with bio-naphtha and eventually some MTG. During the development of R33 it was found that the first charge of R33 had a comparatively high YSI resulting in elevated PN-emissions during engine testing. Following charges showed improved compositions with significant lower YSI values. Two of them are part of this study and named here as R33_4 and R33_5. They both come from deliveries that supply Audi's internal filling stations since mid 2022 and that are used since end of 2022 additionally to fill up new cars before handover to customers at Audi's Ingolstadt customers center [11].

R50 consequently means a renewable content of 50% aiming now to CO_2 savings of roughly 40% compared to pure fossil fuels. This is in our case accompanied simultaneously by the wish to rise the octane level to RON98 which is of high interest for sportscars and historic cars. Since the octane level of bio-naphtha and MTG components is limited it is to expect that some ETG has to be taken to meet the RON98 requirement. This will immediately rise the overall aromatics content as well as the content of heavier aromatics leading to a higher YSI level and a rise of PN-emissions.

Key parameters of the fuels are summarized in Table 1. Pleasant values are marked green, critical values are marked yellow, very critical values are marked red. It is interesting to note that two distillation parameters of MTO are as critical or even worse as those of EU5cert with respect to an often-reported impact of EVAP@150 °C and T90 on PN-emissions. The PN-measurements in the experimental section of the previous paper showed that these parameters were not crucial for MTOs sooting tendency since it outperformed the EU6cert fuel in terms of low PN-emissions. So, MTO should come out to be the low-sooting benchmark in this study.

Regarding the accumulation of critical parameters in Table 1, EU5cert fuel probably is the worst-case fuel in this study. But, it is accompanied directly by the E20_PN fuel which on top shows a reduced volumetric heating value. This property leads to a longer injection duration in order to compensate the lower energy content. This consequently rises the risk of wall-wetting and pool fires which mean both immediate high-grade PN-emissions.

Table 1. Key fuel properties.

		MTO Audi	EU6 cert	R33_5	RON95	E25 Audi	R33_4	RON98	R50	E20_ worst	EU5 cert
RON	-	107	97	96	96	101	95	99	100	104	101
DVPE	kPa	68	58	60	58	56	54	64	59	57	60
Density	kg/m3	722	749	739	749	755	735	742	761	761	754
EVAP@150	%	78	94	93	94	89	89	93	88	86	84
T90	°C	174	125	142	143	153	-	145	158	173	178
FBP	°C	184	191	180	170	195	184	178	195	201	204
Aromatics	%v	0	26	23	31	21	23	31	32	27	32
Aromatics >C8	%v	0	3	6	7	9	9	8	11	13	16
YSI_calc	-	62	82	84	97	85	92	100	103	103	116
Oxygen	%m	2,7	3,5	3,6	3,5	9,4	3,3	2,6	3,6	7,4	1,8
Ether	%v	15	0	0,5	0	0	3,1	14	4	0	0
Ethanol	%v	0	10	10	9	26	7,5	0	8	20	5
LHV (mass.)	MJ/kg	43,2	41,3	41,8	41,9	40,1	42,0	42,4	41,4	39,4	42,0
LHV (vol.)	MJ/L	31,2	30,9	30,9	31,4	30,3	30,9	31,5	31,5	30,0	31,7

In addition to the parameters of standard fuel analysis the sooting tendency of the fuels is expressed by a calculated and simplified Yield Sooting Index (YSI) approach that can be derived either from reformulyzer data or from detailed hydrocarbon analysis (DHA). Details of the simplified YSI calculation method are given in [7], basic reference information about YSI can be found in [12, 13].

2.2 Test Method and Operating Conditions

The Audi engine used for the test bench measurements was a production calibrated turbo charged 4-cylinder inline engine with gasoline direct injection. Table 2 shows the technical data.

Table 2. Audi engine technical data.

Engine	2.0l TFSI EA888 evo4 MLB LK2 EU6AP
Engine Type	4 Cylinder, inline, turbocharged
Bore × Stroke	82,50 mm × 92,80 mm
Displacement	1984 ccm
Injection System	Direct Injection, lateral Position
Injection Pressure	350 bar
Max. Power/1/min	195 kW/5000–6500
Max. Torque/1/min	380 Nm/1600/4300
Gearbox	DL382
Transmission	7-Speed S-Tronic
Load/Testbench	2268 kg

The engine test measurements were carried out with the mounted 7-speed s-Tronic gearbox on a highly dynamic dynamometer with external conditioned fluids as well as controlled fuel, air and inter-cooler temperatures. In this study a temperature of 23 °C was chosen as standard start condition. The experimental setup is shown in Fig. 1.

The gaseous emissions were measured using a AVL i60 iGEM S2, the particulate emissions (PN) by an AVL CPC model AVL489 (23 nm). PN-emissions were measured as engine-out emissions to be able to compare directly the influence of the different fuels on the raw PN-emissions level. Additionally, the PN-emissions were measured a second time in the CVS-tunnel together with the gaseous emissions. This allows to evaluate the effect of the gasoline particulate filter (GPF) on the PN-charateristics of the different fuels and also delivers an equivalent to tailpipe PN-emissions even from the engine dynamometer.

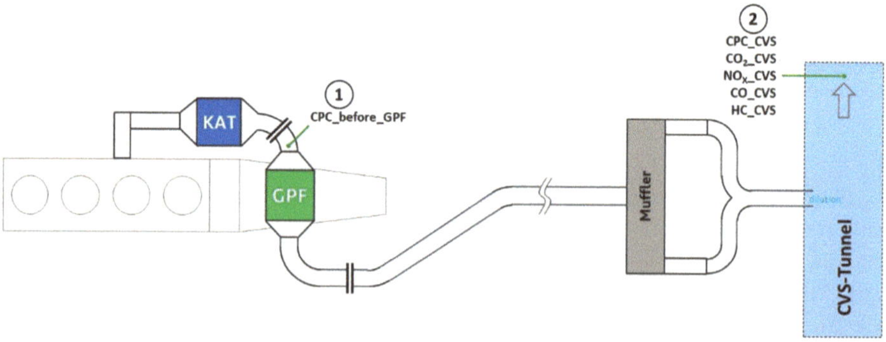

Fig. 1. Experimental setup

Test cycles were the official WLTC as the legal reference and further an Audi-internal real driving emissions cycle (RDE), representing higher engine loads and more dynamic driving. The test cycles themselves were running in a fully automatic mode w/o manual influence. To minimize variations 5 cycles of each type were measured and afterwards averaged. Furthermore, after changing fuel a conditioning program with the EU5 cert fuel was performed to avoid interfering effects from the fuel used before.

3 Results

In this chapter, first the results of the gaseous emissions will be highlighted briefly. Thereafter, the PN-emissions of the fuels will be compared and discussed in more detail.

3.1 Gaseous Emissions

Figure 2 shows the gaseous emissions NOx, CO, HC and NMHC of the different fuels in the WLTC and RDE cycles in comparison to the corresponding legal limits of EU6 AP. Note that our testing is done on a dynamic engine test bench. Thus, the emission results are not directly comparable to homologation measurements which rely on vehicle testing. Nonetheless, the results allow a comprehensive comparison of the influence of the different fuels on their individual emission level.

NOx emissions of all fuels in WLTC are on a very low level. Some differences between the fuels can be seen, all of them not leaving the general level. In principle, the same observation can be made for the RDE cycle, but all NOx emissions are now on a higher level by a factor of 2..3 compared to WLTC due to higher temperatures in the engine caused by the higher loads and accelerations of the RDE cycle. Nonetheless, NOx emissions still remain far below the legal limit. Notice, that a new catalyst was used in this study, so these low emission levels are not very surprising since there also exist stringent legal limits for aged catalysts at 160 tkm.

Following the argumentation above also for CO, the CO emissions remain far below the legal limit, the emissions of the RDE cycle being again on a visibly higher level of about a factor of 1.5 due to the rigorous RDE testing conditions.

HC and NMHC emissions show an almost identical distribution to each other, the NMHCs being slightly higher but still on a very low level. The RDE cycle shows no impact on HC and NMHC emissions for all fuels compared to the WLTC.

All in all, no strong and systematic distinction between the tested fuels is observed throughout all cycles in this study as long as the gaseous emissions are considered.

3.2 PN-emissions

Figure 3 shows the PN-emissions in the WLTC and RDE cycles as engine-out (=raw)-emissions before the GPF. This gives us the opportunity to make visible the direct influence of the different fuels on the PN-emissions without biassing the result by the GPF. Additionally, it is a good equivalent how vehicles without GPF will react on different fuels. The fuels are given following their raw PN-emissions of the WLTC raising from left to right with the EU5cert fuel being the 100% reference. Note that the scale between all measurements is not identical. Instead, the scale is alsways normalized to the emissions of the EU5cert fuel in each cycle and measurement location.

Regarding the raw emissions of the fuels in WLTC, three groups of behavior can be seen. In the first group, PN-emissions are roughly **5 times (!)** less than those of the EU5cert fuel. Within the first group emissions raise from MTO over EU6cert to the R33_5 which represents the most developed and modern fuel with intermediate renewable content in this study. The ascending order of the 3 fuels is clearly visible and corresponds to their ascending YSI ratings from 62..85. The PN-emissions of all fuels of the first group are on a very favorable level.

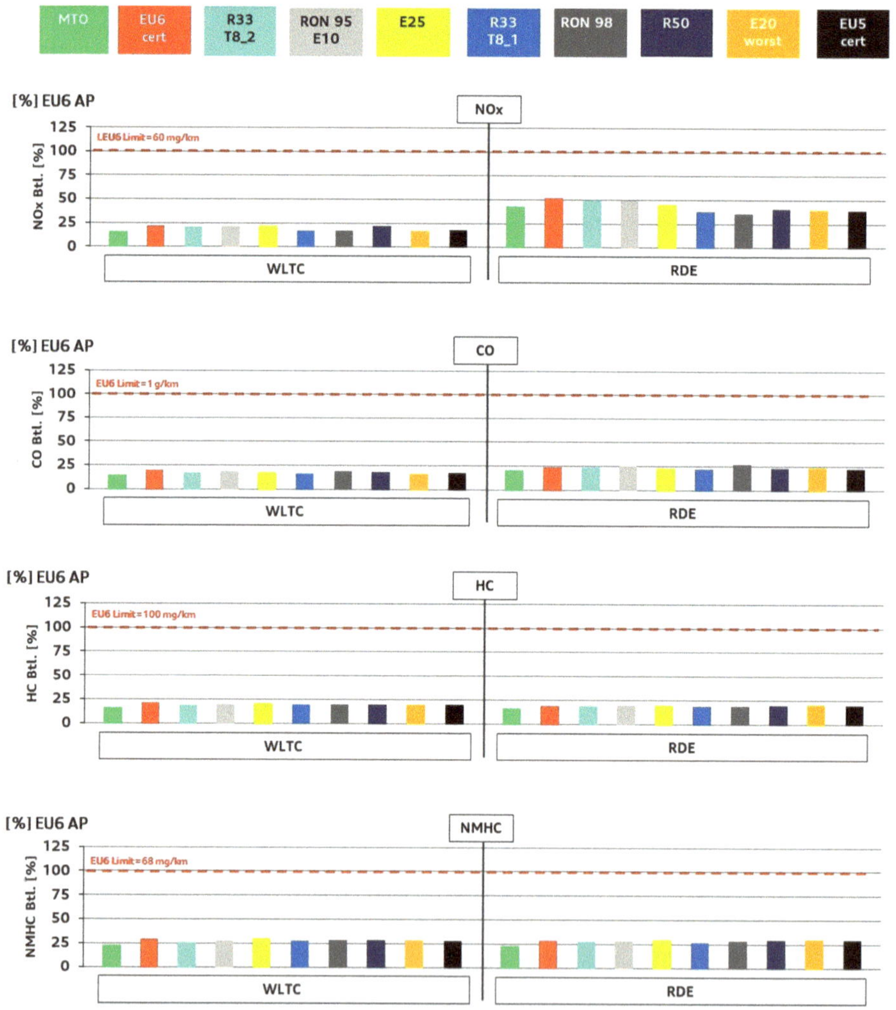

Fig. 2. Audi gaseous emissions of the different fuels in two different test cycles measured on a dynamic engine test bench

In the second group (RON95E10, E25 and R33_4) the PN-emissions level in WLTC is almost constant at a level of a factor of almost 2.5 less compared to EU5_cert. This still means a huge difference and represents an acceptable level of PN-emissions. The corresponding YSI values range from 88..94. The combination of FBP 195 °C and 25% ethanol content in the E25 (s. Table 1) does not seem to play a significant role in WLTC.

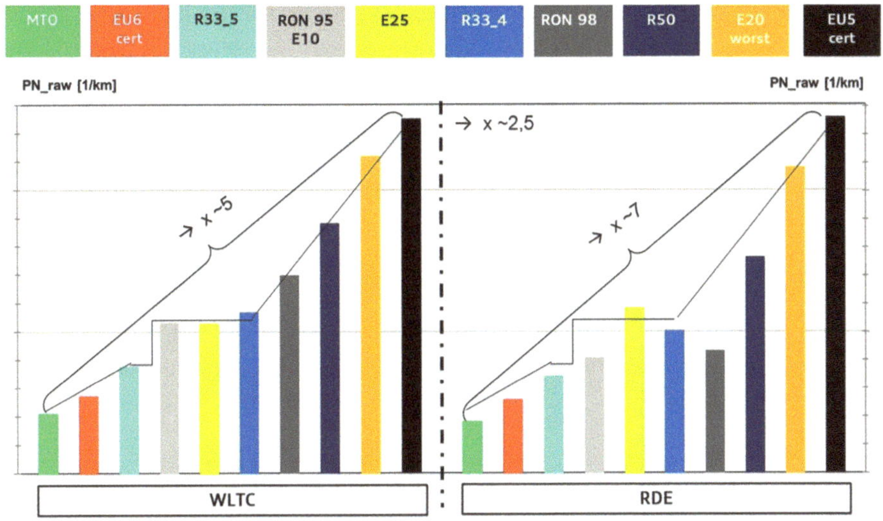

Fig. 3. Engine-out PN-emissions of the different fuels in the two different test cycles measured on a dynamic engine test bench. PN-emissions of EU5cert fuel in the WLTC is defined as 100%

The third group starts with R33_4 and then continues with the RON98 fuel at a PN-emissions level of roughly a factor 2 less than EU5cert which can still be regarded as an acceptable level. PN-emission then raise constantly over R50 and E20_worst to the 100% value of EU5cert. The R50, having a similar YSI as the E20_worst seems to profit from its FBP of 195 °C leading to a factor of 1,4 less PN-emissions than the EU5cert. Obviously, the combination of critical parameters in the especially as a very critical fuel designed E20_worst (high T90, high FBP, high YSI, low volumetric heating value) does not accumulate to disastrous PN-emissions above the EU5cert level but instead, PN-emissions stay distinctly below EU5cert.

Switching to the RDE cycle, the shape of the three WLTC groups is transferred 1:1 to the PN results and makes it possible to study changes in the fuel's PN-behavior under the more challenging conditions of the RDE cycle. Referred to the EU5cert fuel, there is a factor of somewhat more than 2,5 between the raw emissions of WLTC compared to RDE. Within RDE there is a factor of 7 between the lowest (MTO) and the highest (EU5cert) raw PN-emissions.

In the first group of fuels no significant changes occur, but the level of the group seems to be a little bit lower. In the second group, PN-emissions of the RON95E10 stay remarkably below the expected level, which is also true for the RON98 fuel in the third group. These two fuels have the lowest FBP in this study, both staying well below 180 °C. Under the more challenging conditions of the RDE cycle (longer fuel injection events for higher loads, less time to evaporate and complete mixture formation, higher risk of wall-wetting and resulting pool-fires) it is obvious that the low FBP of these two fuels combined with comparatively high volumetric heating values (s. Table 1) help to minimize PN-emissions. To a certain content, this is also true for the R50. In contrast, the E25 fuel having none of these advantages, stays at its relative level and overrides the shape of the WLTC-reference shape slightly. Regarding the E20_worst, its PN-emissions stay again distinctly below the EU5cert.

Figure 4 shows the results of the same cycles in the CVS tunnel, giving a very good equivalent to tailpipe PN-emissions. Again, the results are normalized to the EU-5cert as the 100% PN-reference. The factor between the raw-PN values and the PN-values in the CVS tunnel is a function of the filtration efficiency of the GPF which was once close to 80% in the beginning of the GPF development and is now near 90%. The more interesting question is if there is still a visible distinction between "good" and "critical" fuels or if the GPF equalizes any difference between the fuels. The answer is very clear, the shape of the raw-PN results can be found nearly 1:1 again after the GPF in WLTC with some individual deviations (s. Fig. 4).

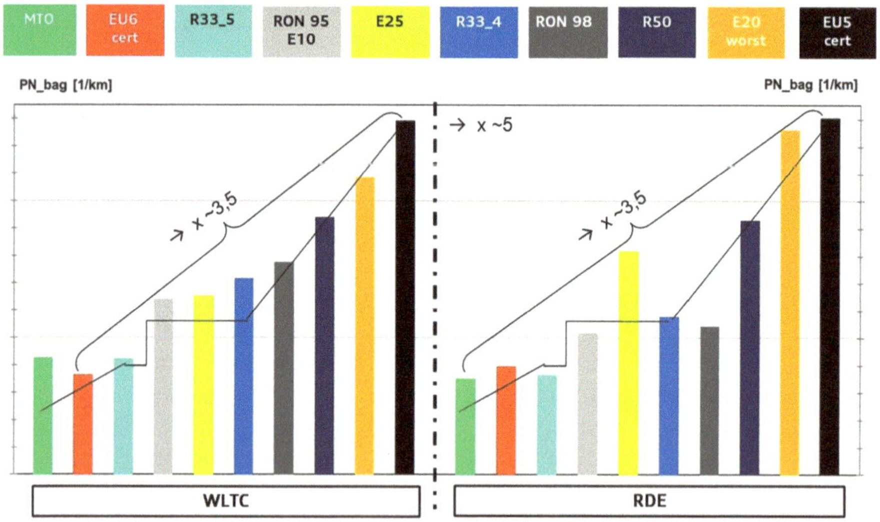

Fig. 4. Tailpipe PN-emissions of the different fuels in the two different test cycles measured on a dynamic engine test bench. PN-emissions of EU5cert fuel in the WLTC is defined as 100%

In detail, the factor between the EU5cert and the fuel with the lowest emissions in WLTC still is at a level of 3.5 in the CVS tunnel. It is interesting to note that in the first group the PN-emissions of MTO apparently pass the GPF easier as expected, leading to an emissions level about two times higher than the reference curve, but still on a low level compared to the other groups. The second group surpasses the reference curve in WLTC to a certain degree as a hole but without changing their order compared to the raw emissions. Beginning with the RON98 the third group follows quite exactly the shape of the reference curve up to the EU5cert fuel.

Switching to the RDE cycle, PN-emissions in the CVS tunnel (Fig. 4) have an overall emissions level about a factor 5 higher than in WLTC in the CVS tunnel. The factor between EU5cert and the fuel with the lowest emissions within the RDE cycle is still at a factor of 3.5 as in WLTC, showing again that the GPF does not equalize the differences between the fuels. As in the raw emissions of the RDE cycle the convenient properties of the two fuels RON95E10 and RON98 at higher loads and accelerations still persist after the GPF in RDE and help again keeping the PN-emissions lower than expected. The three modern fuels with intermediate renewable content (R33 and R50) all confirm their good and partially even very good position (R33_5) regarding the PN-emissions in this study also in the CVS tunnel results in the RDE cycle

In contrast, the PN-emissions of E25 and E20_worst seem to pass the GPF in the RDE cycle better than expected, both now overriding clearly the reference curve, but still staying below the EU5cert fuel. Further investigations could be interesting to confirm and explain this special behavior of the E20..25 fuels in the RDE cycle, but it can be stated that even the E20_worst fuel with its artificial accumulation of critical parameters does not surpass the PN-emissions of the EU5cert fuel from which it was derived.

4 Summary and Outlook

In this study, a set of EN228 compliant fuel candidates with increased renewable content as well as two intermediate ethanol fuels were evaluated concerning their gaseous and particulate (PN-) emissions. Their renewable content ranged from 20..50%. The fuel candidates were benchmarked against market-like fuels (RON95E10, RON98), reference fuels (EU5cert and EU6cert) and a high quality fuel proposal from Audi (MTO) already evaluated and published in a former study.

First, the fuels were screened broadly in typical dynamic test cycles (WLTC, RDE) regarding gaseous and raw PN-emissions as is done very often in such kind of studies. Second, the PN-emissions were simultaneously measured in the CVS tunnel, giving the opportunity to compare the PN-emission characteristics of all fuels before and after the GPF.

The following main findings can be highlighted:

- Dynamic engine cycle test results have shown that the fuels do not exhibit any significant differences in HC, NMHC, NOx and CO tailpipe emissions compared to the legal limits. As in the previous study, approximately neutral behavior on gaseous emissions was concluded.
- Regarding raw PN-emissions as well as PN-emissions in the CVS tunnel, distinct fuel effects could be measured. MTO as high quality fuel shows lowest YSI and in most cases lowest PN emissions, whereas the EU5cert fuel constantly shows the highest YSI and PN emissions and confirmed its position as the fuel with the highest PN-emissions under all conditions.
- The group of EN228 compliant fuels with intermediate renewable content (R33, R50) shows on one hand that very low PN-emissions are possible within this group, even close to the benchmark fuels MTO and EU6cert. On the other hand, aiming for higher renewable content while simultaneously raising RON to 98 makes fuel formulation more complicated and results to date in a higher but not yet critical PN-emissions level.
- Although not being a benchmark concerning the sooting index YSI, the RON95E10 and RON98 fuels in this study surprise with remarkably low PN-emissions in the RDE cycle. Their low final boiling point (FBP) obviously helps to avoid bad mixture formation effects under RDE conditions and should inspire the actual discussions at CEN to lower FBP in the European fuel standard EN228 as it was already addressed by VW at CEN. This measure was already implemented in China's fuel standards in 2016 and represents one of the easiest measures to improve air quality in metropolitan areas concerning PN caused by ICE driven vehicles with and without GPF.
- The fuels with 20..25% ethanol content (E25, E20_worst) do not show unexpected behavior in WLTC but seem to have some disadvantages in the RDE cycle. However, the E25 stays far away from a critical PN level and even the E20_worst with its challenging composition came close to, but did not surpass the PN-emissions of EU5cert under any condition in this study.

It can be highlighted that all tested fuel candidates behave calibration neutral for all tested operating points and conditions. The emission results are promising for drop-in renewable fuel components. Besides the CO_2 reduction potential for the existing vehicle fleet, it could be worked out, that the fuel properties of the partly renewable fuels can be even improved within EN228 limits if the fuel formulation is carefully performed.

On the other side it is evident that fuel compositions like EU5cert with high YSI, high heavy aromatics content, high FBP should be generally avoided in future. This is valid for pure fossil fuels as well as for taking such a kind of fuel as a base fuel for products with an elevated renewable content: fuel quality can and should start to help improve air quality with respect to PN-emissions, in the legacy fleet as well as in newer vehicles equipped with a GPF.

References

1. Kramer, U., Bothe, D., Gatzen, C. et al.: *Future Fuels: FVV Fuels Study IV – The Transformation of Mobility to the GHG-neutral Post-fossil Age,* Final Report, FVV Report no. 1269, 2021.
2. Hartung, S.: *Powertrains of the Future – How We Will Meet our Climate Goals through Technology Neutrality*, 42nd International Vienna Motor Symposium, April 2021.
3. Balthasar F., Mikulic I., Kolbeck A. et al.: *Pathways to Meet Renewable Energy Targets in Transport*, 43rd Vienna Motor Symposium, April 2022
4. Garbe T., et al., *Blue Gasoline – A Solution towards CO_2 neutral mobility for existing and new vehicles with availability already today*, 32nd Aachen Colloquium, Oktober 2023
5. Sarwal R., Kumar S., Mehta A. et al.: *Roadmap for Ethanol Blending in India 2020–25*, Juni 2021
6. Ahlers B.R., *India forms global biofuel alliance at G20 summit*, India Today, September 2023
7. Krämer H., Send M., Gessner M. et al., *Drop-In Renewable Gasoline Fuels for CO_2 Reduction: Evaluation and Demonstration of Functional Potentials*, 10. Internationaler Motorenkongress Baden-Baden, Februar 2023
8. Krämer, H., Borovsky, J., Bulc, A.: *Hochoktaniger paraffinischer Ottokraftstoff innerhalb EN228 E5*. In: Krahl, J., Munack, A., Eilts, P., Bünger, J. (eds.) Fuels Joint Research Group, Kraftstoffe für die Mobilität von Morgen, Band 30, pp.63–70. Cuvillier Verlag (2021).
9. Chaud, B., OPTISOCHEM, *Report on Delivery of First 2G Isobutene Samples*, www.optisochem.eu, 2018.
10. Chaud, B., OPTISOCHEM, *First report on the Delivery of Several Tons of 2G Isobutene, According to Program, with Full Characterization*, www.optisochem.eu, 2019.
11. Winkler J., *Für CO_2-neutrale Produktionsstandorte: nachhaltige Kraftstoffe ab Werk*, https://www.audi-mediacenter.com/de/pressemitteilungen/fuer-co2-neutrale-produktions-standortenachhaltige-kraftstoffe-ab-werk-15098, Audi Media Info, Dezember 2022
12. Das, D.D., St. John, P., McEnally, C.S., Kim, S., Pfefferle, L.D. *Measuring and Predicting Sooting Tendencies of Oxygenates, Alkanes, Alkenes, Cycloalkanes, and Aromatics on a Unified Scale*, Combustion and Flame, 2017.
13. Montgomery, M.J., Das, D.D., McEnally, C.S., Kim, S., Pfefferle, L.D., *Analyzing the Robustness of the Yield Sooting Index as a Measure of Sooting Tendency*, The Combustion Institute, 2018.

Fuel Science:Sustainable Mobility Based on Renewable Resources

Bastian Lehrheuer(✉), Theodoros Kossioris, Paul Muthyala, Marcel Neumann, und Stefan Pischinger

RWTH Aachen University, Forckenbeckstr. 4, 52074 Aachen, Germany
Lehrheuer@tme.rwth-aachen.de

1 Introduction

In 2021, approximately 740 million metric tons of carbon dioxide (CO_2) were emitted in the EU due to the combustion of fuels in road transportation. Passenger cars and motorcycles were responsible for the majority of these emissions, accounting for 64%. Trucks and buses contributed 27%, while light commercial vehicles made up the remaining 10%. Despite efforts to reduce emissions in recent years, progress has been limited. Although improvements in engine technology, exhaust aftertreatment systems, and the use of new fuels such as RON 95 E10 have helped reduce pollutant emissions, these measures have been outweighed by the rising volume of traffic and an increasing number of highly motorized vehicles with relatively high fuel consumption. Between 1990 and 2021, annual CO_2 emissions from road transportation in the EU increased by 21%. The most significant increase was observed in light commercial vehicles, with a 49% rise in CO_2 emissions. CO_2 emissions from trucks and buses grew by 28%, while those from passenger cars increased by 15 Renewable Diesel Fuel. To achieve climate neutrality by 2050, the EU aims to make significant strides with the European Green Deal [EEA23].

A crucial strategy for mitigating GHG emissions and alleviating urban air pollution entails the combination of renewable energy generation with the direct electrification of the transportation sector. However, certain segments of the transportation that are challenging to electrify, such as aviation, shipping, and trucks, may continue to rely on liquid energy carriers in the foreseeable future. For these specific transport sectors, the transition from fossil-based to clean, bio-based, and electricity-based synthetic fuels offers a promising solution to reduce GHG emissions and combat urban air pollution.

Synthetic fuels offer advantages due to their compatibility with existing infrastructure for liquid energy carriers and their ability to indirectly incorporate renewable energy into the transportation sector by utilizing renewable sources for fuel

A. Heintzel (Hrsg.): Int.Engine.Congr. 2024, Proceedings, S. 312–320, 2025.
https://doi.org/10.1007/978-3-658-47966-4_22

production. This can include harnessing renewable energy from regions abundant in wind and solar resources, such as Chile, Australia, North Africa, and Middle Eastern countries.

However, synthetic fuels are facing four main challenges: (1) using mature technologies, (2) compatibility with global fuel standards and elastomer materials, (3) reducing combustion-induced emissions, such as nitrogen oxides (NO_X) and particulate matter (PM), and (4) supporting the transition to net-zero greenhouse gas emissions.

A comprehensive study including the production process & fuel design, material compatibility investigations up to combustion emissions measurements and a subsequent, comparative well-to-wheel Life Cycle Assessment (LCA) reveals that Hydroformylated Fischer-Tropsch (HyFiT) fuels, a combination of two established technologies, Fischer-Tropsch (FT) synthesis and hydroformylation, can effectively tackle all of these challenges simultaneously [Sim24]. This article gives a more detailed impression of the vehicle tests carried out as part of this study and highlights the potential for greenhouse gas and pollutant emission savings on a tank-to-wheel basis.

2 Hydroformylated Fischer-Tropsch (HyFiT) Fuels

The HyFiT-fuel production concept consists of several process units including FT synthesis and hydroformylation. Depending on the FT synthesis, the resulting HyFiT-fuel is either rich in alkanes (C5 to C17) or alcohols (C6 to C11), so that the desired alcohol-alkane ratio of the HyFiT-fuel can be adjusted. The alcohol percentages of 20 and 40 wt% (HyFiT-20% & HyFiT-40%) were selected for the vehicle tests, as these cover a reasonable range of minimum tolerable viscosity with high cetane number at 20 wt% and minimum acceptable cetane number at 40 wt%. The fuel properties of the two mixtures can be found in Table 1. To comply with the lubricity limits of max. 460 μm (EN590), 2000 ppm of R655 have been added. An influence on the other characteristic values could not be observed.

Table 1. Fuel properties of HyFiT fuels with 20 and 40 wt% alcohol content [Sim24]

		HyFiT-20%	HyFiT-40%
Lower Heating Value	$MJ\ kg^{-1}$	42.9	41.5
Derived Cetane Number	–	64	56
Viscosity	mm^2s^{-1}	1.5	2.0
Lubricity (HFRR[1])	μm	568	546
		430[2]	430[2]

[1]High frequency reciprocating rig
[2]Measured with additive: 2000 ppm of R655

3 Vehicle Powertrain Layout and Testing Boundary Conditions

The demonstration vehicle is categorized as a light commercial vehicle (LCV). It is equipped with a 4-cylinder 2.0 L internal combustion engine (ICE) boasting a rated power of 130 kW, featuring high-pressure (HP) exhaust gas recirculation (EGR). The vehicle's exhaust aftertreatment system (EATS) includes a diesel oxidation catalyst (DOC), a diesel particulate filter (DPF) with a selective catalytic reduction (SCR) coating (SCRF), functioning both as DPF and SCR, and an ammonia slip catalyst (ASC). Both the engine and the exhaust aftertreatment system of the vehicle underwent certification in accordance with heavy-duty standards (EU VI D).

For demonstration purposes, the conventional vehicle underwent a transformation into a high voltage full hybrid electric vehicle, completed with an automatic 8-gear transmission. The hybrid powertrain configuration adopted is denoted as P0P2, as illustrated in Fig. 1. Although the electric motor placed in the P0 position for enabling the ICE's On/Off switching, the hybrid powertrain predominantly functions as a P2 hybrid. Essential specifications of the hybrid powertrain are summarized in Table 2. Further details regarding the hybrid demonstrator's configuration and its operational features can be found in [Sch20a] and [Sch20b].

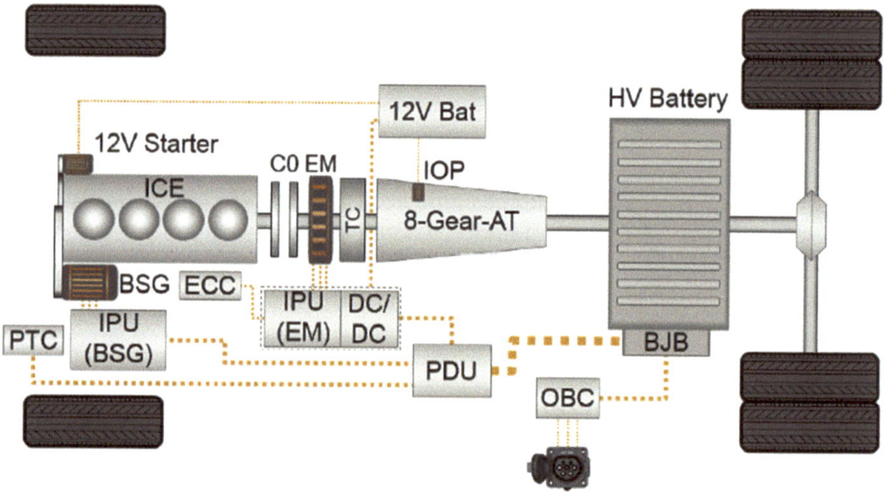

Fig. 1. Powertrain layout of the hybrid electric LCV demonstrator

Table 2. Electric components of the hybrid electric LCV demonstrator

		P0P2-HEV
EM (Electric Machine) Power	kW	60 (cont.)/95 (peak)
EM Type	–	PMSM
Li-Ion Battery Capacity	kWh	14
Nominal System Voltage	V	310

During each test, the total vehicle weight remained constant at 3.3 metric tons. As per regulations, carbon dioxide (CO_2) emissions for vehicles with a total mass below 3.5 metric tons, falling under the N1 category, are certified in accordance with the Worldwide Harmonized Light Vehicle Test Cycle (WLTC) standards [RegEU19]. The WLTC is designed to emulate the typical driving patterns of both passenger cars and vans across Europe. Forming part of the pollutant emissions certification process for light-duty vehicles next to the Real Driving Emissions (RDE) testing requirements, which were introduced within the EU6d-temp regulatory framework, the WLTC is commonly employed for emission calibration.

Given that the vehicle is used as a truck with a speed limit of 120 km/h for safety considerations, the WLTC120 was selected as the appropriate testing cycle for the experimental campaign. All measurements were conducted at an ambient temperature of 23 °C and with a consistent overall vehicle mass of 3.3 metric tons, under charge sustain State of Charge (SoC) conditions. Prior to each test, activities such as the regeneration of the SCRF and its pre-loading with ammonia (NH_3) were carried out. These procedures were performed to establish uniform boundary conditions, ensuring a valid and equitable comparison of the emission measurements across all tests.

4 Engine-Out and Tailpipe Emissions Performance of HyFiT Fuel as Drop-In

Figure 2 provides a summary of the WLTC120 emission measurements for a) carbon dioxide (CO_2), d) particulate matter (PM) and e) & f) nitrogen oxides (NO_X). Vehicle hardware as well as calibration has been kept unchanged characterizing the fuel variation as "drop-in". As previously indicated, all emissions measurements have been adjusted for ΔSoC variations across the entire test cycle to ensure a consistent basis for comparison. Figure 2 b) illustrates the ΔSoC deviations and the corresponding additional CO_2 emissions observed between the initial SoC and the final SoC for each test case. This variation can be attributed, on one hand, to driver influences and, on the other hand, to the fuel change, both of which contribute to a slight difference in power distribution between the Internal Combustion Engine (ICE) and the Electric Motor (EM). Consequently, this results in a minor variance in the amount of battery energy utilized within the system over the entire cycle.

Furthermore, an empirically determined 2 % increase in CO_2 emissions over the entire WLTC120 cycle is factored in for the fossil Diesel case to account for Diesel Particulate Filter (DPF) regeneration events that occur in real-world street operations. To accommodate the less frequent need for regeneration events when utilizing HyFiT fuels, owing to their significantly reduced engine-out soot emissions, the CO_2 increase is proportionally scaled down based on the decrease in engine-out soot emissions observed in the HyFiT cases compared to fossil Diesel (see Fig. 2 c)).

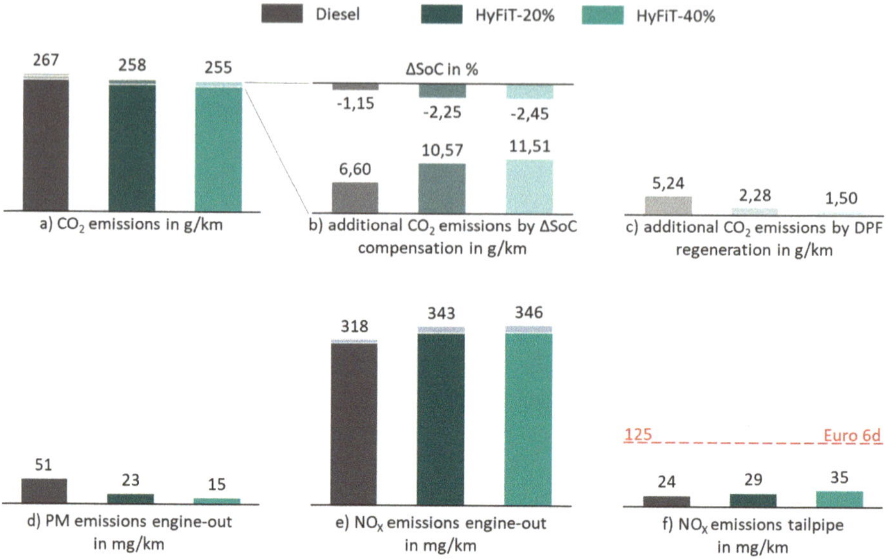

Fig. 2. Cumulative a)-c) carbon dioxide (CO_2), d) particulate matter (PM) and e)-f) nitrogen oxides (NO_X) emissions during WLTC120 of Diesel, HyFiT-20% and HyFiT-40% fuels as Drop-In

Based on the findings presented in Fig. 2, we observe an approximate 3 % reduction in carbon dioxide (CO_2) emissions with HyFiT-20% and about a 5% reduction in CO_2 emissions with HyFiT-40%, as compared to fossil Diesel. This decrease can be attributed to the higher oxygen content of both HyFiT fuels. In terms of engine-out soot emissions, there is an approximately 55 % reduction with HyFiT-20% and an approximately 72 % reduction with HyFiT-40%, again owing to the increased oxygen content.

Regarding nitrogen oxides (NO_X) emissions, we note that engine-out NO_X emissions are approximately 8% higher for HyFiT-20% and approximately 9% higher for HyFiT-40% when compared to fossil Diesel. It is essential to highlight that the influence of the different fuels on tailpipe NO_X emissions appears more pronounced than on engine-out NO_X emissions. This is attributed to the combined effects of both higher engine-out emissions and the delayed exhaust aftertreatment system (EATS) heat-up on the tailpipe NO_X emissions. Consequently, tailpipe NO_X emissions seem to be around 24% higher for HyFiT-20% and approximately 48% higher for HyFiT-40% when compared to fossil Diesel.

Notably, for increased vehicle speed at the end of the WLTC120 a substantial increase in engine-out NO_X emissions followed by a modest increase in tailpipe emissions were observed during the WLTC using HyFiT fuels. This increase can be primarily attributed to the non-optimized urea dozer strategy of the SCRF, which is calibrated based on Diesel fuel properties. However, it's essential to emphasize that EATS optimization for the HyFiT fuels falls outside the scope of the current study and was not further investigated.

The HyFiT fuels have effectively demonstrated their Drop-In compatibility and have shown advantages in terms of reducing both CO_2 and soot emissions. The relatively poorer tailpipe NO_X performance can be attributed to the fact that the internal combustion engine (ICE) and exhaust aftertreatment system (EATS) calibration were not optimized for HyFiT fuels. In a comprehensive vehicle calibration process, both the ICE and the Selective Catalytic Reduction (SCR) control strategy would be adjusted to align with the combustion characteristics of the new fuel. However, it is worth highlighting that the tailpipe NO_X emissions performance of the HyFiT fuels, even though it surpasses that of fossil Diesel, still falls below the current EU6d legislation (125 mg/km).

5 Exploiting the Emission Reduction Potential of HyFiT Fuels

In the following section, the authors aim to showcase the full emission reduction potential achievable by promptly implementing the examined HyFiT fuels as Drop-In substitutes in the existing fleet combined with a potential improvement by an optimized calibration. Through a theoretical investigation, the substantial advantages of HyFiT fuels in reducing engine-out soot emissions by as much as 70% will be leveraged to estimate the potential reduction in overall tailpipe NO_X emissions.

The approach comprises three steps. In the initial step, the emissions performance of both HyFiT and Diesel fuels is compared at specific characteristic operating points (OPs) aligned with the operational profile of the LCV demonstrator during the WLTC. Following this, the observed trends are extrapolated to powertrain level using simulation. In this step, the engine-out emissions recorded during the transient tests with HyFiT fuels are adjusted to model two distinct low-NO_X internal combustion engine (ICE) calibration scenarios, derived from previous Single Cylinder Engine measurements. Finally, the tailpipe emissions are ultimately predicted for the engine-out emissions optimized for low-NO_X, incorporating assumed efficiencies for the individual exhaust aftertreatment system (EATS) components.

As shown in [Neu18], depending on the permissible limit of soot emissions and the calibration of in-cylinder anti-pollution measures, such as high exhaust gas recirculation (EGR) strategies and retarded injection, hydroformylated Fischer-Tropsch fuels can enable indicated specific NO_X levels as low as 0.1 g/kWh. This advantage offered by oxygenated fuels can also be leveraged to mitigate tailpipe NO_X emissions. Nonetheless, it's important to acknowledge that in-cylinder anti-pollution measures come with trade-offs that negatively affect CO_2 emissions. Consequently, such a low-NO_X mode should only be engaged under specific operating conditions where the exhaust aftertreatment system (EATS) cannot achieve high conversion efficiencies, like during warm-up. Here, the low-NO_X mode of HyFiT fuels is explored until the SCRF reaches its designated operating temperature range of approximately 200 °C after 825 seconds. Beyond this point, the SCRF can effectively convert nearly all engine-out NO_X emissions, resulting in virtually negligible tailpipe NO_X emissions.

Two scenarios for an optimized HyFiT engine calibration where modelled:

a) A scenario in which, during the initial 0–825 seconds phase, the engine-out soot emissions of the HyFiT fuels match those of fossil Diesel
b) A scenario in which, during the initial 0-825 seconds phase, the tailpipe NO_X emissions of the HyFit fuels match those of fossil Diesel

The first scenario illustrates the theoretical maximum potential for NO_X emissions reduction, which concurrently results in a maximum increase in engine-out soot emissions, reaching levels akin to those of the fossil Diesel case. In the second scenario, the necessary engine-out NO_X emissions level is estimated to ensure equivalent tailpipe NO_X emissions with fossil Diesel. It's important to note that after approximately 825 seconds, the vehicle's SCR system reaches its desired operating temperature range, rendering low-NO_X measures unnecessary. For the purposes of this study, the emissions level in the phase beyond 825 seconds is assumed to remain constant and equivalent to the measured Drop-In emissions.

Figure 3 provides a comprehensive overview of all emissions results, accounting for the ΔSoC correction and DPF regeneration in all cases. According to the results, both HyFiT fuels exhibit lower CO_2 emissions when compared to fossil Diesel. This trend persists even when the HyFiT-dedicated low-NO_X calibration is implemented, albeit with a minor drawback compared to the pure Drop-In scenario. Likewise, in all cases involving HyFiT fuels, engine-out soot emissions are notably reduced in contrast to the case of fossil Diesel.

Based on the simulation studies, leveraging the full potential for reducing engine-out NO_X emissions with HyFiT fuels can result in a remarkable up to 35 % reduction in tailpipe NO_X emissions over the entire WLTC120 in comparison to fossil Diesel. This reduction can be attributed to the significantly lower level of engine-out NO_X emissions achieved before the Selective Catalytic Reduction Filter (SCRF) light-off is reached, compensating for the limited de-NO_X capacity of the system during this phase. As for tailpipe soot emissions, the DPF effectively filters nearly 100 % of the emitted soot in all cases, resulting in almost negligible tailpipe soot emissions and are therefore not shown here.

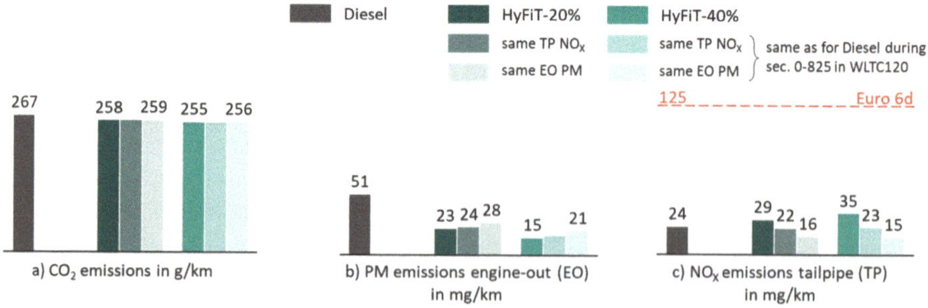

Fig. 3. Cumulative a) carbon dioxide (CO_2), b) particulate matter (PM) and c) nitrogen oxides (NO_X) emissions during WLTC120 of Diesel, HyFiT-20% and HyFiT-40% fuel with optimized calibration for emission reduction

As a result, the HyFiT-dedicated low-NO_X strategies do introduce certain disadvantages related to CO_2 and engine-out soot emissions. Nonetheless, these drawbacks can be offset on one hand by the substantial reduction in tailpipe NO_X emissions and on the other hand by the efficient functioning of the Diesel Particulate Filter (DPF), which ensures zero tailpipe soot emissions.

6 Conclusions

In conclusion, our extensive vehicle measurement campaign has effectively assessed the drop-in compatibility of HyFiT fuels within the WLTC120 framework. Our tests have verified that these E-Fuels can seamlessly enable the same vehicle drivability as fossil Diesel. Furthermore, we identified a noteworthy CO_2 reduction benefit of up to 5%. The standout feature of HyFiT fuels lies in their significant reduction of engine-out soot emissions by a remarkable 50–70%, which can serve as a valuable counterbalance to the downside of increased tailpipe NO_X emissions. It's worth noting that even with the immediate implementation of HyFiT fuels, we observed no violation of the stringent EU6d NO_X limits in the WLTC120, underscoring the neutrality of HyFiT fuels in terms of air quality impact, as defined by regulatory vehicle emissions standards.

However, the HyFiT-fuels produce slightly higher NO_X engine-out emissions of around 8–9%, resulting in higher NO_X tailpipe emissions for HyFiT-20% (29 mg/km) and HyFiT-40% (35 mg/km) than for diesel (24 mg/km).

Utilizing the exceptionally low engine-out soot levels achievable with HyFiT fuels, it becomes possible to substantially reduce tailpipe NO_X emissions to levels significantly below those of fossil Diesel without burdening the Diesel Particulate Filter (DPF). Through the implementation of HyFiT-dedicated low-NO_X strategies, the remarkably low engine-out emissions during these phases can offset the limitations of the Selective Catalytic Reduction (SCR) efficiency until the required temperature is reached. This, in turn, leads to an overall reduction in tailpipe NO_X emissions, particularly in areas characterized by emissions related to the cold exhaust aftertreatment system. In this study, a 35% reduction in tailpipe NO_X emissions over the entire WLTC120 in comparison to fossil Diesel was observed.

Given that HyFiT fuels are drop-in capable, these measures can be readily integrated into the existing vehicle fleet through minor adjustments in the internal combustion engine calibration, provided that the vehicle powertrain hardware can accommodate these adaptations.

Looking ahead to the future of vehicle fleets, the insights from our study can be pivotal in shaping the EU7 technology package for vehicle manufacturers. An optimized HyFiT-dedicated low-NOx strategy has the potential to downsize both the deNOx unit and the DPF, resulting in cost savings and improved ICE efficiency by reducing EATS-induced backpressure. Furthermore, the exceptional capability of HyFiT fuels to limit engine-out NOx emissions during cold starts can obviate the need for additional heating devices in the EATS, such as electrical heaters.

In conclusion, the benefits of HyFiT fuels extend beyond their contribution to achieving “net-zero” CO_2 emissions. They also hold promise for enhancing air quality impacted by the transportation sector. Consequently, HyFiT fuels, and fuels based on renewable resources in general, should be recognized as a technology option for the next generation of vehicles. Future research efforts should focus on enhancing powertrain models and conducting further vehicle testing. This entails developing adaptable powertrain models capable of accurately predicting pollutant emissions and fuel consumption for a range of carbon advanced fuels, as well as integrating ICE models and EATS components into a comprehensive powertrain simulation approach. Moreover, comprehensive testing campaigns with vehicle demonstrators should address the low-NOx capabilities of ICEs powered by renewable fuels in real driving conditions and at ambient temperatures below 0 degrees Celsius. Additionally, test campaigns featuring EU7 demonstrators using new fuels can provide valuable insights into the powertrain system and uncover potential challenges and opportunities, thereby contributing to a more holistic evaluation.

Acknowledgements. This work was funded by the Deutsche Forschungsgemeinschaft (DFG, German Research Foundation) under Germany´s Excellence Strategy—Cluster of Excellence 2186 “The Fuel Science Center”—ID: 390919832.

Reference

[EEA23] European Environment Agency (EEA): Greenhouse Gas Emissions by Source Sector., Eurostat Data Code: env_air_gge, https://ec.europa.eu/eurostat/databrowser/bookmark/1ffd5837-18ba-4e68-9634-1eaf2d6d4023?lang=de), last check: 31.05.2023.

[Sim24] Simon Völker, Niklas Groll, Marvin Bachmann, Leonard Müller, Marcel Neumann, Theodoros Kossioris, Paul Muthyala, Bastian Lehrheuer, Marius Hofmeister, Andreas Vorholt, Katharina Schmitz, Stefan Pischinger, Walter Leitner, and André Bardow: Towards carbon-neutral and clean combustion with hydroformylated Fischer-Tropsch fuels, (submitted)

[Sch20a] Schaub et al., Hybrid Powertrain Concepts for Light Commercial Vehicles, 29th Aachen Colloquium Sustainable Mobility 2020

[Sch20b] Schaub et al., Electrified efficiency – diesel hybrid powertrain concepts for light commercial vehicles, ATZ, Internationaler Motorenkongress 2020

[RegEU19] Regulation (EU) 2019/631 of the European Parliament and of the Council of 17 April 2019

[Neu18] Neumann, D., Pischinger, S., Zubel, M., Heuser, B., Thenert, K., Leitner, W., Schönen, M., Schaub, J., and Jörg, C. (2018). Power-to-Liquids - Kompensation variierender E-Fuel Zusammensetzungen mittels Digital Rate Shaping /Power-to-Liquids - Compensation of Varying E-Fuel Compositions via Digital Rate Shaping. In 39. Internationales Wiener Motorensymposium 26.-27. April 2018, B. Geringer and H.-P. Lenz, eds. (: VDI Verlag), pp. 230–258

Building Industrial Capacity for Advanced Biofuels to Meet the 2030 Targets

Dina Bacovsky[1(✉)], Doris Matschegg[1], Andrea Sonnleitner[1], Ioannis Tsiropoulos[2], David Chiaramonti[3], Theodor Goumas[4], und George Vourliotakis[4]

[1] Graz, Austria
{dina.bacovsky,doris.matschegg, andrea.sonnleitner}@best-research.eu
[2] E3-Modelling SA, Graz, Austria
tsiropoulos@e3modelling.com
[3] Politecnico di Torino, Graz, Austria
david.chiaramonti@polito.it
[4] EXERGIA SA, Graz, Austria
{t.goumas,g.vourliotakis}@exergia.gr

1 Introduction

Biofuels are one of the main means to reduce greenhouse gas (GHG) emissions from the transport sector. The European Union (EU) has adopted regulations to promote their use, and in doing so distinguishes between different categories. While the use of biofuels produced from food/feed feedstocks is promoted only up to 7% of road transport energy use, so-called advanced biofuels are more actively promoted through a separate sub-target. But these advanced biofuels are more difficult and costlier to produce and technologies for their production are not yet fully commercialized. Will it be possible to build the required industrial production capacity to meet the 2030 targets for advanced biofuels?

This paper is based on the study "Development of outlook for the necessary means to build industrial capacity for drop-in advanced biofuels" [1] carried out for the European Commission in 2023[1]. This paper, however, reflects the views of the authors of this paper, and the European Commission shall not be liable for any consequences stemming from this paper.

The goal of this study is to investigate and analyze factors contributing to the development of advanced and sustainable biofuels production within the relevant European policy and regulatory framework. Its primary goal is to facilitate the timely market entry of advanced biofuels and endorse technological solutions through reliable findings to achieve the 2030 and 2050 targets. This study takes into account the

[1] The full study is available for download here: https://data.europa.eu/doi/10.2777/679307.

A. Heintzel (Hrsg.): Int.Engine.Congr. 2024, Proceedings, S. 321–333, 2025.
https://doi.org/10.1007/978-3-658-47966-4_23

evolving EU policy context for sustainable alternative fuels, which is undergoing significant revision and forms a core component of both the Fit for 55 package and the REPowerEU plan. The Consortium consisted of EXERGIA SA (Leader), Politecnico di Torino (POLITO), BEST—Bioenergy and Sustainable Technologies GmbH, BTG Biomass Technology Group BV, Wageningen Institute for Environment and Climate Research (WENR), and E3-Modelling S.A.

The study was executed along six tasks, focusing on: the analysis of demand potential of alternative to fossil fuels for transport using specialized models for the EU (Task 1), the analysis of the resource potential for drop-in advanced biofuels collecting updated data for feedstock availability in the EU (Task 2), the analysis of the potential for industrial capacity of advanced biofuels based on the industry's perspective for the expansion plans and capacity of production (Task 3), synthesis of the outlook for industrial capacity with a focus on the identification of gaps between demand/supply/feedstock availability (Task 4), the analysis of socio-economic impacts, including GHG emissions and costs, GDP, and employment generation (Task 5), and the organization of a workshop to collect feedback from competent thematic and industry experts on the project findings (Task 6).

2 EU Regulation and Targets

2.1 Renewable Energy Directive

Biofuels production and consumption in the EU is mainly driven by the Renewable Energy Directive. It was first published in 2009, setting a target for each Member State to reach a share of energy from renewable sources in all forms of transport of at least 10% of the final consumption of energy in transport in 2020.

RED II was published in 2018, setting the target for 2030 at 14%. But in the light of the Green Deal objectives, the EU has stepped up ambitions, and the revised Renewable Energy Directive now requires Member States to set national targets to either achieve a share of 29% renewable energy in transport or a 14.5% GHG emissions reduction in the overall transport fuel mix.

The Renewable Energy Directive furthermore introduced sustainability criteria for biofuels and distinguishes between different types of biofuels for which specific subtargets or limits apply.

- The contribution of biofuels that are produced from food and feed crops is limited to 7% of final consumption of energy in the transport sector in each Member State.
- Advanced biofuels are defined as fuels produced from feedstocks listed in Annex IX Part A of the Directive. This list includes lignocellulosic feedstocks (e.g., agricultural and forestry residues, such as wheat straw/corn stover/bagasse, wood-based biomass), non-food crops (e.g., grasses, miscanthus, algae), or industrial waste and residue streams. Formerly, there was a specific sub-target solely for advanced biofuels and biogas for transport, but in the revised RED this sub-target also includes renewable fuels of non-biological origin (often referred to as e-fuels) while it is raised to 5.5% in 2030.

- The contribution of biofuels and biogas produced from Annex IX Part B feedstocks (namely used cooking oil and animal fats) is limited to 1.7% of the energy content of transport fuels.
- Biofuels produced from feedstocks that do not fall into any of the above categories are not limited or covered by a specific sub-target.

All biofuels produced from feedstocks listed in Annex IX Part A or B can be double-counted for the purpose of reaching the target for the share of renewables in transport. In addition to biofuels, renewable electricity applied in electric vehicles or other transport is counted towards the target.

Given (a) the limits for biofuels produced from food and feed crops and those produced from used cooking oil and animal fats, and (b) the uncertainties related to the pace of the widespread adoption of electromobility and the associated concerns with respect to critical material availability and competitiveness of the EU in the sector, achieving the 2030 target for renewable energy in the transport sector largely relies on advanced biofuels and biogas produced from Annex IX Part A feedstocks. Production technologies for these, however, still need to be demonstrated and commercialized.

2.2 Shipping and Aviation

Additional demand for biofuels is expected to come from two new regulations dedicated to the shipping and aviation sectors. The final regulations stipulate as follows, while the assessment carried out under [1] is based on the provisional regulations as available early 2023.

ReFuel EU Aviation was published in October 2023 and sets obligations for aviation fuel suppliers to include minimum shares of Sustainable Aviation Fuels (SAF) in all aviation fuel supplied. The minimum share starts at 2% in 2025, reaches 6% in 2030 and further increases in steps up to 70% in 2050. Biofuels produced from food and feed crops are not eligible as SAF.

FuelEU Maritime was published in September 2023 and sets limits on the GHG intensity of energy used on board of ships above 5000 gross tonnage. The limit starts at –2% as compared to the fossil reference in 2025, reaches -6% in 2030 and further increases in steps up to -80% in 2050.

3 Biofuel production pathways

Four biofuel production pathways have reached market maturity and are widely deployed in Europe so far. These include:

- Biodiesel (FAME) obtained from the (trans)esterification of fatty acids
- Ethanol obtained from the fermentation of sugar and starch crops
- Hydrotreated vegetable oil (HVO, often also referred to as renewable diesel) obtained from the hydrotreatment of vegetable oils
- Biomethane obtained from anaerobic digestion and subsequent upgrading of the resulting biogas.

Depending on the feedstock used, the resulting biofuels are either conventional (food/feed-based), or advanced (Annex IX Part A-based), or Annex IX Part B-based.

Advanced biofuels production pathways include the following:

- Fermentation of lignocellulosic feedstock to ethanol
- Gasification and synthesis to methanol/DME/ammonia
- Gasification and subsequent methanation
- Gasification and subsequent Fischer Tropsch synthesis
- Pyrolysis
- Hydrothermal liquefaction
- Lignin depolymerisation
- Alcohol-to-Jet (ATJ) and Methanol-to-Jet (MTJ)

Technologies for the production of these advanced biofuels are at Technology Readiness Levels (TRL) of 7 to 9, i.e. at various stages of demonstrating their technical and economic feasibility. Since they are not yet fully mature, current production volumes are limited.

Our analysis furthermore includes biodiesel (FAME) and HVO produced from cover crops. This is a new category of biomass feedstock that might be included in a future revision of the feedstock list in Annex IX Part A of the Renewable Energy Directive.

4 Current Biofuel Production and Use

4.1 Commercial Technologies

In 2019, biofuels provided a 5.83% of road transport energy in the EU-27, and electricity 0.08%; accordingly, fossil fuels still provided more than 94% of road transport energy. Within the biofuels share, biodiesel (FAME) and HVO together provided some 82%, ethanol 17% and biomethane 1% [2].

Latest estimates for EU-27 biofuel production include some 9.9 million t/y of biodiesel (FAME) production[2]; 5.1 million t/y of HVO production[3]; another 5.1 million t/y of ethanol production[4]; and around 3.8 billion m^3/y of biomethane production[5]. Installed capacities for biodiesel (FAME) and ethanol may be higher since facilities are sometimes idle or do not operate on full capacity when feedstock prices are high [1].

[2] Combined FAME and HVO production of 15 million t/y according to EBB interview conducted 21.03.2023.

[3] HVO production capacity estimated according to EU Biofuels Barometer 2020 and ETIP Bioenergy database on production facilities.

[4] ePURE members produced 5.57 billion litres in 2020 and constitute around 85% of EU production capacity, according to ePURE interview conducted 03.03.2023.

[5] 37 TWh of biomethane production in EU in 2021, according to EBA interview conducted 14.03.2023.

The majority of the EU liquid biofuels production is based on food/feed crops. Data for 2017 shows that only around 3% of ethanol production in EU-27 and the UK was produced from lignocellulosic material or other feedstocks listed in Annex IX Part A and thus qualified as advanced biofuel; the vast majority was based on sugars and cereals. For Biodiesel (FAME), some 75% were based on virgin oils (i.e. food/ feed crops), while some 20% were based on used cooking oil and around 5% on animal fats. For HVO, used cooking oil was used for almost 30% of production and animal fats for around 10%, the remainder again was based on virgin oils. [2]

Current EU production capacity of advanced biofuels is estimated at 300,000 t/y[6], and production capacity of biomethane from anaerobic digestion at 3.8 billion m^3 in 2021[7]. These small amounts will certainly fall short of meeting the 5.5% target imposed for advanced biofuels combined with RFNBOs in 2030 [1].

4.2 Technologies Under Demonstration

A set of technologies that is capable of producing advanced biofuels (i.e. from Annex IX Part A feedstocks) is currently under small- or large-scale demonstration and needs to prove their technical and economic feasibility. Based on the demoplants database maintained by IEA Bioenergy Task 39 in collaboration with ETIP Bioenergy [3], and acknowledging that data may be incomplete, we estimate around 200,000 t/y of advanced ethanol, and around 100,000 t/y of pyrolysis oil production capacity. Gasification and subsequent production of biomethane (synthetic natural gas, SNG) is not widely demonstrated yet (2,000 t/y of installed and operational production capacity), gasification and subsequent synthesis to methanol or DME stands at 600 t/y, and gasification and subsequent Fischer Tropsch fuel production was demonstrated at 8,000 t/y. The largest operational hydrothermal liquefaction production capacity amounts to 1,400 t/y. So far there are no alcohol-to-jet fuel facilities in Europe [1, 3].

The largest operational European demonstration facilities for these advanced biofuels are listed in **Table 1** below.

6 Based on the demoplants database available at https://www.demoplants.best-research.eu.

7 37 TWh of biomethane production in EU in 2021, according to EBA interview conducted 14.03.2023.

Table 1. Largest operational advanced biofuels facilities in Europe

Company	Location	TRL	Product	Capacity
Clariant[a]	Podari, Romania	8 (first of a kind)	Ethanol	50,000 t/y
AustroCel Hallein	Hallein, Austria	8 (first of a kind)	Ethanol	30,000 t/y
Versalis/Eni	Crescentino, Italy	9 (commercial)	Ethanol	25,000 t/y
Domsjoe Fabriker	Ornskoldsvik, Sweden	9 (commercial)	Ethanol	19,000 t/y
Borregaard Industries	Sarpsborg, Norway	9 (commercial)	Ethanol	15,800 t/y
St1	Kajaani, Finland	6–7 (demonstration)	Ethanol	8,000 t/y
Total	Dunkirk, France	6–7 (demonstration)	FT liquids	8,000 t/y
Advanced Biofuels Solutions	Swindon, UK	8 (first of a kind)	Biomethane (SNG)	1,500 t/y
Karlsruhe Institute of Technology	Eggenstein-Leopolds-hafen, Germany	6–7 (demonstration)	DME	608 t/y
Empyro (Twence)	Hengelo, Netherlands	9 (commercial)	Bio-oil	24,000 t/y
Green Fuel Nordic	Lieksa, Finland	9 (commercial)	Bio-oil	24,000 t/y
Pyrocell	Gavle, Sweden	9 (commercial)	Bio-oil	24,000 t/y
Silva Green Fuel	Tofte, Norway	6–7 (demonstration)	Bio-crude	1,400 t/y

[a]As was announced on 6th December 2023, the Clariant facility is now closed

Note that currently only the pyrolysis oil produced at Pyrocell is upgraded to transport fuel within Preem´s refinery in Lysekil, while the pyrolysis oil produced at the Empyro and the Green Fuel Nordic plants is being used as heating oil. However, because of the addition of hydrogen during the upgrading of fast pyrolysis bio oil (FPBO) to finished transport fuel, the yield from FPBO to finished fuel is around 100%

Current *production capacities* for biofuels are summarized in **Table 2** below. Note that for conventional biofuels, capacities are higher than actual production.

Table 2. Estimated European biofuel production capacities in 2023

Fuel/pathway	2023 production capacity
FAME (all feedstock categories)	12 million t/y
HVO/HEFA (all feedstock categories)	5.1 million t/y
Conventional ethanol	5.8 million t/y
Biomethane from anaerobic digestion	3.8 billion m^3
Advanced ethanol	200,000 t/y
ATJ	–
Gasification + methanol/DME	600 t/y
Gasification + SNG	2,000 t/y
Gasification + FT—diesel	8,000 t/y
Pyrolysis—bio-oil	100,000 t/y
HTL—biocrude	1,400 t/y

5 Achieving the 2030 Target for Advanced Biofuels

5.1 Advanced Biofuels Demand to Achieve the 2030 Target

The first Task of the "Development of outlook for the necessary means to build industrial capacity for drop-in advanced biofuels" study [1] aimed at quantifying the potential demand for advanced biofuels in the transport sector within the policy framework of the legislative package Fit for 55 (July 2021) and the targets outlined in the REPowerEU communication. 5 scenarios and 4 sensitivity variants were considered, accounting for two sets of basic drivers: limited deployment of technologies such as electric vehicle battery technology, alternative fuels infrastructure, electrolyzers and direct air capture; and equal contribution of sectors, including road transport, to the Effort Sharing Regulation target.

The assessment finds that under scenario assumptions biofuel consumption increases notably from current levels, predominantly originating from road transport. Overall biofuel consumption is projected to reach between 32 and 40 Mtoe in 2030, depending on the scenario. Of these, conventional biofuels are expected to range from 8 to 12 Mtoe. Advanced biofuels (i.e. from Annex IX Part A feedstocks) are expected to provide 15 to 19 Mtoe, and biofuels derived from waste oil and fats (i.e. from Annex IX Part B feedstocks) 9 to 10 Mtoe [1].

5.2 Capability of the Biofuels Industry to Expand Production Capacities Under Current Market Conditions

Task 3 in the above-mentioned study [1] was dedicated to analyze and evaluate the industrial capacity for supplying advanced biofuels and biomethane for transport, particularly those fuels derived from feedstocks mentioned in Annex IX Part A of the Renewable Energy Directive. The analysis reviewed existing infrastructure, technical expertise and stakeholder capabilities, both in current operations and prospective value chains.

A first assessment considered *current market conditions.* It was based on data collected through:

- An online survey sent to 140 technology developers and advanced biofuels producers,
- Interviews with 11 industry associations and major stakeholders in the field, and
- Estimates of 14 industry experts and academics on specific production technologies.

The resulting data was structured along the biofuel production pathways, condensed into a table, and discussed with 6 high-level experts from industry and academia with a broad view on biofuels.

The experts concluded that many advanced biofuels technologies are not yet commercially available, and large-scale demonstrations are needed to further develop these technologies as well as to create appropriate supply chains for so far unused feedstocks. However, investments in such large-scale demonstrations are very costly and risky and do not appear very attractive to investors, who can choose from a wide range of investment opportunities also outside the biofuels sector. At current state of technology development, CAPEX requirements and market demand, the experts envisaged the following merit order of investments into biofuels facilities:

- Anaerobic digestion with upgrading to biomethane, and HVO/HEFA: both already well-known technologies whose products can serve more than one market
- Second-generation (2G) ethanol, gasification for the production of biomethane, gasification for the production of biomethanol/DME/ammonia, and pyrolysis: higher CAPEX requirements, technologies not yet fully proven
- Alcohol-to-jet (ATJ): mandated shares just picking up from 2025 onwards, requires waste-and-residues-based alcohol as a feedstock; gasification for the production of Fischer-Tropsch Synthetic Paraffinic Kerosene (FT-SPK): very high CAPEX requirements); and Hydrothermal Liquefaction (HTL): technology still under development
- RFNBOs: very high fuel production costs due to large renewable energy demand

The resulting estimate for 2030, under current market conditions, includes 15 Mtoe of biomethane production through anaerobic digestion and 3.4 Mtoe of advanced biofuels. This can potentially satisfy the projected policy-driven 2030 demand, but includes the assumption that all biomethane produced would be available to the transport sector. Yet, biomethane has many different applications and not all is expected to be available to the transport sector. Europe does not even have a large enough fleet capable of running on natural gas or biomethane right now. The experts estimated that only 10% of biomethane from anaerobic digestion and 5% of biomethane from gasification plus methanation would be available to the transport sector.

Figure 1 below displays these values as well as the further projection to 2050, when more advanced biofuel technologies will have reached maturity.

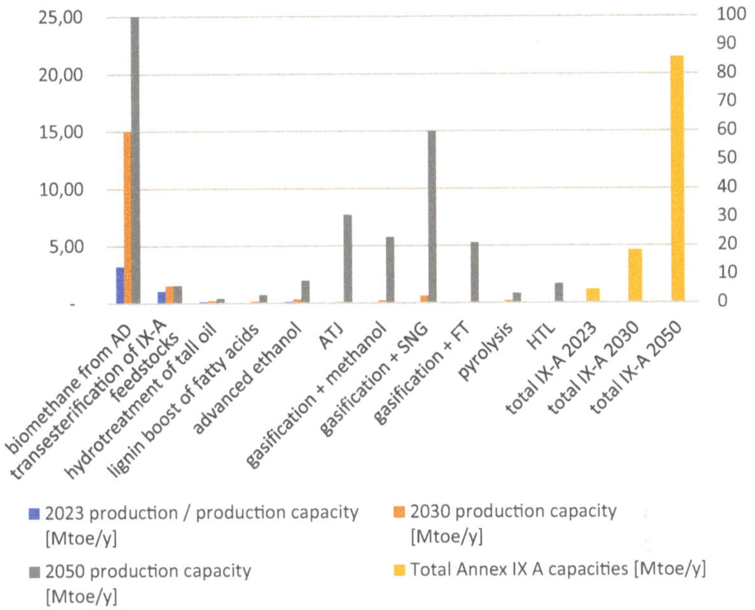

Fig. 1. Estimated evolution of biofuel and biomethane production capacities from Annex IX A feedstocks in Europe under current market conditions, converted to Mtoe/y. Left side: distinct biofuel pathways; right side: total Annex IX A capacities in 2023/2030/2050

5.3 Capability of the Biofuels Industry to Expand Production Capacities Under Favorable Market Conditions

Since current market conditions might not trigger the investments into demonstration facilities needed to further develop advanced biofuels technologies, within the study "Development of outlook for the necessary means to build industrial capacity for drop-in advanced biofuels" we also assessed the *possibility of production capacity expansion under entirely favorable market conditions*. Even if clients are ready to invest, capacity expansion cannot happen overnight, since for some technologies the construction, erection and start-up of large-scale facilities takes up to four years, and also engineering capacity of technology providers is limited by the availability of knowledgeable staff.

Thus, a second assessment of the capability of the biofuels industry to expand production capacities was undertaken. Again, data was collected from industry, this time through structured interviews with leading technology providers for nine biofuel pathways. These are hydrotreatment; cellulosic ethanol; ATJ; gasification to methanol; gasification to SNG; gasification to FT; pyrolysis; HTL; and lignin depolymerization. The analysis comprised the following steps:

Step 1: Together with the interviewees from each company, the author assessed the currently installed capacity in Europe, the largest operational facility globally, the size of a future commercial facility, the lead time until fully operational, and the number of facilities that the company can support per year.

Step 2: Based on these numbers, the author calculated the increase in capacity in the coming years (starting 2024) that could contribute to the volumes needed in 2030, and from that the resulting 2030 capacity of the specific technology provider.

Step 3: Considering that for most pathways there are also competitors with similar capabilities, a "competitor´s factor" was used to estimate European capacity in 2030.

Step 4: 2050 capacity was estimated by the use of a multiplier. The standard factor used was 5 (corresponding to 4 years lead time); it was 7 for technologies with a ramp-up of size ahead of 2030, and 10 for technologies just starting at commercial scale in 2030.

Step 5: Resulting 2030 and 2050 capacities were trimmed by feedstock availability, as provided through Task 2 within the study (see [1], high biomass mobilization scenario). However, trimming proved necessary only for the production of HVO based on cover crops (this is a feedstock category that could become part of Annex IX Part A in the future); all other pathways did not exceed related feedstock availability.

Table 3, **Table 4** and **Fig. 2** provide further details on the methodology and the results.

Table 3. Data collection and calculation example for possible capacity expansion by 2030 for a single technology provider. This refers to steps 1 and 2 in the above

Data Description	Data Value	Source
Largest operational facility globally	30 000 t/y	Company + Author
Size of a future commercial facility	180 000 t/y	Company + Author
Lead time until fully operational	4 years	Company + Author
Number of facilities that the company can support by year	3	Company + Author
Average capacity of one facility installed in 2024 and resulting overall capacity	30 000 t/y → 90 000 t/y	Author
Average capacity of one facility installed in 2025 and resulting overall capacity	90 000 t/y → 270 000 t/y	Author
Average capacity of one facility installed in 2026 and resulting overall capacity	180 000 t/y → 540 000 t/y	Author
Average capacity of one facility installed in 2027—2030	Facilities built in these years cannot contribute to 2030 operational capacity because of 4 years lead time	Author
Resulting 2030 capacity of this company	**900 000 t/y**	Author

Table 4. Estimated possible capacity expansion by 2030 and 2050 by pathways. Provides the results of steps 2, 3 and 4 as described above

Technology pathway	Resulting 2030 capacity of company [Mtoe/y]	Factor for competitors doing the same	Total 2030 capacity of pathway [Mtoe/y]	Factor for 2050	Total 2050 capacity of pathway [Mtoe/y]
Hydrotreatment of Annex IX A feedstocks	0.7	1	0.7	3	1.7
Lignin depoly-merisation	2.3	1	2.3	5	11.7
Cellulosic ethanol	1.0	5	5.0	7	34.7
ATJ	0.9	3	2.8	7	19.8
Gasification + methanol	0.3	5	1.4	7	9.7
Gasification + SNG	3.0	5	14.9	7	104.6
Gasification + FT	1.6	5	8.2	10	81.8
Pyrolysis	1.0	5	5.1	7	35.5
HTL	0.3	3	0.8	10	7.6

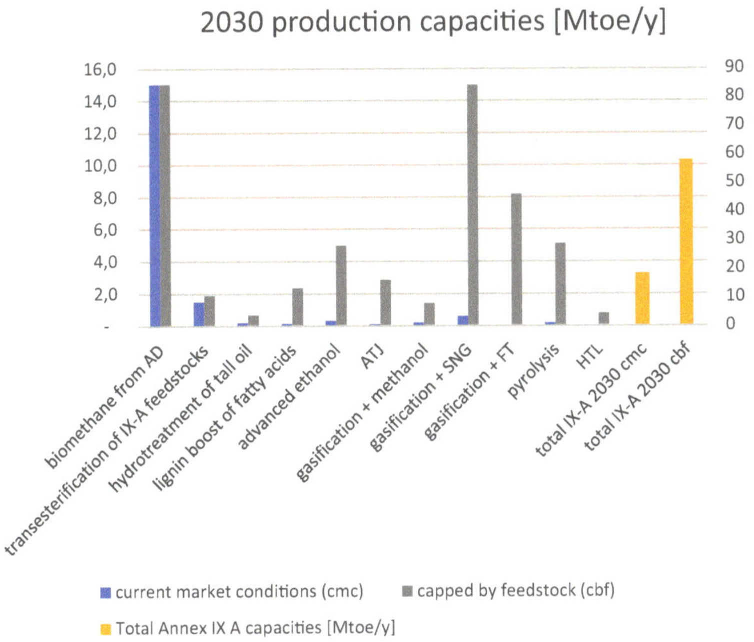

Fig. 2. Comparison of projected 2030 production capacities for advanced biofuels and biomethane from Annex IX A feedstocks by pathway under current market conditions and under favorable market conditions, after capping by feedstock availability

Results show that in the case of favorable market conditions, capacity expansion could be much faster than under current market conditions, since quite a number of technology providers is at the edge of commercialisation of their technologies and ready to support multiple projects at a time. Total advanced biofuels and biomethane reach 58.0 Mtoe in 2030, by far exceeding the projected demand. However, again—despite the many different applications of biomethane and related demand from other sectors than transport—all biomethane is attributed to the transport sector. If allocating only 10% of biomethane from AD and 5% of biomethane from gasification to the transport sector, advanced biofuels are projected to reach 28.8 Mtoe and biomethane in transport 1.5 Mtoe. These amounts even exceed the projected policy-driven demand of 15 to 19 Mtoe.

6 Summary and Conclusions

Task 1 of the study [1] projects a policy-driven demand for advanced biofuels and biomethane (i.e. biofuels produced from Annex IX Part A of the Renewable Energy Directive) of 15 to 19 Mtoe by 2030.

The assessment under Task 3 of the study [1] estimates the potential related capacity expansion *under current market conditions* to reach 18.4 Mtoe by 2030. If only considering 10% of biomethane from anaerobic digestion and 5% of biomethane from gasification to be available to the transport sector, this potential falls to 4.3 Mtoe, which is below the projected policy-driven demand.

Under the assumption that *market conditions are absolutely favorable,* the technical possible capacity expansion, capped by feedstock availability, could reach 58 Mtoe by 2030. Again, if considering only 10% of biomethane from anaerobic digestion and 5% of biomethane from gasification, this potential falls to 30.3 Mtoe. This is however still sufficient to meet the projected policy-driven demand.

We thus conclude that reaching the projected policy-driven demand for advanced biofuels in 2030 is possible, at least from a technological perspective. It does, however, not happen if left to market forces alone. The development of viable production technologies takes years to decades, and investments in demonstration facilities are an absolute must to allow for the required technological learning. Bold incentives that create real business opportunities to investors will be needed to stimulate such deployment.

The potential feedstock base of lignocellulosic materials is much larger than the potential of oil crops; consequently, bringing technologies that are based on lignocellulosic materials to market maturity will be essential for achieving high volumes of biofuel production in the long term.

The multitude of technology providers, particularly those based in Europe, is an asset, increasing the chances of reaching the advanced biofuels targets even if some of the facilities to be installed will likely fail economically. Society must be prepared to lose some money through failures for the benefit of developing and deploying the best-performing technologies.

References

1. European Commission, Directorate-General for Research and Innovation, Georgiadou, M., Goumas, T., Chiaramonti, D., *Development of outlook for the necessary means to build industrial capacity for drop-in advanced biofuels—Final report*, Georgiadou, M.(editor), Goumas, T.(editor), Chiaramonti, D.(editor), Publications Office of the European Union, 2024, https://data.europa.eu/doi/org/10.2777/679307
2. EEA Report No 02/2022. Transport and environment report 2021. Decarbonising road transport — the role of vehicles, fuels and transport demand
3. Demoplants database maintained by IEA Bioenergy Task 39 in collaboration with ETIP Bioenergy, available at https://www.demoplants.best-research.eu, accessed in August 2023

Importance of Green Methanol in Sustainable Mobility and Contribution of Oxidative Fluorination

Henrik Schuster, Lukas D. Ernst, and Ingo Krossing(✉)

Materials Research Center, University Freiburg, Freiburg, Germany
{henrik.schuster,lukas.ernst,}@fmf.uni-freiburg.de,
krossing@uni-freiburg.de

1 Importance of Green Methanol and Its Synthesis

1.1 Motivation

Since the pre-industrial age, the concentration of carbon dioxide (CO_2) on earth has increased by 140 ppm, from 280 ppm to over 420 ppm, according to the Keeling curve, see Fig. 1 [1]. It is projected that, without intervention, the concentration of CO_2 will increase to 500 ppm by 2050.[2] The Paris Agreement of 2015 aims to address global climate change by limiting the increase in the global average temperature since the pre-industrial age to well below 2 °C or to 1.5 °C [3]. Achieving the target poses a significant challenge due to the current anthropogenic CO_2 emissions of 36 Gt per year. To ensure consistent CO_2 neutrality, it's crucial to balance emissions, storage, and recycling. Cumulative emissions should remain below 1440 Gt between 2000 and 2050, with annual emissions decreasing accordingly to meet the set target [2]. It's worth noting that CO_2 in the environment contributes to the greenhouse effect, causing it to absorb more energy radiating from the earth's surface, along with other greenhouse gases in the stratosphere, resulting in warming of the earth's climate [4].

CO_2 is present in the environment due to various sources, with a significant amount originating from the extraction and use of fossil fuels such as oil and natural gas. Other sources of CO_2 include biomass and emissions from industry and infrastructure. To address this issue, a practical and effective approach would involve implementing a closed-loop CO_2 cycle, comprising carbon capture, storage of the gas, and utilization.

This is known as the CCUS-process (Carbon Capture and Utilization and Storage) process [2].

A. Heintzel (Hrsg.): Int.Engine.Congr. 2024, Proceedings, S. 334–344, 2025.
https://doi.org/10.1007/978-3-658-47966-4_24

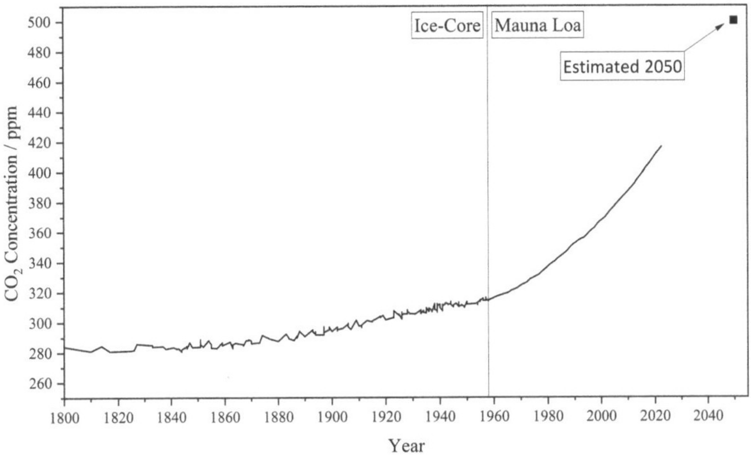

Fig. 1. The keeling curve clearly demonstrates the steady rise in CO_2 concentration per year to the present day. The data before 1958 was obtained through ice-core examination and has since been collected from the Mauna Loa Observatory in Hawaii [1].

This method has shown great potential in reducing carbon emissions and promoting sustainable energy practices. One promising method for CO_2 storage, along with the storage of excess energy, is referred to as *Power-to-Liquid*. This process involves extracting CO_2 from the atmosphere and converting it with hydrogen to produce liquid fuels. The hydrogen used in this process is derived from the electrolysis of water, which can be powered by surplus energy from wind, water, and solar power. Chemicals produced *via* this green method, such as Green Methanol, are called green chemicals. Methanol is a viable synthetic product in this process [5, 6]. It can be utilized both as a fuel and as a precursor for further products, such as formaldehyde, biodiesel, and dimethyl ether, with steadily increasing demand, see Fig. 2 [7].

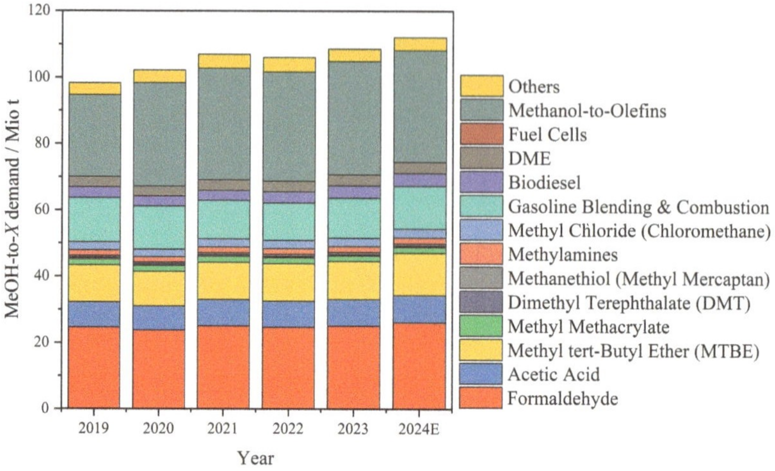

Fig. 2. Yearly global supply and demand of products made from methanol (MeOH) from 2018 until the estimated data of 2023 [7].

1.2 Methanol Synthesis from Heterogeneous Catalysis

Currently, methanol is synthesized from synthesis gas (carbon monoxide CO, hydrogen H_2 with traces of CO_2), see Eq. (1), which is obtained through non-sustainable coal gasification or methane steam reforming pathways [8]. The motivation for sustainable methanol synthesis is to redirect the heterogeneous catalyzed methanol synthesis from synthesis gas towards the recycling of the greenhouse gas CO_2, see Eq. (2). It should be noted that the CO_2-hydrogenation is less exothermic than CO-hydrogenation (ARH^0 = – 49.4 kJ mol^{-1} *vs.* – 90.8 kJ mol^{-1}) and provides an equivalent of water with each methanol molecule [9, 10]. The water moistens the catalyst's surface and deactivates the active sites. Increasing the catalysis temperature promotes the desorption of water, however, it also supports the endothermic reverse water-gas shift (rWGS) reaction, which converts the feed gas CO_2 and H_2 into CO and additional water [11–13].

$$CO + 2\,H_2 \rightleftharpoons CH_3OH \qquad \Delta RH^0 = -90.8\ \text{kJ mol}^{-1} \tag{1}$$

$$CO_2 + 3\,H_2 \rightleftharpoons CH_3OH + H_2O \quad \Delta RH^0 = -49.4\ \text{kJ mol}^{-1} \tag{2}$$

$$CO_2 + H_2 \rightleftharpoons CO + H_2O \qquad \Delta RH^0 = +42.1\ \text{kJ mol}^{-1} \tag{3}$$

The calculated methanol productivity Y_{Me}, MeOH- and CO-selectivity S_i of CO_2-hydrogenation and CO_2-conversion, shown in Fig. 3, emphasize the issue of high temperature for methanol synthesis. With increasing temperature, the yield and selectivity of MeOH decrease while more CO is produced [14].

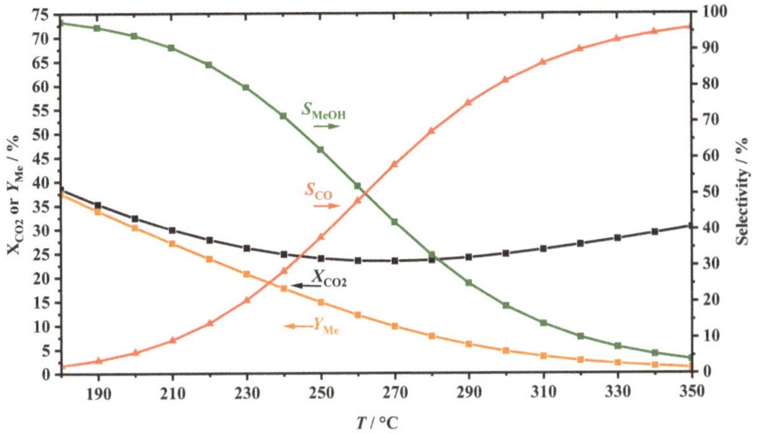

Fig. 3. Thermodynamic equilibrium values in dependence of temperature T at 40 bar for CO_2-hydrogenation to methanol. Calculated by D. Kubas et al [14].

Furthermore, the conversion rate relies on the catalyst loading, which is quantified by the parameter known as gas hourly space velocity (GHSV), representing the flow of feed gas per unit volume of catalyst and time.

In practical terms, comparing catalysts, particularly with the literature, proves to be challenging due to the multitude of parameters involved such as temperature, pressure, GHSV, reactor, and particle diameter. Ensuring maximum reproducibility of all condi-tions is paramount for catalysis research.

Industrially, heterogeneous catalysts used for methanol synthesis are based on $Cu/ZnO/Al_2O_3$. In this process, copper (Cu) is the active species on which the methanol synthesis takes place. Zinc oxide (ZnO) is a geometric spacer that disperses the Cu units and reduces the activation energy of CO_2 chemisorption with defect sites VO2+ [15]. This phenomenon is also referred to as Cu/Zn-synergism. The third component, Alumina (Al_2O_3), is used for long-term stability to achieve batch lifetimes of three to four years [16–19]. For CO_2-hydrogenation Cu/ZnO catalysts are also suitable, but Al_2O_3 has a high water affinity, leading to a "drowning" of the catalysts surface and reconstruction of ZnO, which consequently results in a quick deactivation of the catalyst [12, 20].

2 Research at Freiburg Materials Research Center

At Freiburg Materials Research Center, the catalyst systems are precipitated from the metal nitrates with sodium carbonate Na_2CO_3 using the so-called co-precipitation method. The Cu/ZnO model system is precipitated at a temperature of 40 °C at pH 6.5 within 40 min [15, 21].

2.1 Oxidative Fluorination of Methanol Catalysts

To achieve catalysts that can be used to produce methanol at lower temperatures, while being more water-suitable, existing catalyst systems, based on the Cu/ZnO synergism, are modified at the Materials Research Center in Freiburg, Germany, to increase MeOH productivity and selectivity. The working group of Prof. Ingo Krossing has developed catalysts in which the oxygen of the CuO/ZnO precatalyst is replaced with fluorine using oxidative fluorination, whereby the following reactions take place on the oxides and carbonates of the precipitated and calcined precatalyst, Eqs. (4) and (5) [13, 22]. M represents the metal component.

$$2\,MO_x + 2yF_2 \rightarrow 2\,MOx - yF_2y + yO_2 \tag{4}$$

$$2\,M(CO_3)_x + 2yF_2 \rightarrow 2\,M(CO_3)x - yF_2y + yO_2 + 2yCO_2 \tag{5}$$

These reactions are highly exothermic and take place at room temperature. A mild fluorination of 1.2 wt-% F (determined by ion chromatography) is sufficient to significantly increase the catalytic activity. When the fluorinated precatalysts react with hydrogen, firstly to activate and prepare the active elemental Cu^0 before catalysis, and secondly by reaction with the hydrogen-containing feed gas $CO_2/3\,H_2$, defects are generated in the zinc oxide lattice ZnO_{1-x}, which capture electrons from the H_2 and activate CO_2 for methanol synthesis, see Fig. 4. S. Fehr was able to show the increase of electron-filled defect sites in diffuse reflection infrared spectroscopy and quantified the defect sites via N_2O-chemisorption and X-ray Photoelectron Spectroscopy. With a higher amount of ZnO_{1-x} defect sites, the activity of CO_2-conversion to methanol increases [15, 23]. While in the untreated Cu/ZnO system (CZ-0) CO is still predominantly formed from the rWGS, the MeOH rate and MeOH selectivity increase with a higher degree of fluorination, so that from about 5 wt% F of the catalyst the same amount of MeOH is formed as CO, see Fig. 5.

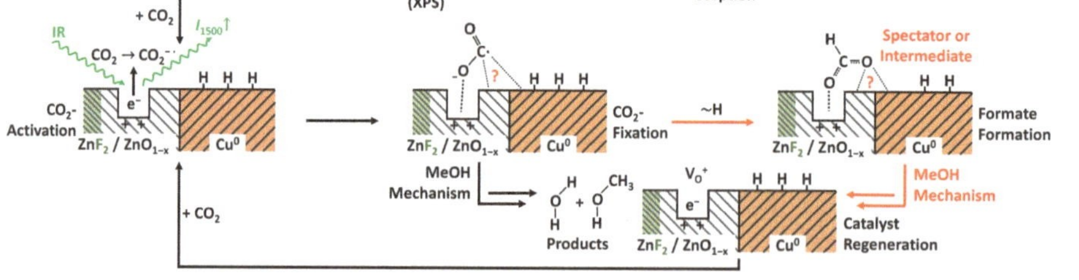

Fig. 4. Catalytic sequence of the process of oxidative fluorination. This mechanism was established by S. Fehr *et al.* from the Krossing Group [15, 23].

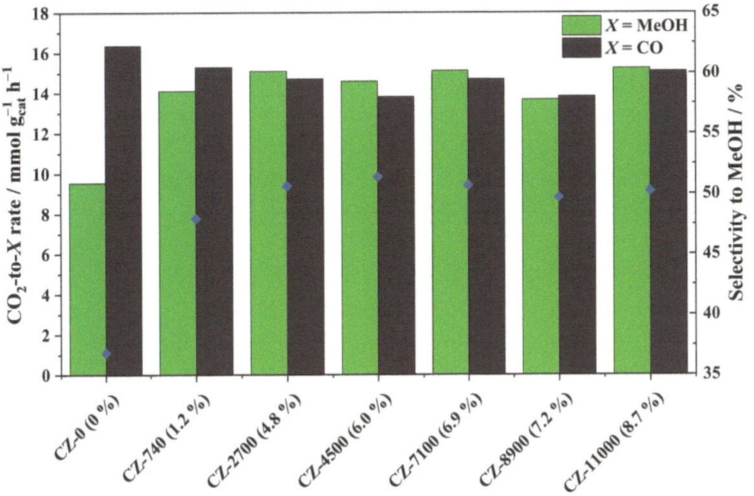

Fig. 5. Formation rate for MeOH (green) and CO (anthracite) from $CO_2/3\ H_2$ for chosen fluorinated Cu/ZnO-catalysts (CZ-*x*) at 250 °C, 40 bar and a GHSV of 19.8 L $Kg_{cat^{-1}}$ $h-^{1}$. *x* corresponds to the theoretical degree of fluorination in µmolF $g_{cat^{-1}}$. In parentheses is the determined fluorination amount in wt-% of the unreduced precatalyst. The blue diamonds show the selectivity for MeOH formation. With an increasing degree of fluorination, methanol rate and selectivity increase considerably.

Due to the high hazard associated with elemental fluorine, the industrial implementation of fluorination using F_2 is challenging. H. Schuster from Krossing Group in the Freiburg Materials Research Center is therefore investigating alternative fluorinating agents that replicate the positive effects of fluorination while being safe to use. In the coming years, publications can be expected that demonstrate the new type of fluorination.

2.2 Ternary Methanol Catalysts

As mentioned above, the industrially used Cu/ZnO/Al_2O_3 catalyst contains a third component, Al_2O_3, which ensures the longterm stability of the system. For methanol synthesis, the Freiburg Materials Research Center also investigates other ternary components that have their maximum effect on the conversion of CO_2 to methanol. A very promising system is the Cu/ZnO/ZrO_2 system, in which zirconia ZrO_2 is used as the third component [13, 14], V. Dybbert investigated the application of oxidative fluorination to the Cu/ZnO/ZrO_2 system and found that the zirconia contained therein binds the fluorine, depriving the Cu/ZnO system that is responsible for methanol synthesis of the positive effect of fluorine. The activity of the catalyst then decreased significantly [13].

L. D. Ernst investigated ternary systems that achieve positive effects in methanol synthesis through oxidative fluorination. This is especially noticeable in the iron-containing Cu/ZnO/FeO_x systems. Due to the Fisher-Tropsch (FT) active species Fe, Cu/ZnO/FeO_x-catalysts also produce methane and other higher alkanes. Fluorination suppresses the FT-activity of alkane synthesis and increases the selectivity for methanol formation [24]. In further studies, L. D. Ernst worked on ternary magnesium-containing Cu/ZnO/MgO systems. Figure 6 shows a variation in synthesis parameters, such as magnesium content, synthesis temperature, and application of one-hour aging after precipitation. The most promising system was 1.0 Mg_70_1, which was precipitated with a composition of 3.75 (Cu) : 1.73 (Zn) : 1.0 (Mg) at 70 °C and pH 7.0 and then aged for one hour. It is used as the system for further syntheses of a Cu/ZnO/MgO catalyst and will be referred to as CZMg. The nonfluorinated catalyst showed the highest activity and after treatment with fluorine, it further rose to the level of the above-mentioned Cu/ZnO/ZrO_2 benchmark catalyst system.

The degree of fluorination of the CZMg system was further varied and subjected to a long-term measurement of 14 days' time-on-stream, see Fig. 7. The MeOH formation increases by oxidative fluorination and, above all, the long-term stability is controlled by the degree of fluorination. It is also interesting to mention that during the first two days, a restructuring processes in the system led to an increase in activity. This is particularly noticeable for the CZMg-0 and CZMg-2700 (−0/−2700 refers to μmol applied F per gram catalyst) catalysts. The CZZ catalyst was also included as a reference in the long-term stability study. It does not show an increase in activation, but on the contrary shows a small decrease in MeOH productivity. In the long term, CZMg-2700 and CZZ have similar catalytic activity.

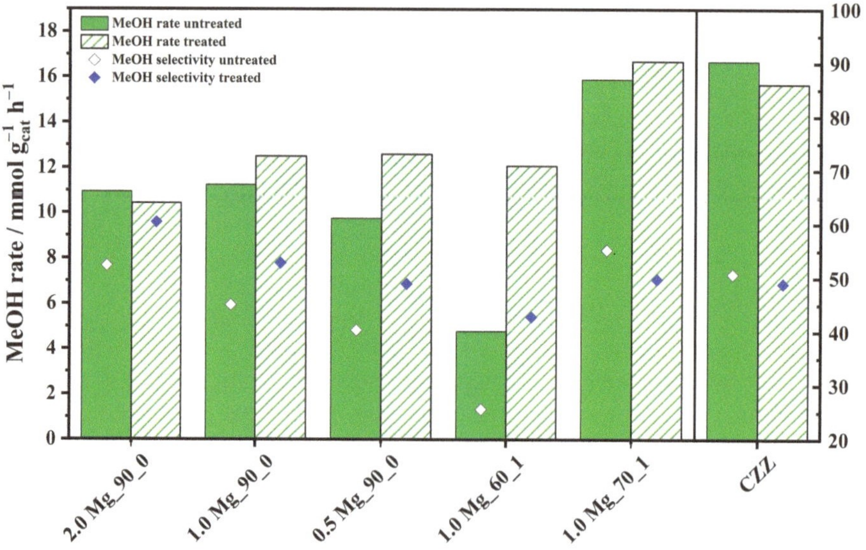

Fig. 6. MeOH formation rate of untreated (green) and fluorinated (lined, 1.3 wt% F) Cu/ZnO/MgO-catalysts with CO_2/3 H_2 at 250 °C, 40 bar and a GHSV of 19.8 L $Kg_{cat^{-1}}$ h^{-1}. In the nomenclature *x* Mg_*y*_*z*, *x* corresponds to the metal content of magnesium used in the synthesis 3.75 (Cu) : 1.73 (Zn) : x (Mg). *y* is the synthesis temperature in °C and *z* indicates whether the precipitation synthesis was followed by a one-hour ageing. The fluorinated 1.0 Mg_70_1-Catalyst is the first catalyst, featuring the same MeOH rate as the reference Cu/ZnO/ZrO_2-System (CZZ) [25].

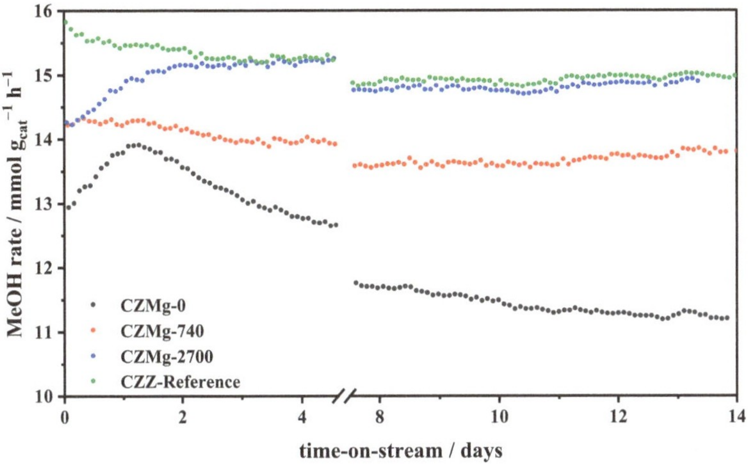

Fig. 7. Long-term stability testing of chosen CZMg-catalysts and the reference CZZ for 14 days. The methanol rate is displayed against the time-on-stream. The measurement was carried out in $CO_2/3\ H_2$ as feed gas at 250 °C, 40 bar, and a GHSV of 19.8 L $Kg_{cat^{-1}}$ $h-^{1}$. It was technically not possible to obtain measurements from days 5 to 7. Fluorinated CZMg-2700 and CZZ show especially after three days similar catalytic activity.

The use of the alternative fluorination agent described above for the fluorination of ternary catalyst sytems is currently being tested at the Freiburg Materials Research Center. Considering the results presented, it is recommended to look for future publications that focus on further investigations and the extension of the methods presented.

3 Conclusion

The research conducted at the Freiburg Materials Research Center marks a significant contribution in sustainable methanol production. Central to this is the development and testing of oxidative fluorinations in the synthesis of Cu/ZnO-containing catalysts. This approach enhances the efficiency and selectivity of CO_2 conversion to green methanol, a possible step in sustainable energy and mobility solutions.

Particularly noteworthy is the exploration of magnesium-containing ternary Cu/ZnO/MgO systems (CZMg), which have shown promising results. The CZMg system exhibited high catalytic activity, especially after fluorination, reaching levels comparable to the well-known Cu/ZnO/ZrO_2 system. Long-term studies revealed that methanol formation in these systems increases with the degree of fluorination, importantly maintaining stability over extended periods. This investigation is relevant for industrial applications, as it demonstrates both the effectiveness and sustainability of these catalysts.

The introduction of alternative fluorination agents addresses the safety concerns associated with elemental fluorine, showcasing a commitment to sustainable and safe chemical practices. This innovation aligns with the broader goal of reducing carbon emissions and promoting green chemistry, particularly in the context of *Power-to-Liquid* processes using green hydrogen.

Future research, as indicated, will likely focus on further investigations and the extension of the proposed methods. These developments are important for scaling up the process for industrial application, optimising the reaction conditions and integrating these processes into existing systems. The ongoing work at the Freiburg Materials Research Centre not only contributes to green chemistry, but also helps to create more environmentally friendly approaches to chemical production and energy generation.

References

1. C. D. Keeling, S. C. Piper, R. B. Bacastow, M. Wahlen, T. P. Whorf, M. Heimann, H. A. Meijer, in *A History of Atmospheric CO_2 and Its Effects on Plants, Animals, and Ecosystems* (Eds.: I.T. Baldwin, M.M. Caldwell, G. Heldmaier, R.B. Jackson, O.L. Lange, H.A. Mooney, E.-D. Schulze, U. Sommer, J.R. Ehleringer, M. Denise Dearing, T.E. Cerling), Springer, New York, NY, **2005**, pp. 83–113.
2. T. P. Senftle, E. A. Carter, *Acc. Chem. Res.* **2017**, *50*, 472–475.
3. H. L. van Soest, M. G. J. Elzen, D. P. van Vuuren, *Nat. Commun.* **2021**, *12*, 2140.
4. S. Bakan, E. Raschke, *promet* **2002**, 283485–94.
5. P. Schmidt, V. Batteiger, A. Roth, W. Weindorf, T. Raksha, *Chem. Ing. Tech.* **2018**, *90*, 127–140.
6. P. Schmidt, W. Weindorf, A. Roth, V. Batteiger, F. Riegel, *Power-to-Liquids: Potentials and Perspectives for the Future Supply of Renewable Aviation Fuel,* Umweltbundesamt, Dessau-Roßlau, **2016**.
7. "Methanol Price and Supply/Demand," can be found under https://www.methanol.org/methanol-price-supply-demand/, **2024**.
8. G. Bozzano, F. Manenti, *Prog. Energy Combust. Sci.* **2016**, *56*, 71–105.
9. J. Abu-Dahrieh, D. Rooney, A. Goguet, Y. Saih, *Chem. Eng. J.* **2012**, *203*, 201– 211.
10. 10. G. Bonura, C. Cannilla, L. Frusteri, A. Mezzapica, F. Frusteri, *Catal. Today* **2017**, *281*, 337–344.
11. Y. A. Daza, J. N. Kuhn, *RSC Adv.* **2016**, *6*, 49675–49691.
12. M. B. Fichtl, D. Schlereth, N. Jacobsen, I. Kasatkin, J. Schumann, M. Behrens, R. Schlögl, O. Hinrichsen, *Appl. Catal., A* **2015**, *502*, 262–270.
13. V. Dybbert, S. M. Fehr, F. Klein, A. Schaadt, A. Hoffmann, E. Frei, E. Erdem, T. Ludwig, H. Hillebrecht, I. Krossing, *Angew. Chem. Int. Ed.* **2019**, *58*, 12935– 12939.
14. D. Kubas, M. Semmel, O. Salem, I. Krossing, *ACS Catal.* **2023**, *13*, 3960–3970.
15. S. M. Fehr, K. Nguyen, I. Krossing, *ChemCatChem* **2022**, *14*, e202101500.
16. M. Behrens, D. Brennecke, F. Girgsdies, S. Kißner, A. Trunschke, N. Nasrudin, S. Zakaria, N. F. Idris, S. B. A. Hamid, B. Kniep, R. Fischer, W. Busser, M. Muhler, R. Schlögl, *Appl. Catal., A* **2011**, *392*, 93–102.
17. S. Zander, E. L. Kunkes, M. E. Schuster, J. Schumann, G. Weinberg, D. Teschner, N. Jacobsen, R. Schlögl, M. Behrens, *Angew. Chem. dt. Ed.* **2013**, *125*, 6664–6669.
18. T. Fujitani, J. Nakamura, *Catal. Lett.* **1998**, *56*, 119–124.

19. C. Tisseraud, C. Comminges, T. Belin, H. Ahouari, A. Soualah, Y. Pouilloux, A. Le Valant, *J. Catal.* **2015**, *330*, 533–544.
20. T. Lunkenbein, F. Girgsdies, T. Kandemir, N. Thomas, M. Behrens, R. Schlögl, E. Frei, *Angew. Chem. Int. Ed.* **2016**, *55*, 12708–12712.
21. M. Behrens, R. Schlögl, *Z. anorg. allg. Chem.* **2013**, *639*, 2683–2695.
22. J. Eicher, H. Hillebrecht, V. Dybbert, E. Frei, M. Hill, A. Schaadt, I. Krossing, T. Ludwig, *Catalyst Comprising Fluorinated Metal Oxide, Manufacture Process and Hydrogenation Process*, **2016**, WO2016066823A1.
23. S. M. Fehr, K. Nguyen, C. Njel, I. Krossing, *ACS Catal.* **2021**, *11*, 13223–13235.
24. L. D. Ernst, I. Krossing, *8. Internationaler Motorenkongress 2021*, Springer Vie-weg, Wiesbaden, **2021**, pp. 579–587.
25. L. D. Ernst, S. Polinski, M. Günthel, W. Marquart, S. Raseale, N. Fischer, I. Krossing, *to be published*.

Autorenverzeichnis

A. Heintzel (Hrsg.): Int.Engine.Congr. 2024 Proceedings, S. 345–346, 2025.
https://doi.org/10.1007/978-3-658-47966-4

Kolibri 360 GmbH
Ferdinand-Jühlke-Straße 7
99095 Erfurt, Deutschland
produktsicherheit@kolibri360.de